Textbooks in Telecommunication Engineering

Era of the Internet and Network Science/Engineering

Series Editor

Tarek S. El-Bawab, Ph.D., Boston, MA, USA

Telecommunication and networks have evolved to embrace all aspects of our everyday life, including smart cities and infrastructures, healthcare, banking and businesses, manufacturing, space and aviation, meteorology and climate change, oceans and marine life, Internet of Things, defense, homeland security, education, research, social media, entertainment, and many others. Network applications and services continue to expand, virtually without limits. Therefore, specialized telecommunication and network engineering programs are recognized as a necessity to accelerate the pace of advancement in this field, and to prepare a new generation of engineers for imminent needs in our modern life. These programs need curricula, courses, labs, and textbooks of their own.

The IEEE Communications Society's Telecommunication Engineering Education (TEE) movement, led by Tarek S. El-Bawab- the editor of this Series, resulted in recognition of this field of engineering by the Accreditation Board for Engineering and Technology (ABET) on November 1, 2014. This Springer Series was launched to capitalizes on this milestone. The Series goal is to produce high-quality textbooks to fulfill the education needs of telecommunication and network engineering, and to support the development of specialized undergraduate and graduate curricula in this regard. The Series also supports research in this field and helps prepare its scholars for global challenges that lay ahead. The Series have published innovative textbooks in areas of network science and engineering where textbooks have been rare. It is producing high-quality volumes featuring innovative presentation media, interactive content, and online resources for students and professors.

Book proposals are solicited in all topics of telecommunication and network engineering including, but not limited to: network architecture and protocols; traffic engineering; network design, dimensioning, modeling, measurements, and analytics; network management and softwarization; cybersecurity; synchronization and control; applications of artificial intelligence in telecommunications and networks; applications of data science in telecommunications and networks; network availability, reliability, protection, recovery and restoration; wireless communication systems; cellular technologies and networks (through 5G, 6G, and beyond); satellite and space communications and networks; optical communications and networks; heterogeneous networks; broadband access and free-space optical communications; MSO/cable networks; storage networks; optical interconnects; and data centers; social networks; transmission media and systems; switching and routing (from legacy to today's paradigms); network applications and services; telecom economics and business; telecom regulation and policies; standards and standardization; and laboratories.

Proposals of interest shall be for textbooks that can be used to develop university courses, either in full or in part. They should include recent advances in the field while capturing whatever fundamentals that are necessary for students to understand the bases of the topic and appreciate its evolution trends. Books in this series will provide high-quality illustrations, examples, end-of-chapters' problems/exercises and case studies.

This series is indexed in Scopus. For further information and to submit proposals, please contact the Series Editor, Dr. Tarek S. El-Bawab, telbawab@ieee.org; or Mary James, Executive Editor at Springer, mary.james@springer.com.

Lin Zhou • Lin Bai

Low-latency Communication

Theory and Application

Springer

Lin Zhou
Southern University of Science and Technology
Shenzhen, China

Lin Bai
Beihang University
Beijing, China

ISSN 2524-4345 ISSN 2524-4353 (electronic)
Textbooks in Telecommunication Engineering
ISBN 978-3-032-06626-8 ISBN 978-3-032-06627-5 (eBook)
https://doi.org/10.1007/978-3-032-06627-5

This Springer imprint is published by the registered company Springer Nature Switzerland AG
The registered company address is: Gewerbestrasse 11, 6330 Cham, Switzerland

Dedicated to our family and cats.

Preface

In this book, we aim to comprehensively introduce the theoretical foundations for low-latency communications and its applications, by covering recent advances from the information theoretic literature in the past decades and by presenting various applications. This book is partitioned into two parts: theoretical foundations and applications. In the theoretical foundation part, we cover topics including source and channel coding for point-to-point cases and multiuser cases, illustrating the information theoretical characterizations for the performance of optimal codes. In the applications part, we cover topics including multi-input multi-output (MIMO) communication, multi-connectivity, user scheduling, and covert communication.

This book is targeted for senior undergraduate students, graduate students, professional researchers, and engineers who are interested in low-latency communications. The theoretical foundation part can be designed as either a high-level undergraduate course or a graduate course that teaches information theoretical aspects of low-latency communications. The applications part can be designed as an advanced graduate course that specializes to the practical communication systems with multiple transmitting and receiving antennas, where we care transmission reliability, user scheduling for base stations, and physical layer security. For professional researchers who are interested in future directions of low-latency communication analyses, the foundational part presents elementary techniques to conduct research in this domain, and the applications demonstrate examples of how the tools can be used. For engineers working in the domain of communications, the performance benchmarks can be used as the design guidelines, analogous to how the Shannon capacity has been used to evaluate the performance of codes despite its asymptotic nature.

To our best knowledge, this book is the first to cover theoretical foundations of low-latency communication, introducing the cutting-edge advances from the information theory society in the past decade. We hope this book benefits readers, especially people outside the domain of low-latency communication theory, by providing an easy-accessible yet in-depth introduction of related results with proofs, examples, and exercises.

Beijing, China
December, 2024

Lin Zhou
Lin Bai

Acknowledgement The authors would like to thank Changhong Liu, Yiming Wang, Zhuangfei Wu, Jinpeng Xu, and Lina Zhu for their help in preparing materials, generating plots, and proofreading parts of this book.

The authors acknowledge Prof. Tarek El-Bawab and the publication staff of Springer for providing us this great opportunity to contribute a book in the series of Textbooks in Telecommunication Engineering and for their kind help throughout the process of preparation, writing, review, and publication of this book.

Declarations

L. Zhou was supported by the National Natural Science Foundation of China under Grant 62201022, the Beijing Municipal Natural Science Foundation under Grant 4232007, and Fundamental Research Funds for the Central Universities.

L. Bai was supported by the National Natural Science Foundation of China under Grant U22B2008.

Contents

Part II Applications

Acronyms

AAN	Additive arbitrary noise
AMS	Arbitrary memoryless source
AWGN	Additive white Gaussian noise
AO	Alternative optimization
AOIUS	AO iterative user group selection algorithm
BC	Broadcast channel
CLT	Central limit theorem
CDI	Channel distribution information
CSI	Channel state information
CSIR	CSI at the receiver
CSIT	CSI at the transmitter
CSITR	CSI at both the transmitter and the receiver
DF	Decode and forward
DMC	Discrete memoryless channel
DMS	Discrete memoryless source
FBL	Finite block length
GIUS	Greedy iterative user scheduling algorithm
GMS	Gaussian memoryless source
i.i.d.	independent and identically distributed
i.n.i.d.	independent and non-identically distributed
JEP	Joint excess-distortion probability
JD	Joint decoding
JNN	Joint nearest neighbor
JSCC	Joint source-channel coding
LBUS	Lower-bound based user scheduling algorithm
LD	Large deviations
MAC	Multiple access channel
MC	Multi-connectivity
MIMO	Multi-input multi-out
NN	Nearest neighbor
NOMA	Non-orthogonal multiple access

noCSI	No CSI at either the transmitter or the receiver
OMA	Orthogonal multiple access
P2P	Point to point
RAC	Random access channel
SEP	Separate excess-distortion probability
SIC	Successive interference cancellation
SINR	Signal-to-interference and noise ratio
SNR	Signal-to-noise ratio
SR	Successive refinement
SSCC	Separate source-channel coding

Part I
Theory

Chapter 1
Introduction

1.1 Motivation

Low latency is an inherent nature of practical communication systems. In 5G, a typical scenario is *ultra-reliable and low-latency communication (URLLC)* [1], which finds diverse applications including automatic driving, Tactile internet, and internet of things [2]. In upcoming 6G, the latency tolerable is set to be 0.1 millisecond, which is one hundred-th of the tolerable latency of 5G [3]. To achieve stringent latency requirements, short blocklength codes are used [4].

The theoretical foundation of communication systems was founded by Shannon [5]. In this seminal paper, for the first time ever, Shannon demonstrated that mathematical models could be used to study communication systems. In particular, Shannon showed that all communication problems are either data compression or data transmission. Data compression, also known as *source coding*, aims to reduce redundancy of an information source. Data transmission, also known as *channel coding*, aims to transmit a message reliably over a noisy channel. In terms of functions, source coding helps improve transmission efficiency while channel coding helps combat noise over the channel. In terms of mathematical models, source coding can also be understood as data transmission over a noiseless channel. Using above mathematical formulations, Shannon proved well-known theoretical benchmarks for source and channel coding. In particular, Shannon showed that as the blocklength increases, the error probability of channel coding vanishes if the message rate is less than the capacity of the noisy channel, while the error probability of source coding vanishes if the compression rate is greater than the entropy of the source. These two results, also known as the channel and source coding theorems, guide the design of practical codes in the past several decades. Specifically, LDPC code [6] and Polar code [7] are shown to be capacity achieving for data transmission, and both codes are standardized in 5G.

Over the past several decades, Shannon's results have been generalized to multiuser communication, providing further insights into how the practical commu-

L. Zhou, L. Bai, *Low-latency Communication*, Textbooks in Telecommunication Engineering, https://doi.org/10.1007/978-3-032-06627-5_1

nication systems should be designed. The collective efforts have resulted in several classical textbooks in the information theory domain, including [8–10]. In particular, [8] is an excellent textbook for beginners of information theory, introducing the basic concepts and related results for point-to-point (P2P) communication systems. In contrast, [9] is an advanced textbook focusing on multiuser communication systems, while [10] is a further technical textbook for researchers interested in refined analyses of communication systems, which characterize how the error probability behaves as a function of the coding rate.

Although insightful, all above studies are in nature of *Shannon theory*, assuming infinite blocklength to obtain exact theoretical benchmarks. The assumption of infinite blocklength enables the theoretical analyses and allows for elegant formulas as performance benchmarks. However, infinite blocklength leads to infinite latency, which contradicts stringent low-latency requirements of practical communication systems and calls for theoretical studies of low-latency communication.

1.2 Literature Review

To fill the above gap between theory and practice, in the past one and a half decades, researchers have refined Shannon theory to account for finite blocklength for both P2P and multiuser communication. Remarkably, only 5 years after Shannon's paper, Yushkevich [11] pioneered the theoretical study of low-latency communication, characterizing the so-called *second-order asymptotics* for P2P source coding, which provides a much tighter approximation to the performance of optimal codes with short blocklengths than Shannon's coding theorems. Subsequently, in 1962, Straseen [12] rediscovered the result in [11] and generalized the result to P2P channel coding. The above results were rediscovered by Hayashi [13, 14] and also by Polyanskiy, Poor, and Verdú for channel coding [15]. In particular, the seminal papers [14, 15] have now been used as building blocks to study low-latency wireless communication systems. The above results have been generalized to multiuser communication, with channel coding summarized in [16] and source coding summarized in [17]. These theoretical results lay the foundation of low-latency communications.

This section introduces the history of theoretical analyses for low-latency communication, which is known as *finite blocklength (FBL) analysis* or second-order asymptotics.

1.2.1 Source Coding

Source coding aims to reduce redundancy in source sequences as much as possible so that subsequent transmission can be efficient. *Lossless source coding* is the simplest problem in information theory. In this problem, given a discrete memoryless source sequence of a certain blocklength, one aims to compress the sequence and recover the sequence perfectly. The performance metric is the error probability that quantifies the probability of imperfect reconstruction.

Shannon [5] proved that as the blocklength tends to infinity, the minimal compression rate that ensures vanishing error probability equals the entropy of the source. In other words, if one has an infinitely long sequence to compress, the minimal number of bits required for each source symbol is at least *the entropy* to ensure that the error probability decays to zero as the blocklength increases. Shannon's result is now commonly known as *first-order asymptotics* since it basically characterizes the first dominant term in the compression rate as a function of the blocklength when the blocklength increases. Subsequently, in 1953, Yushkevich [11] refined Shannon's result by showing that when the blocklength is finite, there is an additional dominant rate that scales in the order of reciprocal of the square root of the blocklength and is a function of the both error probability and the variance of the source. Such result is named second-order asymptotics [13] since the second dominant term for the minimal compression rate is characterized as a function of the blocklength. It is now well known that second-order asymptotics provides good approximation to FBL performance of optimal codes. Recently, Kontoyiannis and Verdú [18] and Chen et al. [19] obtained even tighter results by characterizing the third dominant term and bounded remainders.

Another important problem of data compression is *lossy source coding*, where one aims to reproduce a sequence by tolerating imperfections. The imperfection level is measured by a function called *distortion function*, and the value of imperfection between the source sequence and its reproduced version is named distortion. There are performance criteria for lossy source coding given a distortion level: *expected distortion* and the *excess-distortion probability*. Specifically, the expected distortion criterion checks whether the average distortion between the source and reproduced sequence is within the given distortion level, while the excess-distortion probability quantifies the probability that the distortion between the source and the reproduced sequence is above the given distortion level.

Lossy source coding is important for two reasons. Firstly, lossy source coding allows lower compression rates and thus further improves transmission efficiency. Secondly, lossy source coding applies to both discrete and continuous source sequences. Lossy source coding is also known as quantization [20] or rate-distortion. The theoretical study of lossy source coding is again pioneered by Shannon [21]. Analogously to lossless source coding, Shannon characterized the first-order asymptotic term of the minimal compression rate as the blocklength increases. Specifically, Shannon's result implies that when the blocklength tends to infinity, the expected distortion criterion can be satisfied if and only if the compression rate is above a value named the *rate-distortion function* that depends on both the distortion level and the source distribution.

Theoretical analysis of low-latency lossy source coding is quite challenging. This is partially due to the adoption of the expected distortion criterion, which is vastly different from the error probability for lossless source coding. Potentially motivated by the results of the lossless case, the excess-distortion probability was later used, under which the second-order asymptotics was finally derived in the 2010s by Ingber and Kochman [22] and by Kostina and Verdú [23]. The result in [22] holds for any *discrete memoryless source (DMS)* and *Gaussian memoryless source (GMS)*, while the result in [23] holds for any discrete and almost *arbitrary memoryless source (AMS)*. In particular, beyond providing second-order asymptotics, upper and lower bounds on the excess-distortion probability of optimal codes were provided [23].

1.2.2 Channel Coding

Channel coding aims to transmit a message reliably over a noisy channel that distorts transmitted signals. Channel coding is the most fundamental problem in information theory and is widely studied even now. The performance metric is the error probability that quantifies the probability that the transmitted message is decoded incorrectly. Shannon's famous channel coding theorem states that as the blocklength tends to infinity, the maximal message rate equals the *capacity* of the noisy channel to guarantee that the error probability vanishes. Again, the above result is first-order asymptotics that characterizes the first dominant term in the maximal message rate as the blocklength increases. The second-order asymptotics for channel coding was first derived by Strassen [12], who showed that there exists an additional term beyond capacity, which scales in the order of the reciprocal of the blocklength and is a function of both the error probability and the variance of the noisy channel. Strassen's result was rediscovered by two award-winning papers by Hayashi [14] and by Polyanskiy et al. [15]. In particular, the authors of [15] provided a sequence of upper and lower bounds on the error probability of optimal codes when the blocklength is finite, using which second-order asymptotics were derived for noisy channels with both discrete and continuous noise, including the *additive white Gaussian noise (AWGN)* channel. Subsequently, the above second-order asymptotics was further refined by Tomamichel and Tan [24, 25] by characterizing the third-order term. Very recently, Yavas et al. [26] provided another analysis of the third-order term and demonstrated the tightness of their bounds for blocklengths of even tens. All above results hold for discrete-time channels. Furthermore, the above result has been generalized to the continuous-time Possian channel by Sakai et al. [27].

The second-order asymptotics has also been generalized to channels with fading, which models wireless communications. There are two types of fading models [28]: *block fading* and *quasi-static fading*. The random fading is also known as *channel state information (CSI)*. In block fading, the CSI spans all possible values in the transmission of a message, while in quasi-static fading, the CSI remains unchanged during the transmission of the message. For block fading, Polyanskiy and Verdú [29] pioneered the study of second-order asymptotics, which was further generalized to the case with multiple transmit and receive antennas. For quasi-static fading, Yang et al. [30] derived tight bounds on the maximal message rates when multiple antennas are used at the transmitter and the receiver, where CSI is either available or unavailable at the transmitter and receiver. A critical take-home message from [30] is that the second-order term equals to zero. For single antenna channels, Yang et al. [31] compared the second-order asymptotics of block and quasi-static fading. The results in [30, 31] imply that for low-latency communication, quasi-static fading is the correct fading model. Finally, the result in [30] was generalized to the case of massive number of transmit and receiver antennas by Hoydis et al. [32], where the number of antennas scales with the blocklength.

1.2.3 Joint Source-Channel Coding

Any practical communication task is *joint source-channel coding (JSCC)*, where one aims to send a source over a noisy channel reliably. Thus, source coding is required to reduce redundancy of the source sequence and channel coding is required to combat channel noise. When one applies source coding and channel coding separately in a sequential manner, the coding scheme is known as *separate source-channel coding (SSCC)*. In fact, it was shown by Shannon [21] that SSCC is optimal as the blocklength tends to infinity. In other words, when the latency is infinite, in order to achieve given distortion levels, using SSCC and JSCC achieves the same transmission rate in terms of source symbols per channel use. Shannon's result was proved for lossy source coding, which includes the lossless case as a special case when one uses *Hamming distortion measure* and considers the excess-distortion probability. Over the past several decades, Shannon's result guides the wide application of SSCC.

However, when low latency is required, SSCC is strictly inferior [33, 34]. Specifically, the second-order asymptotics for JSCC was first derived by Wang et al. [33] for transmitting a discrete source over a channel with discrete noise. The authors of [33] bounded the performance gap between SSCC and JSCC and showed that the gap exists in the second dominant term in the transmission rate as the blocklength increases. Subsequently, Kostina and Verdú [34] derived upper and lower bounds on the excess-distortion probability of JSCC for arbitrary source and channel when the blocklength is finite, using which second-order asymptotics were derived for transmitting an AMS over any memoryless channel under mild conditions, including transmitting a GMS over an AWGN channel. The results above have been generalized to the case of multi-connectivity by Zhou et al. [35], where transmission reliability is shown to be significantly improved by having several independent transmission channels under quasi-static fading.

1.2.4 Multiterminal Communication

In practical communication systems, there are usually multiple users who wish to communicate simultaneously. Correspondingly, the source and channel coding problems and theoretical benchmarks for the P2P case have been generalized to various settings of multiterminal communication. When multiple users are involved, the theoretical benchmarks are a region instead of a point. The first-order asymptotics characterizes the asymptotic limit of the coding rates when the blocklength tends to infinity. For source coding, the coding rates are compression rates in terms compression bits required per source symbol, while for channel coding, the coding rates are message rates in terms of transmission bits achieved per channel use. To characterize second-order asymptotics, one needs to fix a rate pair on the boundary of the first-order asymptotics and bound how fast the compression rates approach the boundary rate pairs as the blocklength increases under a fixed non-vanishing error probability. Thus, both first and second-order asymptotics for multiterminal problems require a multidimensional analysis and are significantly more challenging. In the following, several multiterminal source and channel coding problems are introduced, which have been studied under low-latency constraints.

A typical multiterminal source coding problem is the *Slepian-Wolf problem* [36], where two transmitters aim to compress their source sequences to a common receiver. Since two users are involved, the theoretical benchmarks concern the relationship among rates of both users, error probability, and the block length. As the block length tends to infinity, Slepian and Wolf [36] characterized the set of rate pairs that ensure the error probability vanishes. The result of [36] is Shannon theoretical and generalized the result of Shannon [5] for lossless source coding. This first-order asymptotic result was recently refined by Tan and Kosut [37] and by Nomura and Han [38], who derived second-order asymptotics that relates the compression rates of both users to the blocklength and the error probability.

Shortly after [37] , another multiterminal source coding problem, the *Gray-Wyner problem* [39], was studied under the low-latency constraint. The Gray-Wyner problem concerns the transmission of a pair of correlated source to two different receivers over a noiseless channel. In this problem, three encoders are used to compress the two correlated sources. The first-order asymptotics for both the lossless and lossy cases were derived by Gray and Wyner [39], and this is why the problem is named. For second-order asymptotics, the lossless case was solved by Watanabe [40], while the lossy case was solved by Zhou et al. [41]. A combination of lossless and lossy source coding problem was proposed by Fu and Yeung [42], where one aims to transmit a source to three receivers, two receivers apply lossy reconstruction of the source while the remaining receiver applies lossless reconstruction. The bounds on second-order asymptotics were derived in [43]. There are two special cases where the above bounds are tight: the *successive refinement (SR) problem* [44], where the lossless recovery is removed, and the Kaspi problem [45, Theroem 1], where one aims to transmit a source to two receivers that apply lossy reconstruction with different distortion levels. For the SR problem, the second-order asymptotics was first derived by No et al. [46] under separate performance requirements for two receivers. Subsequently, Zhou et al. [47] derived corresponding results under the joint performance criterion for both decoders and showed that the joint criterion is more suitable for multiterminal source coding problems.

Two typical multiterminal channel coding models are the *multiple access channel (MAC)* [48–50] and the *broadcast channel (BC)* [51, 52], which model the uplink and downlink communication links of practical multiuser communication systems, respectively. In a MAC, multiple users aim to transmit messages to a common receiver, while in a BC, a single transmitter sends messages to several users. Compared with the source coding counterpart, multiterminal channel coding is more challenging, and most results for low-latency communication are achievability analyses without optimality guarantees. The only exception is the Gaussian MAC with two users and degraded message sets by Scarlett and Tan [53], where one of the transmitters sends messages of both users of the classical two-user MAC. For a general Gaussian MAC, MolavianJazi and Laneman [54] pioneered the study of low-latency communication and obtained achievable second-order asymptotics. Recently, the result of [54] was refined by Yavas, Kostina, and Effros, who obtained further refined third-order asymptotics and also considered a more practical setting of a random access channel (RAC), where the number of transmitters is uncertain. The second-order asymptotics for Gaussian Broadcast channel was recently studied by Tuninetti et al. [55] and by Lin et al. [56].

1.2.5 Mismatched Communication

In all theoretical analyses above, one assumes that the source distribution and the noise distributions are known so that codes can be designed, optimized, and analyzed. This way, the designed codes are optimal for particular source and noise distributions, and this kind of communication is named the matched case. However, in many practical applications, obtaining the exact knowledge of source and noise distributions is challenging. Thus, it is imperative to propose coding schemes ignorant of exact source and noise distribution and ensure good performance. In this case, due to the loss of distribution knowledge, the designed codes are not matched to any particular source or noise and are named mismatched communication. In other words, one aims to use a particular coding scheme for arbitrary source and noise distributions.

Towards this goal, Csiszár and Körner [57] and Hui [58] initiated the study of mismatched communication over a discrete noise channel, where a fixed decoding rule is used regardless of the noise distribution and one can optimize the codebook distribution to achieve the largest possible transmission rate. Bounds on the transmission rate for discrete noise channels were provided in [57, 58], which were subsequently generalized to continuous noise channels by Kaplan and Shamai [59] and by Ganti et al. [60]. The low-latency performance of mismatched communication was analyzed by Scarlett et al. [61]. Recently, a comprehensive survey for this line of mismatched communication is available in [62].

In another line of mismatched communication, Lapidoth [63] proposed to use optimal codes of data transmission over an AWGN channel to send a message over an *additive arbitrary noise (AAN)* channel. In this line of studies, both the codebook distribution and the decoding rule are fixed. Lapidoth proved a first-order asymptotic result in the Shannon theoretic sense, showing that the maximal achievable message rate to ensure vanishing error probability equals the capacity of an AWGN channel with the same *signal to noise ratio (SNR)* as the blocklength tends to infinity. Analogously, Lapidoth [64, Theorem 3] proposed to use optimal codes of lossy source coding for a GMS to compress an arbitrary memoryless source and derived the first-order asymptotic result. The above results are *ensemble tight* [65], which implies the analyses of the performance of the particular code are rigorous and any sub-optimality of coding performance results from the mismatch between the used coding scheme and the source/noise distribution.

Lapidoth's results have also been generalized to account for low-latency communication constraints. In particular, Scarlett et al. [66] refined Lapidoth's result for mismatched channel coding by deriving second-order asymptotics, generalizing [15] from AWGN to AAN channels. The corresponding second-order asymptotics for lossy source coding and JSCC were derived by Zhou et al. [67, 68]. All above results are also ensemble tight. Very recently, advances in the achievability part of second-order asymptotics have been made for mismatched multiterminal communications [69, 70]. Specifically, Bai et al. [69] studied mismatched SR, Wang et al. [70] studied mismatched MAC and RAC, and Wu et al. [71] studied mismatched BC. In all studies, the codes and analyses for mismatched P2P communication [66, 67] have been generalized and combined with multidimensional analyses for the corresponding matched cases [46, 47, 54, 72].

1.3 Scope and Significance

This textbook aims to introduce recent advances in theoretical analyses of low-latency communications from the information theory society and demonstrate applications of these techniques to wireless communications and physical layer security. The book is divided into two parts: theory and applications. In the theory part, FBL analyses for P2P communication will be introduced, including source coding, channel coding and JSCC. Subsequently, the focus shifts to mismatched communication, introducing FBL analyses for both P2P and multiterminal settings. In the applications part, it will be illustrated how FBL analyses can be used in multiple-input multiple-output (MIMO) communication, lossy multi-connectivity, user scheduling, and covert communication. All contents will be presented with problem formulation, theorem statement, illustrating examples, and corresponding exercises. For most contents, either proof sketches or detailed proof steps will be provided.

This book fits well with the series of "Textbook in Telecommunications Engineering", because it covers topics on theoretical analyses of practical low-latency and ultra-reliable communication systems, which form the basis of the current 5G, B5G, and future 6G communications. Using this book, graduate and senior undergraduate students could learn the theoretical foundations and analytical tools of practical communication systems, while engineers in the communication domain could access accurate theoretical benchmarks to guide project designs. In terms of courses, this book could be used in part for a series of courses including communication theory, digital communications, principles of communications, wireless communications, advanced topics in information theory, source coding, channel coding, and multiuser information theory.

This book is the first textbook that systematically introduces theoretical benchmarks of low-latency communications and their applications to wireless communications beyond the classical but asymptotic Shannon theory. Traditionally, when talking about theoretical analyses for communications, one uses asymptotic notions such as capacity or entropy as the performance benchmark. However, in practical communications, low latency is necessary. It is thus impossible to use infinite blocklength codes that were used implicitly by Shannon to derive source/channel coding theorems. This way, asymptotic Shannon theory cannot provide accurate benchmarks for practical communication theorems despite its simplicity and elegance. To resolve the above issues, recent advances in information theory have developed tools to analyze the performance of optimal codes when the blocklength is short and finite. This book aims to introduce such results in detail, addressing practical scenarios where the additive noise is not necessarily Gaussian, and the distribution of the information source is not necessarily available. Furthermore, the book discusses applications of the information theoretical benchmarks to wireless communications systems, where quasi-static fading appears. Compared with existing monographs on low-latency communication and textbooks on information theory or wireless communications, this book complements the literature via its timely contents. The most related monographs are [16, 17]. The monograph [16] focuses on second-order asymptotics for matched P2P communications, with emphasis on channel coding. On the other hand, the monograph [17] focuses on matched source coding, and the only mismatched case covered is lossy source coding. This book complements [16, 17] by focusing on mismatched communication in both the P2P and multiterminal cases and by considering practical wireless communication systems with fading. A classical textbook for wireless communications is [28] by Tse and Viswanath, while two classical textbooks in information theory are [8, 9]. These three books set theoretical foundations for communication systems. The only deficiency is that all books are Shannon theoretical, most of whose results hold perfectly only when the latency is infinite with rare exceptions. This book complements the three classical textbooks by introducing recent theoretical analyses of wired and wireless communications under low-latency constraints.

1.4 Organization

The rest of the book is organized as follows. In Chap. 1.5, notations, definitions of information theoretical quantities, and mathematical tools are presented.

Chapter 2 presents FBL analyses of P2P matched communication, covering topics including lossless and lossy source coding, channel coding, and joint source-channel coding. This chapter sets the theoretical foundation for low-latency communications and builds upon [12–15, 22, 23, 33, 34]. A take-home message is that FBL analyses provide a tight approximation to the performance of optimal codes with short blocklengths, while classical Shannon theoretical results are inaccurate for low-latency communication.

Chapter 3 presents FBL analyses for P2P mismatched channel coding, where optimal codes for data transmission over an AWGN channel are used to transmit a message over an additive arbitrary noise channel. Two kinds of code books including i.i.d. Gaussian codebooks and spherical codebooks are studied and compared in terms of low-latency communication performance. This chapter builds up on [63, 66]. A take-home message is that although both i.i.d. Gaussian and spherical codebooks achieve the same infinite latency performance, the low-latency performance of spherical codebooks is significantly better.

Chapter 4 presents FBL analyses for P2P mismatched JSCC. A novel code is proposed by combining codes for mismatched channel coding [63, 66] and source coding [64, 67] via the idea of unequal error protection [73]. The coding scheme can be used to transmit an arbitrary memoryless source over an additive arbitrary noise channel, and the low-latency performance of such a coding scheme is analyzed. This chapter builds up on [68]. A take-home message is that separate source-channel coding is optimal when the latency is infinite for mismatched communication, while the low-latency performance of mismatched JSCC is much better.

Chapter 5 presents FBL analyses for mismatched successive refinement. Successive refinement is a classical multiterminal lossy source coding problem and finds diverse applications in image and video compression. A mismatched coding scheme is constructed by generalizing the codes for the P2P case and adopting the idea of layered encoding. The coding scheme can be used to compress an arbitrary memoryless source into several copies so that different users can reconstruct a different version of the source with particular tolerable distortion levels. This chapter builds up on [47, 69]. A take-home message is that i.i.d. Gaussian codebooks have better low-latency performance than spherical codebooks.

Chapter 6 presents FBL analyses for mismatched multiterminal transmission including MAC, RAC, and BC. MAC and RAC are models for uplink multiuser communication, while BC models downlink multiuser communication. Mismatched codes are proposed using the P2P case and the idea of either successive interference cancellation (SIC) or joint decoding (JD). The low-latency performance of all three problems is presented. This chapter builds up on [70, 71]. One take-home message exists for each problem: for MAC, JD has better low-latency performance than SIC; for RAC, JD has better performance regardless of latency; for BC, both JD and SIC achieve the same low-latency performance.

Chapter 7 presents FBL analyses of P2P MIMO communication with quasi-static fading. MIMO models practical communication scenarios where both the transmitter and the receiver employ multiple antennas, while quasi-static fading models channels for low-latency wireless communications. The channel state information (CSI) is a random variable that specifies how quasi-static fading affects the transmission of channel inputs. This chapter presents low-latency performance of optimal codes of the matched case for all four possibilities of CSI availability and builds up on [30]. A take-home message is that the asymptotic notion of outage probability and outage capacity is still valid performance benchmarks for low latency wireless communication since the second-order asymptotics equals zero for quasi-static fading channels.

Chapter 8 presents FBL analyses of P2P lossy multi-connectivity with quasi-static fading. Multi-connectivity models the case where multiple independent transmission links are available between the transmitter and the receiver, which is a special case of MIMO where all antennas transmit the same channel input and the CSI matrix is full rank. This chapter considers JSCC and thus generalizes Chap. 7, although for a slightly restricted setting. The main result of this chapter is the low-latency performance for lossy multi-connectivity and builds up on [35]. A take-home message is that two independent links improve transmission reliability significantly while the performance gain of further increasing the number of links degrades.

Chapter 9 presents an application of FBL results for user scheduling in a MAC. User scheduling is a typical problem in multiuser communication, especially for base stations or WIFI. The goal is to select users with good channel quality so that the overall performance is maximized. Under the quasi-static fading model, this chapter presents an attempt in user scheduling, where CSI is either perfectly known or only known from its distribution. Scheduling algorithms for both cases are presented and numerically simulated to demonstrate its convergence and good performance. This chapter builds up [74]. A take-home message is that low-complexity well designed scheduling algorithms can achieve close to optimal performance in terms of outage capacity and outage probability.

Chapter 10 presents FBL analyses of covert communication. Covert communication [75–78] is a new branch in physical layer security, where the communication behavior is required to be ignorant to malicious unauthorized users. In this line of studies, the famous square root law states that given a certain blocklength, the maximal number of bits that can be transmitted reliability and covertly scales in the order of the square root of the blocklength. This chapter presents low-latency performance benchmark for covert communication and builds up on [75, 76, 79]. A take-home message is that for covert communication, the traditional first-order asymptotics equals zero, and the traditional second-order asymptotics becomes the first dominant term.

1.5 Notation, Definitions, and Mathematical Tools

1.5.1 Notation

The set of real numbers, non-negative real numbers, and natural numbers are denoted by $\mathbb{R}$, $\mathbb{R}_+$, and $\mathbb{N}$, respectively. For any two natural numbers $(a, b) \in \mathbb{N}^2$, let $[a : b]$ denote the set of all natural numbers between a and b (inclusive) and let $[a]$ denote $[1 : a]$. For any integers $(a, b) \in \mathbb{N}^2$, let $\mathbf{0}_a$ denote the length-a vector of all zeroes and use $\mathbf{1}_{a,b}$ to denote the $a \times b$ matrix of all ones. Furthermore, for any two integers $(a, b) \in \mathbb{N}^2$ such that $a < b$, let $[b]_a$ denote the set of vectors that contain exactly a unique elements of $[b]$ and let $\mathcal{F}(a, b)$ the set of bijective mappings from $[b]_a$ to $[b]_a$. For any real number $a \in \mathbb{R}$, define $|a|^+$ as $\max\{a, 0\}$. Given any positive integer $m \in \mathbb{N}$, let $\mathbb{R}_+^m \subset \mathbb{R}^m$ be the nonnegative orthant of the m-dimensional Euclidean space, and let $\mathbb{R}_\geq^m \subseteq \mathbb{R}_+^m$ be defined as $\mathbb{R}_\geq^m := \{x^n \in \mathbb{R}_+^m : x_1 \geq \cdots \geq x_m\}$.

Random variables and their realizations are in capital (e.g., X) and lower case (e.g., x), respectively. All sets (e.g., alphabets of random variables) are denoted in calligraphic font (e.g., $\mathcal{X}$). The complement of $\mathcal{X}$ is denoted as $\mathcal{X}^{\mathrm{c}}$. Let $X^n := (X_1, \ldots, X_n)$ be a random vector of length-n and $x^n = (x_1, \ldots, x_n)$ be a particular realization. Let $\|x^n\| = \sqrt{\sum_{i\in[n]} x_i^2}$ denote the ℓ_2 norm of a vector $x^n \in \mathbb{R}^n$. Given two sequences x^n and y^n, the *quadratic distortion measure* (squared Euclidean norm) is defined as $d(x^n, y^n) := \frac{1}{n}\|x^n - y^n\|^2 = \frac{1}{n}\sum_{i\in[n]}(x_i - y_i)^2$. For any $a \in (0, 1)$, the binary entropy function is $H_{\mathrm{b}}(p) = -p \log p - (1 - p) \log(1 - p)$. Given two n-dimensional vectors A^n and B^n, $A^n \geq B^n$ means that for each $i \in [n]$, $A_i \geq B_i$. For matrices, upper case letters of two special fonts are used to denote deterministic matrices (e.g., Y) and random matrices (e.g., $\mathbb{Y}$). Let $\mathrm{tr}(\mathsf{A})$ and $\det(\mathsf{A})$ be the trace and determinant of the matrix A, respectively, and let $\mathrm{span}(\mathsf{A})$ be the subspace spanned by the column vectors of A. The Frobenius norm of a matrix A is denoted by $\|\mathsf{A}\|_{\mathsf{F}} := \sqrt{\mathrm{tr}(\mathsf{A}\mathsf{A}^{\mathrm{H}})}$. The notation $\mathsf{A} \succeq \mathbf{0}$ means that the matrix A is positive semi-definite. Let I_a be the identity matrix of size $a \times a$, and designate by $\mathsf{I}_{a,b}(a > b)$ the $a \times b$ matrix containing the first b columns of I_a.

The set of all probability distributions on an alphabet $\mathcal{X}$ is denoted by $\mathcal{P}(\mathcal{X})$ and the set of all conditional probability distribution from $\mathcal{X}$ to $\mathcal{Y}$ is denoted by $\mathcal{P}(\mathcal{Y}|\mathcal{X})$. Given $P \in \mathcal{P}(\mathcal{X})$, let $\mathrm{supp}(P)$ denote the support of distribution P, i.e., $\mathrm{supp}(P) = \{x \in \mathcal{X} : P(x) > 0\}$. Given a conditional distribution $P_{Y|X} \in \mathcal{P}(\mathcal{Y}|\mathcal{X})$ and $x \in \mathcal{X}$, let $P_{Y|x}$ denote the conditional distribution $P_{Y|X}(\cdot|x)$. Given $P \in \mathcal{P}(\mathcal{X})$ and $V \in \mathcal{P}(\mathcal{Y}|\mathcal{X})$, let $P \times V$ denote the joint distribution induced by P and V.

The standard asymptotic notations are used, such as $\Theta(\cdot)$, $O(\cdot)$ and $o(\cdot)$ (cf. [80]). Let $\mathbb{1}(\cdot)$ be the indicator function, and let $\log(\cdot)$ be with base e unless otherwise stated. For any two sequences $\{a_n\}_{n\in\mathbb{N}}$ and $\{b_n\}_{n\in\mathbb{N}}$, $a_n \sim b_n$ means $\lim_{n\to\infty} a_n/b_n = 1$. For any $(x, \mu, \sigma) \in \mathbb{R}^2 \times \mathbb{R}_+$, let $\mathcal{N}(x; \mu, \sigma^2)$ denote the probability density function (pdf) of a Gaussian random variable with mean μ and variance σ^2. The distribution of a circularly-symmetric complex Gaussian random vector with covariance matrix A is denoted by $\mathcal{CN}(\mathbf{0}, \mathsf{A})$. Given any $t \in \mathbb{R}$, let $\Phi(t) := \int_{-\infty}^{t} \mathcal{N}(x; 0, 1)\mathrm{d}x$ and $\mathrm{Q}(t) := \int_{t}^{\infty} \mathcal{N}(x; 0, 1)\mathrm{d}x$ be the cumulative distribution function (cdf) and complementary cdf of the standard Gaussian. Let $\Phi^{-1}(\cdot)$ and $\mathrm{Q}^{-1}(\cdot)$ be the inverse of $\Phi(\cdot)$ and $\mathrm{Q}(\cdot)$, respectively. Given any integer $k \in \mathbb{N}$, vector $\mathbf{x} = (x_1, \ldots, x_k) \in \mathbb{R}^k$, $\boldsymbol{\mu} = (\mu_1, \ldots, \mu_k) \in \mathbb{R}^k$ and positive semidefinite matrix $\boldsymbol{\Sigma} \in \mathbb{R}^{k \times k}$, let $\Psi_k(\mathbf{x}; \boldsymbol{\mu}, \boldsymbol{\Sigma})$ denote the multivariate generalization of the Gaussian cumulative distribution function (cdf), i.e., $\Psi(\mathbf{x}; \boldsymbol{\mu}, \boldsymbol{\Sigma}) = \int_{-\infty}^{x_1} \cdots \int_{-\infty}^{x_k} \mathcal{N}(\mathbf{x}; \boldsymbol{\mu}; \boldsymbol{\Sigma})\, \mathrm{d}\mathbf{x}$, where $\mathcal{N}(\mathbf{x}; \boldsymbol{\mu}; \boldsymbol{\Sigma})$ is the pdf of a k-variate Gaussian vector $\mathbf{x}$ with mean vector $\boldsymbol{\mu}$ and covariance matrix $\boldsymbol{\Sigma}$.

1.5.2 Definitions

To smoothly present the results in this monograph, necessary information theoretical definitions are recalled in this section. Given any distribution $P_X \in \mathcal{P}(\mathcal{X})$ defined on a finite alphabet $\mathcal{X}$, the entropy is defined as

$$H(X) = H(P_X) := \sum_{x \in \mathrm{supp}(P_X)} -P_X(x) \log P_X(x). \tag{1.1}$$

Note that the notation $H(X)$ is used in classical textbooks as [8] and the notation $H(P_X)$ that clarifies the dependence of the entropy on the distribution is used in [10]. Both notations for the entropy and other information theoretical quantities are used interchangeably. Specifically, the distribution dependence version $H(P_X)$ is used when one needs to specify the distribution of a random variable, while $H(X)$ is used when the distribution of the random variable is clear. Analogously, given a joint probability mass function (PMF) $P_{XY} \in \mathcal{P}(\mathcal{X} \times \mathcal{Y})$ defined on a finite alphabet $\mathcal{X} \times \mathcal{Y}$, the joint entropy is defined as

$$H(X, Y) = H(P_{XY}) = \sum_{(x,y) \in \mathrm{supp}(P_{XY})} -P_{XY}(x, y) \log P_{XY}(x, y), \tag{1.2}$$

and the conditional entropy of Y given X is defined as

$$H(Y|X) = H(P_{Y|X}|P_X) = \sum_{(x,y) \in \mathrm{supp}(P_{XY})} -P_{XY}(x, y) \log P_{Y|X}(x, y), \tag{1.3}$$

where $(P_{Y|X}, P_X)$ are the induced conditional and marginal distributions of P_{XY}. The conditional entropy $H(P_{Y|X}|P_X)$ of X given Y is defined similarly.

Furthermore, the mutual information that measures dependence of two random variables (X, Y) with distribution P_{XY} is defined as

$$I(X; Y) = I(P_X, P_{X|Y}) = H(P_X) - H(P_{X|Y}|P_Y), \tag{1.4}$$

where $P_{X|Y}$ is also induced by P_{XY}. Note that mutual information $I(X;Y)$ is symmetric so that $I(X;Y) = I(Y;X)$. Similar to the definition of entropy, $I(X;Y)$ and the distribution dependence version $I(P_X, P_{X|Y})$ are used interchangeably. Analogously, given the joint distribution P_{XYZ} of three random variables (X, Y, Z) defined on a finite alphabet $\mathcal{X} \times \mathcal{Y} \times \mathcal{Z}$, define the conditional mutual information $I(X;Y|Z)$ as

$$I(X;Y|Z) = I(P_{X|Z}, P_{X|YZ}|P_Z) = H(P_{X|Z}|P_Z) - H(P_{X|YZ}|P_{YZ}), \tag{1.5}$$

where all distributions are induced by the joint distribution P_{XYZ}.

Another critical quantity is the Kullback-Leibler (KL) divergence, also known as the relative entropy. Given any two distributions (P_X, Q_X) defined on the finite alphabet $\mathcal{X}$, the KL divergence $D(P_X \| Q_X)$ is defined as

$$D(P_X \| Q_X) = \sum_{x \in \mathrm{supp}(P_X)} P_X(x) \log \frac{P_X(x)}{Q_X(x)}. \tag{1.6}$$

Note that $D(P_X \| Q_X)$ measures closeness of two distributions P_X and Q_X and equals zero if and only if $P_X = Q_X$.

For any two distributions P_{XY} and Q_{XY} defined on a finite alphabet $\mathcal{X} \times \mathcal{Y}$, the KL divergence $D(P_{XY} \| Q_{XY})$ is defined similarly; when the marginal distributions $P_X = Q_X$, the conditional KL divergence is defined as

$$D(P_{Y|X} \| Q_{Y|X}|P_X) = \sum_{x \in \mathrm{supp}(P_X)} P_X(x) D(P_{Y|X}(\cdot|x) \| Q_{Y|X}(\cdot|x)). \tag{1.7}$$

Similar to KL divergence, total variation distance (TVD) is also a common measurement of the closeness of two distributions. Given any two distributions (P_X, Q_X) defined on the finite alphabet $\mathcal{X}$, the TV distance $\mathrm{TV}(P_X, Q_X)$ is defined as

$$\mathrm{TV}(P_X, Q_X) = \frac{1}{2} \sum_{x \in \mathcal{X}} |P_X(x) - Q_X(x)|. \tag{1.8}$$

1.5.3 Mathematical Tools

This section presents mathematical tools used to derive second-order asymptotics, which are essentially the generalization of central limit theorems. Fix any integer $n \in \mathbb{N}$ and an alphabet $\mathcal{X}$. Let $X^n = (X_1, \ldots, X_n) \in \mathcal{X}^n$ be a collection of n i.i.d. random variables with zero mean and variance σ^2 and let the normalized sum of these n random variables be

$$S_n := \frac{1}{n} \sum_{i \in [n]} X_i. \tag{1.9}$$

First recall the weak law of large numbers [81], which states that the normalized sum S_n converges in probability to its mean.

Theorem 1.1 (The Weak Law of Large Numbers) *For any positive real number $\delta \in \mathbb{R}_+$,*

$$\lim_{n \to \infty} \Pr\{S_n > \delta\} = 0. \tag{1.10}$$

In the proofs of theorems, the Markov inequality and Fano's inequality are usually used.

Theorem 1.2 (Markov Inequality) *For any non-negative real number $\theta \in \mathbb{R}_+$ and any positive real number t,*

$$\Pr\{S_n > t\} \le \frac{\mathbb{E}[\exp(\theta S_n)]}{\exp(t\theta)}. \tag{1.11}$$

Theorem 1.3 (Fano's Inequality) *Let $\mathcal{X}$ be finite. Fix an alphabet $\mathcal{Y}$ and any three random variables $(X, \hat{X}, Y) \in \mathcal{X}^2 \times \mathcal{Y}$ that satisfy the Markov chain $X - Y - \hat{X}$. It follows that*

$$1 + \Pr\{X \neq \hat{X}\} \log |\mathcal{X}| \geq H(X|Y). \tag{1.12}$$

In ensemble converse proof of mismatched communication, the strong large deviations theorem [82, Theorem 3.7.4] will be applied. Recall that $X^n = (X_1, \ldots, X_n)$ is a sequence of i.i.d. random variables. Given any real number $\lambda \in \mathbb{R}$, define log moment generating function

$$\Lambda(\lambda) := \log \mathbb{E}[\exp(\lambda X_1)], \tag{1.13}$$

$$\Lambda^*(\lambda) := \sup_{s \geq 0} \{s\lambda - \Lambda(s)\}. \tag{1.14}$$

Theorem 1.4 (Strong Large Deviations) *There exists a sequence positive real numbers $\{\eta_n\}_{n \in \mathbb{N}}$ in the order of $\Theta\big(n^{-\frac{1}{2}}\big)$ such that*

$$\lim_{n \to \infty} \frac{\Pr\left\{\frac{1}{n} \sum_{i \in [n]} X_i \geq \lambda\right\}}{\eta_n \exp(-n\Lambda^*(\lambda))} = 1. \tag{1.15}$$

The Berry-Esseen Theorem for i.i.d. random variables [83, 84], a refined version of the central limit theorem, is usually applied to refined asymptotics including second-order and third-order. Assume that the third absolute moment of X_1 is finite, i.e., $T := \mathbb{E}[|X_1|^3] < \infty$.

Theorem 1.5 (The Berry-Esseen Theorem) *For each $n \in \mathbb{N}$,*

$$\sup_{t\in\mathbb{R}} \left| \Pr\left\{ S_n \geq t\sqrt{\frac{\sigma^2}{n}} \right\} - \mathrm{Q}(t) \right| \leq \frac{T}{\sigma^3\sqrt{n}}. \tag{1.16}$$

The Berry-Esseen theorem states that the probability that the normalized sum S_n deviates from its mean by a sequence which scales as $\Theta\left(\frac{1}{\sqrt{n}}\right)$ is well approximated by the same probability for a standard normal variable, with the difference in the order of $O\left(\frac{1}{\sqrt{n}}\right)$ that depends on the variance σ^2 and the third absolute moment T. The assumption that T is finite is satisfied for DMS. It is the mathematical theorem that one applies in the analysis of second-order asymptotics for source and channel coding problems that involve a single encoder.

For certain problems, independent but not identically distributed (i.n.i.d.) random variables are used. Let $X^n = (X_1, \ldots, X_n)$ be a sequence of random variables, where each random variable X_i has zero mean, variance $\sigma_i^2 := \mathbb{E}[X_i^2] > 0$, and finite third-absolute moment $T_i := \mathbb{E}[|X_i|^3]$. Define the average variance and third-absolute moment as follows:

$$\sigma^2 := \frac{1}{n} \sum_{i\in[n]} \sigma_i^2, \tag{1.17}$$

$$T := \frac{1}{n} \sum_{i\in[n]} T_i \tag{1.18}$$

The Berry-Esseen theorem for i.n.i.d. random variables states as follows.

Theorem 1.6 (General Version of the Berry-Esseen Theorem) *For each $n \in \mathbb{N}$,*

$$\sup_{t\in\mathbb{R}} \left| \Pr\left\{ S_n \geq t\sqrt{\frac{\sigma^2}{n}} \right\} - \mathrm{Q}(t) \right| \leq \frac{6T}{\sigma^3\sqrt{n}}. \tag{1.19}$$

To derive low-latency performance for multiterminal communication problems with multiple encoders, the following multivariate generalization of the Berry-Esseen theorem [85]. Fix any integer $d \in \mathbb{N}$. For each $i \in [n]$, let $\mathbf{X}_i = (X_{i,1}, \ldots, X_{i,d})$ be a d-dimensional random vector with zero mean vector and covariance matrix $\mathbf{\Sigma}$ and let $\mathbf{t} = (t_1, \ldots, t_d) \in \mathbb{R}^d$ be any deterministic vector. Given any integer $n \in \mathbb{N}$, let the normalized sum vector be $\mathbf{S}_n := \frac{1}{\sqrt{n}} \sum_{i\in[n]} \mathbf{X}_i$. Let the third absolute moment of $\mathbf{X}_1$ be $T := \mathbb{E}[\|\mathbf{X}_1\|^3]$.

Theorem 1.7 (Vector Version of the Berry-Esseen Theorem) *For each $n \in \mathbb{N}$, it follows that*

$$\sup_{\mathbf{t}\in\mathbb{R}^d} \left| \Pr\{\mathbf{S}_n \leq \mathbf{t}\} - \Psi_d(\mathbf{t}; \mathbf{0}_d, \mathbf{\Sigma}) \right| \leq \frac{K(d)T}{\sqrt{n}}, \tag{1.20}$$

where $>$ refers to elementwise comparison and $K(d)$ is a constant depending on the dimension d only (see [86, 87] for explicit bounds).

To derive low-latency performance for mismatched communication, the following Berry-Esseen theorem for functions by MolavianJazi and Laneman [54, Prop. 1] is used to bound error probabilities. Fix any integer $l \in \mathbb{N}$. For each $i \in [n]$, let $\mathbf{X}_i = (X_{i,1}, \ldots, X_{i,d})$ be a d-dimensional random vector generated from any given pdf. Furthermore, let $\mathbf{t} = (t_1, \ldots, t_d) \in \mathbb{R}^d$ be any vector and let $\mathbf{f}(\mathbf{t}) = (f_1(\mathbf{t}), \ldots, f_l(\mathbf{t})) : \mathbb{R}^d \to \mathbb{R}^l$ be a l-component vector-function with continuous second-order partial derivatives in a neighborhood of $\mathbf{t} = \mathbb{E}[\mathbf{X}_1]$ whose Jacobian matrix $\mathbf{J} = \{J_{i,j}\}_{i\in[l],j\in[d]}$ satisfies

$$J_{i,j} := \left.\frac{\partial f_i(\mathbf{t})}{\partial t_j}\right|_{\mathbf{t}=\mathbb{E}[\mathbf{X}_1]}. \tag{1.21}$$

Assume that the third absolute moment of the random vector is finite, i.e., $\mathbb{E}[\|\mathbf{X}_1 - \mathbb{E}[\mathbf{X}_1]\|^3] < \infty$, and assume that the covariance matrix $\mathsf{Cov}[\mathbf{J}\mathbf{X}_1]$ is positive definite.

Theorem 1.8 (Function Version of the Berry-Esseen Theorem) *There exists a finite positive constant $B \in \mathbb{R}_+$ such that, for any convex Borel-measurable set $\mathcal{D} \in \mathbb{R}^l$ and for all $n \in \mathbb{N}_+$,*

$$\left| \Pr\left\{ \sqrt{n}\left(\mathbf{f}\left(\frac{1}{n}\sum_{t=1}^{n}\mathbf{X}_t\right) - \mathbf{f}(\mathbb{E}[\mathbf{X}_1])\right) \in \mathcal{D}\right\} - \Pr\{Z^l \in \mathcal{D}\} \right| \leq \frac{B}{\sqrt[4]{n}}, \tag{1.22}$$

where $Z^l \sim \mathcal{N}(\mathbf{0}_l, \mathbf{J}\,\mathsf{Cov}[\mathbf{X}_1]\mathbf{J}^{\mathrm{T}})$ is a l-dimensional Gaussian random vector.

1.6 Exercises

1.1 Why is low-latency desired in practical communication systems? What is the requirement of codes in low-latency communication?

1.2 Explain why Shannon theory is inaccurate for practical communications including 5G and 6G.

1.3 What is mismatched communication? Why is mismatched communication important?

1.4 Fix any $p \in (0, 1)$. Let $\mathcal{X} = \mathcal{Y} = \{0, 1\}$, $P_X = [1/2, 1/2]$ and $P_{Y|X} = p\mathbb{1}(Y = X) + (1 - p)\mathbb{1}(Y \neq X)$. Calculate the entropy $H(P_X)$, $H(P_Y)$, the conditional entropy $H(P_{X|Y})$, $H(P_{Y|X})$ and the mutual information $I(P_X, P_{X|Y})$.

1.5 Prove the Markov inequality in Theorem 1.2.

1.6 Fix $n \in \mathbb{N}$ and $p \in (0, 1)$. Let $X^n = (X_1, \ldots, X_n)$ be a sequence of i.i.d. random variables generated from the Bernoulli distribution with $P_X(1) = p$. Let $S_n = \frac{1}{n}\sum_{i\in[n]}(X_i - \mathbb{E}[X_i])$. Calculate upper and lower bounds of $\Pr\{S_n \geq t\sqrt{\frac{\sigma^2}{n}}\}$ using the Berry-Esseen theorem in Theorem 1.5 and specify the third-absolute moment T.

References

1. P. Popovski, Č. Stefanović, J.J. Nielsen, E. de Carvalho, M. Angjelichinoski, K.F. Trillingsgaard, A.-S. Bana, Wireless Access in Ultra-Reliable Low-Latency Communication (URLLC). IEEE Trans. Commun. **67**(8), 5783–5801 (2019)
2. N.H. Mahmood, I. Atzeni, E.A. Jorswieck, O.L. Alcaraz López, Ultra-reliable low-latency communications: foundations, enablers, system design, and evolution towards 6G. Found. Trends Commun. Inf. Theory **20**(5–6), 512–747 (2023)
3. K.B. Letaief, W. Chen, Y. Shi, J. Zhang, Y.-J.A. Zhang, The roadmap to 6G: AI empowered wireless networks. IEEE Commun. Mag. **57**(8), 84–90 (2019)
4. M.C. Coşkun, G. Durisi, T. Jerkovits, G. Liva, W. Ryan, B. Stein, F. Steiner, Efficient error-correcting codes in the short blocklength regime. Phys. Commun. **34**, 66–79 (2019)
5. C.E. Shannon, A mathematical theory of communication. Bell Syst. Techn. J. **27**(1), 379–423 (1948)
6. R. Gallager, Low-density parity-check codes. IRE Trans. Inf. Theory **8**(1), 21–28 (1962)
7. E. Arikan, Channel polarization: A method for constructing capacity-achieving codes for symmetric binary-input memoryless channels. IEEE Trans. Inf. Theory **55**(7), 3051–3073 (2009)
8. M.T.C.A.J. Thomas, A. Thomas, *Elements of Information Theory* (Wiley-Interscience, Hoboken, 2006)
9. A. El Gamal, Y.-H. Kim. *Network Information Theory* (Cambridge University Press, Cambridge, 2011)
10. I. Csiszár, J. Körner, *Information Theory: Coding Theorems for Discrete Memoryless Systems* (Cambridge University Press, Cambridge, 2011)
11. A.A. Yushkevich, On limit theorems connected with the concept of entropy of Markov chains. Uspekhi Matematicheskikh Nauk **8**(5), 177–180 (1953)
12. V. Strassen, Asymptotische Abschätzungen in Shannon's informationstheorie, in *Transactions of the Third Prague Conference on Information Theory etc*, Czechoslovak Academy of Sciences (1962), pp. 689–723
13. M. Hayashi, Second-order asymptotics in fixed-length source coding and intrinsic randomness. IEEE Trans. Inf. Theory **54**(10), 4619–4637 (2008)
14. M. Hayashi, Information spectrum approach to second-order coding rate in channel coding. IEEE Trans. Inf. Theory **55**(11), 4947–4966 (2009)
15. Y. Polyanskiy, H.V. Poor, S. Verdú, Channel coding rate in the finite blocklength regime. IEEE Trans. Inf. Theory **56**(5), 2307–2359 (2010)
16. V.Y.F. Tan, Asymptotic estimates in information theory with non-vanishing error probabilities. Found. Trend. Commun. Inf. Theory **11**(1–2), 1–184 (2014)

17. L. Zhou, M. Motani, Finite blocklength lossy source coding for discrete memoryless sources. Found. Trend. Commun. Inf. Theory **20**(3), 157–389 (2023)
18. I. Kontoyiannis, S. Verdú, Optimal lossless data compression: Non-asymptotics and asymptotics. IEEE Trans. Inf. Theory **60**(2), 777–795 (2014)
19. S. Chen, M. Effros, V. Kostina, Lossless source coding in the point-to-point, multiple access, and random access scenarios. IEEE Trans. Inf. Theory **66**(11), 6688–6722 (2020)
20. R.M. Gray, D.L. Neuhoff, Quantization. IEEE Trans. Inf. Theory **44**(6), 2325–2383 (1998)
21. C.E. Shannon, Coding theorems for a discrete source with a fidelity criterion. IRE Nat. Conv. Rec **4**, 142–163 (1959)
22. A. Ingber, Y. Kochman, The dispersion of lossy source coding, in *IEEE Data Compression Conference* (2011), pp. 53–62
23. V. Kostina, S. Verdú, Fixed-length lossy compression in the finite blocklength regime. IEEE Trans. Inf. Theory **58**(6), 3309–3338 (2012)
24. M. Tomamichel, V.Y.F. Tan, A tight upper bound for the third-order asymptotics for most discrete memoryless channels. IEEE Trans. Inf. Theory **59**(11), 7041–7051 (2013)
25. V.Y.F. Tan, M. Tomamichel, The third-order term in the normal approximation for the AWGN channel. IEEE Trans. Inf. Theory **61**(5), 2430–2438 (2015)
26. R.C. Yavas, V. Kostina, M. Effros, Third-order analysis of channel coding in the small-to-moderate deviations regime. IEEE Trans. Inf. Theory **70**(9), 6139–6170 (2024)
27. Y. Sakai, V.Y.F. Tan, M. Kovačević, Second- and third-order asymptotics of the continuous-time Poisson channel. IEEE Trans. Inf. Theory **66**(8), 4742–4760 (2020)
28. D.N.C. Tse, P. Viswanath, *Fundamentals of Wireless Communication* (Cambridge University Press, Cambridge, 2005)
29. Y. Polyanskiy, S. Verdú, Scalar coherent fading channel: Dispersion analysis, in *IEEE International Symposium on Information Theory Proceedings* (2011), pp. 2959–2963
30. W. Yang, G. Durisi, T. Koch, Y. Polyanskiy, Quasi-static multiple-antenna fading channels at finite blocklength. IEEE Trans. Inf. Theory **60**(7), 4232–4265 (2014)
31. W. Yang, G. Durisi, T. Koch, Y. Polyanskiy, Block-fading channels at finite blocklength, in *Proceedings of the ISWCS 2013; The Tenth International Symposium on Wireless Communication Systems*, Ilmenau (2013), pp. 1–4
32. J. Hoydis, R. Couillet, P. Piantanida, The second-order coding rate of the MIMO quasi-static Rayleigh fading channel. IEEE Trans. Inf. Theory **61**(12), 6591–6622 (2015)
33. D. Wang, A. Ingber, Y. Kochman, The dispersion of joint source-channel coding, in *49th Annual Allerton Conference on Communication, Control, and Computing* (2011), pp. 180–187
34. V. Kostina, S. Verdú, Lossy joint source-channel coding in the finite blocklength regime. IEEE Trans. Inf. Theory **59**(5), 2545–2575 (2013)
35. L. Zhou, A. Wolf, M. Motani, On lossy multi-connectivity: finite blocklength performance and second-order asymptotics. IEEE J. Sel. Areas Commun. **37**(4), 735–748 (2019)
36. D. Slepian, J.K. Wolf, Noiseless coding of correlated information sources. IEEE Trans. Inf. Theory **19**(4), 471–480 (1973)
37. V.Y.F. Tan, O. Kosut, On the dispersions of three network information theory problems. IEEE Trans. Inf. Theory **60**(2), 881–903 (2014)
38. R. Nomura, T.S. Han, Second-order Slepian-Wolf coding theorems for non-mixed and mixed sources. IEEE Trans. Inf. Theory **60**(9), 5553–5572 (2014)
39. R. Gray, A. Wyner, Source coding for a simple network. Bell Syst. Tech. J. **53**(9), 1681–1721 (1974)
40. S. Watanabe, Second-order region for Gray-Wyner network. IEEE Trans. Inf. Theory **63**(2), 1006–1018 (2017)
41. L. Zhou, V.Y.F. Tan, M. Motani, Discrete lossy Gray-Wyner revisited: second-order asymptotics, large and moderate deviations. IEEE Trans. Inf. Theory **63**(3), 1766–1791 (2017)
42. F.W. Fu, R.W. Yeung, On the rate-distortion region for multiple descriptions. IEEE Trans. Inf. Theory **48**(7), 2012–2021 (2002)
43. L. Zhou, M. Motani, Non-asymptotic converse bounds and refined asymptotics for two source coding problems. IEEE Trans. Inf. Theory **65**(10), 6414–6440 (2019)

44. B. Rimoldi, Successive refinement of information: characterization of the achievable rates. IEEE Trans. Inf. Theory **40**(1), 253–259 (1994)
45. A. Kaspi, Rate-distortion function when side-information may be present at the decoder. IEEE Trans. Inf. Theory **40**(6), 2031–2034 (1994)
46. A. No, A. Ingber, T. Weissman, Strong successive refinability and rate-distortion-complexity tradeoff. IEEE Trans. Inf. Theory **62**(6), 3618–3635 (2016)
47. L. Zhou, V.Y.F. Tan, M. Motani, Second-order and moderate deviation asymptotics for successive refinement. IEEE Trans. Inf. Theory **63**(5), 2896–2921 (2017)
48. C.E. Shannon, Two-way communication channels. in *Proceedings of the Fourth Berkeley Symposium on Mathematical Statistics and Probability*, vol. 1 (University of California Press, 1961), pp. 611–645
49. R. Ahlswede, Multiway communication channels, in *IEEE ISIT*, Tsahkadsor (1971), pp. 23–51
50. H. Liao, Multiple access channels. Ph.D. dissertation, Dept. Elec. Eng., Univ. of Hawaii, 1972
51. T. Cover, Broadcast channels. IEEE Trans. Inf. Theory **18**(1), 2–14 (1972)
52. P. Bergmans, Random coding theorem for broadcast channels with degraded components. IEEE Trans. Inf. Theory **19**(2), 197–207 (1973)
53. J. Scarlett, V.Y.F. Tan, Second-order asymptotics for the Gaussian MAC with degraded message sets. IEEE Trans. Inf. Theory **61**(12), 6700–6718 (2015)
54. E. MolavianJazi, J.N. Laneman, A second-order achievable rate region for Gaussian multi-access channels via a central limit theorem for functions. IEEE Trans. Inf. Theory **61**(12), 6719–6733 (2015)
55. D. Tuninetti, P. Sheldon, B. Smida, N. Devroye, On second order rate regions for the static scalar gaussian broadcast channel. IEEE J. Sel. Areas Commun. **41**(7), 1982–1999 (2023)
56. P.-H. Lin, S.-C. Lin, P.-W. Chen, M.A. Mross, E.A. Jorswieck, Second order rate regions of gaussian broadcast channels under heterogeneous blocklength constraints. IEEE Trans. Commun. **72**(2), 801–814 (2024)
57. I. Csiszár, J. Körner, Graph decomposition: a new key to coding theorems. IEEE Trans. Inf. Theory **27**(1), 5–12 (1981)
58. J. Hui, Fundamental issues of multiple accessing. Ph.D. dissertation, MIT (1983)
59. G. Kaplan, S. Shamai, Information rates and error exponents of compound channels with application to antipodal signaling in a fading environment. Arch. Elek. Über. **47**(4), 228–239 (1993)
60. A. Ganti, A. Lapidoth, E. Telatar, Mismatched decoding revisited: General alphabets, channels with memory, and the wide-band limit. IEEE Trans. Inf. Theory **46**(7), 2315–2328 (2000)
61. J. Scarlett, A. Martinez, A. Guillên i Fàbregas, Mismatched decoding: error exponents, second-order rates and saddlepoint approximations. IEEE Trans. Inf. Theory **60**(5), 2647–2666 (2014)
62. J. Scarlett, A. Guillén i Fàbregas, A. Somekh-Baruch, A. Martinez, Information-theoretic foundations of mismatched decoding. Found. Trends Commun. Inf. Theory **17**(2–3), 149–401 (2020)
63. A. Lapidoth, Nearest neighbor decoding for additive non-Gaussian noise channels. IEEE Trans. Inf. Theory **42**(5), 1520–1529 (1996)
64. A. Lapidoth, On the role of mismatch in rate distortion theory. IEEE Trans. Inf. Theory **43**(1), 38–47 (1997)
65. R. Gallager, The random coding bound is tight for the average code (corresp.). IEEE Trans. Inf. Theory **19**(2), 244–246 (1973)
66. J. Scarlett, V.Y.F. Tan, G. Durisi, The dispersion of nearest-neighbor decoding for additive non-Gaussian channels. IEEE Trans. Inf. Theory **63**(1), 81–92 (2017)
67. L. Zhou, V.Y.F. Tan, M. Motani, Refined asymptotics for rate-distortion using Gaussian codebooks for arbitrary sources. IEEE Trans. Inf. Theory **65**(5), 3145–3159 (2019)
68. L. Zhou, V.Y.F. Tan, M. Motani, The dispersion of mismatched joint source-channel coding for arbitrary sources and additive channels. IEEE Trans. Inf. Theory **65**(4), 2234–2251 (2019)
69. L. Bai, Z. Wu, L. Zhou, Achievable refined asymptotics for successive refinement using gaussian codebooks. IEEE Trans. Inf. Theory **69**(6), 3525–3543 (2023)

70. Y. Wang, L. Bai, Z. Wu, L. Zhou, Achievable Second-Order Asymptotics for MAC and RAC with Additive Non-Gaussian Noise. arXiv: 2410.10312 (2024)
71. Z. Wu, L. Bai, J. Xu, L. Zhou, M. Motani, The dispersion of broadcast channels with degraded message sets using Gaussian codebooks. arXiv: 2410.17540 (2024)
72. R.C. Yavas, V. Kostina, M. Effros, Gaussian multiple and random access channels: finite-blocklength analysis. IEEE Trans. Inf. Theory **67**(11), 6983–7009 (2021)
73. Y.Y. Shkel, V.Y.F. Tan, S.C. Draper, Unequal message protection: asymptotic and non-asymptotic tradeoffs. IEEE Trans. Inf. Theory **61**(10), 5396–5416 (2015)
74. L. Zhu, L. Bai, L. Zhou, J. Choi, Efficient user scheduling for uplink hybrid satellite-terrestrial communication. IEEE Trans. Wirel. Commun. **22**(3), 1885–1899 (2023)
75. B.A. Bash, D. Goeckel, D. Towsley, Limits of reliable communication with low probability of detection on AWGN channels. IEEE J. Sel. Areas Commun. **31**(9), 1921–1930 (2013)
76. L. Wang, G.W. Wornell, L. Zheng, Fundamental limits of communication with low probability of detection. IEEE Trans. Inf. Theory **62**(6), 3493–3503 (2016)
77. M.R. Bloch, Covert communication over noisy channels: a resolvability perspective. IEEE Trans. Inf. Theory **62**(5), 2334–2354 (2016)
78. Q. Zhang, M. Bakshi, S. Jaggi, Covert communication with polynomial computational complexity. IEEE Trans. Inf. Theory **66**(3), 1354–1384 (2020)
79. X. Yu, S. Wei, S. Huang, X. Zhang, The First and Second Order Asymptotics of Covert Communication over AWGN Channels. arXiv:2305.17924 (2023)
80. T. Cormen, C. Leiserson, R. Rivest, C. Stein, *Introduction to Algorithms*, 2nd edn. (McGraw-Hill Science/Engineering/Math, Columbus, 2003)
81. W. Feller, Law of large numbers for identically distributed variables. Introd. Probab. Theory Appl. **2**, 231–234 (1971)
82. R.R. Bahadur, R.R. Rao, On deviations of the sample mean. Ann. Math. Statist. **31**(4), 1015–1027 (1960)
83. A.C. Berry, The accuracy of the Gaussian approximation to the sum of independent variates. Trans. Am. Math. Soc. **49**(1), 122–136 (1941)
84. C.G. Esseen, *On the Liapounoff Limit of Error in the Theory of Probability* (Almqvist & Wiksell, Stockholm, 1942)
85. F. Gotze, On the rate of convergence in the multivariate CLT. Ann. Probab. **19**, 724–739 (1991)
86. V. Bentkus, On the dependence of the Berry-Esseen bound on dimension. J. Stat. Plann. Infer. **113**, 385–402 (2003)
87. M. Raič, A multivariate Berry–Esseen theorem with explicit constants. Bernoulli **25**(4A), 2824–2853 (2019)

Chapter 2
Theoretical Foundations

This chapter sets the theoretical foundation of low-latency communications. Low latency is inherent in communication systems, while Shannon theory is asymptotic by providing accurate theoretical benchmarks only when the latency (blocklength) is infinite. Despite this mismatch, Shannon theory has been used to guide the system design in the past till 5G. In the past one and a half decades, motivated by the use of short blocklength codes to achieve low latency, the information theory society has rediscovered and developed tools to study finite blocklength (FBL) performance of optimal codes. This chapter presents such advances for the point-to-point case, including lossless and lossy source coding, channel coding, and joint source-channel coding (JSCC). For all problems, the FBL performance deviates from the asymptotic Shannon theoretical results as a function of the blocklength and the error probability. FBL analysis shows that Shannon's separation theorem fails at low latency: JSCC achieves significantly better performance than separate source-channel coding (SSCC). The proof techniques for FBL analysis are no longer typicality arguments pioneered by Shannon and used by a sequence of follow-up studies. Instead, the information spectrum method and novel non-asymptotic bounds are combined with the Berry-Esseen theorem to yield desired results. The proof in this chapter emphasizes these new bounds and tools.

2.1 Lossless Source Coding

Source coding aims to reduce redundancy from the information source so that efficient storage and transmission can be subsequently used. Depending on whether exact recovery is required, source coding can be categorized into lossless source coding and lossy source coding. Lossy source coding is also known as quantization or rate-distortion. In lossless source coding, the source sequence should be recovered exactly from its compressed version, while in lossy source coding, the

L. Zhou, L. Bai, *Low-latency Communication*, Textbooks in Telecommunication Engineering, https://doi.org/10.1007/978-3-032-06627-5_2

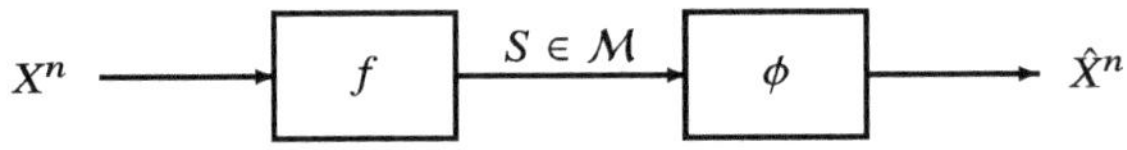

Fig. 2.1 System model for lossless source coding

reproduced source sequence can be different from the source sequence but should be within a certain distortion level under a given distortion function. In this section, the second-order asymptotics for lossless source coding is presented.

Shannon's well-known source coding theorem [1] states that entropy is the minimal compression rate to compress any discrete memoryless source (DMS) in lossless source coding under the condition that the error probability vanishes to zero as the blocklength tends to infinity. Note that infinite blocklength means infinite latency. Subsequently, Yushkevich [2] pioneered the study of low-latency lossless source coding, which was refined by Strassen [3] and rediscovered by Hayashi [4]. In particular, Hayashi coined the name of second-order coding rate and second-order asymptotics. As will be illustrated subsequently, the results of Shannon, Yushkevich, and Strassen characterize the first, second, and third dominant terms, respectively, for the finite blocklength rate of optimal codes for lossless source coding.

2.1.1 Problem Formulation

Let $\mathcal{X}$ be a finite alphabet and let $P_X \in \mathcal{P}(\mathcal{X})$ be an arbitrary probability mass function (PMF). Fix two integers $(n, M) \in \mathbb{N}^2$ and define the set of integers $\mathcal{M} := [M] = \{1, \ldots, M\}$. Assume that $X^n = (X_1, \ldots, X_n) \in \mathcal{X}^n$ is generated i.i.d. from P_X. As shown in Fig. 2.1, the task of lossless source coding is to compress X^n into an index $S \in \mathcal{M}$ and recover X^n perfectly from the compressed version S. A code specifies how the compression (encoding) and the recovery (decoding) are done. The performance metric of any code is the error probability that quantifies the probability of the error event, where the reproduced source sequence using the compressed version is not the source sequence X^n. Thus, there is a tradeoff among the blocklength n, the codebook size M, and the error probability. The theoretical study characterizes this tradeoff under various assumptions on the blocklength and error probabilities.

Definition 2.1 An (n, M)-code for lossless source coding consists of an encoder $f : \mathcal{X}^n \to \mathcal{M}$, and a decoder $\phi : \mathcal{M} \to \mathcal{X}^n$.

Consistent with literature, one can use $\hat{X}^n$ to denote the reproduced source sequence using the decoder, i.e., $\hat{X}^n = \phi(f(X^n))$.

To evaluate the performance of any (n, M)-code, two metrics are considered: efficiency and accuracy. The efficiency is measured via the compression rate $R_n := \frac{\log M}{n}$ and the accuracy is measured via the following error probability

$$\mathrm{P}_{\mathrm{e},n} := \Pr\{\hat{X}^n \neq X^n\}. \tag{2.1}$$

Thus, the study of lossless source coding is to characterize the tradeoff between the compression rate R_n and the error probability $\mathrm{P}_{e,n}$ given any blocklength $n \in \mathbb{N}$.

A trivial case is no compression when $M \geq |\mathcal{X}|^n$ and thus $R_n > \log |\mathcal{X}|$. This way, each source sequence can be mapped into a unique index, and the error probability will be exactly zero. However, any practical code should compress the source sequence and thus requires $M < |\mathcal{X}|^n$. This means that several source sequences will be mapped into the same index and the error probability will be non-zero. Fix any positive real number $\varepsilon \in (0, 1)$. The theoretical benchmark is the minimal compression rate given any blocklength n, subject to an error probability of at most ε, i.e.,

$$R^*(n, \varepsilon) := \inf\{R_n : \exists \text{ an } (n, M)\text{−code s.t. } \mathrm{P}_{e,n} \leq \varepsilon\}. \tag{2.2}$$

Theoretical studies of lossless source coding concern characterizing $R^*(n, \varepsilon)$. It follows from [1–3] that

$$R^*(n, \varepsilon) = a + \frac{b}{\sqrt{n}} + \frac{c \log n}{n} + O\left(\frac{1}{n}\right), \tag{2.3}$$

where $(a, b, c) \in \mathbb{R}^3$ are the coefficients for the first, second, and third dominant terms. Shannon [1] bounded the first-order term $\lim_{\varepsilon \to 0} \lim_{n \to \infty} R^*(n, \varepsilon)$ with vanishing error probability explicitly and show that $a = H(P_X)$ when $\varepsilon \to 0$. Yushkevich [2] bounded the first two terms for any $\varepsilon \in (0, 1)$, showing that $a = H(P_X)$ regardless of ε and characterized b as a function of the source distribution and the error probability. Strassen [3] further showed that $c = -\frac{1}{2}$. Finally, Kontoyiannis and Verdú [5, Section 5] and Chen, Kostina, and Effros [6, Theorem 5] bounded the $O\left(\frac{1}{n}\right)$ term explicitly.

Throughout the book, the second-order asymptotics that characterizes the first two dominant terms in the coding rate for various problems will be introduced, while the further refined analysis characterizing third-order terms and above is discussed when it is available for particular problems. Note that third-order asymptotics is challenging and only available for a small subset of communication problems.

2.1.2 Second-Order Asymptotics

In this section, the characterization of $R^*(n, \varepsilon)$ is presented with proofs, up to second dominant terms.

Given any $x \in \mathcal{X}$, define the *entropy density* as

$$\imath(x|P_X) := -\log P_X(x). \tag{2.4}$$

Note that the entropy $H(P_X)$ defined in (1.1) is the average of the entropy density random variable, i.e.,

$$H(P_X) = \mathbb{E}_{P_X}[\imath(X|P_X)]. \tag{2.5}$$

To present second-order asymptotics, the following *dispersion function* is needed:

$$V(P_X) := \mathrm{Var}[\imath(X|P_X)], \tag{2.6}$$

which is the variance of the entropy density random variable. Recall that $\mathrm{Q}^{-1}(\cdot)$ is the inverse function of the *complementary Gaussian cdf.*

Theorem 2.1 *For any* $\varepsilon \in (0, 1)$, *if the source dispersion function* $V(P_X)$ *is positive, the minimal compression rate* $R^*(n, \varepsilon)$ *satisfies*

$$R^*(n, \varepsilon) = H(P_X) + \frac{\sqrt{V(P_X)}\mathrm{Q}^{-1}(\varepsilon)}{\sqrt{n}} + O\left(\frac{\log n}{n}\right), \tag{2.7}$$

where the remainder term satisfies $-\frac{1}{2}\log\frac{\log n}{n} + O\left(\frac{1}{n}\right) \leq O\left(\frac{\log n}{n}\right) \leq O\left(\frac{1}{n}\right)$.

Shannon's source coding theorem is recovered from (2.7) of Theorem 2.1 with a stronger converse argument by letting $n \to \infty$, i.e., for any $\varepsilon \in (0, 1)$,

$$\lim_{n\to\infty} R^*(n, \varepsilon) = H(P_X). \tag{2.8}$$

In other words, the asymptotic minimal compression rate equals entropy, the first-order statistic of the entropy random variable, regardless of the error probability ε. Such a result is known as *strong converse*. Strong converse refines the weak converse argument of Shannon by showing that any code with an asymptotic rate less than entropy has error probability of one when the blocklength tends to infinity, while weak converse could only show that error probability is bounded away from zero under the same condition.

For simplicity, one can use $R_{\mathrm{soa}}(n, \varepsilon)$ to denote the first two terms in (2.7), i.e.,

$$R_{\mathrm{soa}}(n, \varepsilon) = H(P_X) + \frac{\sqrt{V(P_X)}\mathrm{Q}^{-1}(\varepsilon)}{\sqrt{n}}. \tag{2.9}$$

The formula $R_{\mathrm{soa}}(n, \varepsilon)$ is known as second-order asymptotics since it collects asymptotic expansion of the finite block length rate $R^*(n, \varepsilon)$ up to the second dominant term. Note that the first-order term $H(P_X)$ is the average of the entropy density random variable, while the second-order term depends on both the variance of the entropy density random variable and the error probability. The same logic generalizes to other problems discussed in this book, where the first two dominant terms depend on first- and second-order statistics of certain density random variables. In Fig. 2.2, for a Bernoulli source with parameter, with error probability

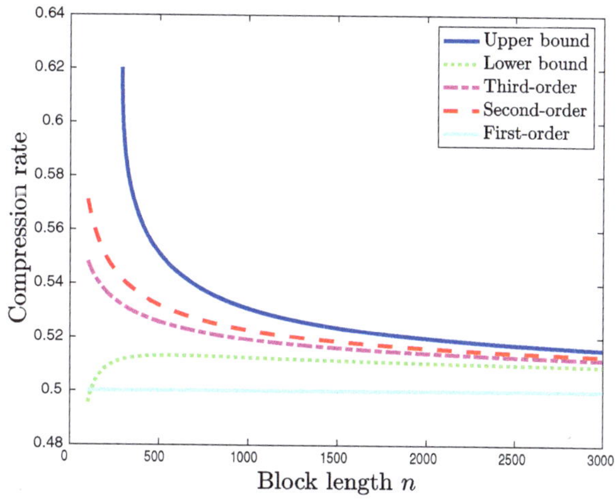

Fig. 2.2 Plot of upper and lower bounds, and first-second and third-order asymptotics for the minimal compression rate $R^*(n, \varepsilon)$ when $\varepsilon = 0.1$ for a Bernoulli source with $P_X = [0.2, 0.8]$

$\varepsilon = 0.15$, the second-order asymptotic approximation $R_{\rm soa}(n, \varepsilon)$ is plotted and compared with its finite blocklength upper and lower bounds in (2.29) and (2.49), respectively. It is observed that $R_{\rm soa}(n, \varepsilon)$ well approximates $R^*(n, \varepsilon)$ while the asymptotic Shannon bound $H(P_X)$ is quite away from $R^*(n, \varepsilon)$, illustrating the accuracy of second-order asymptotics and the inaccuracy of Shannon theory for low-latency communication.

The second-order asymptotics relates the blocklength, the compression rate, and the error probability. Theorem 2.1 presents bounds on the minimal compression rate $R^*(n, \varepsilon)$ as a function of the error probability ε and the blocklength n. This relationship can be also transformed to derive bounds on the error probability as a function of the blocklength and the compression rate. Using $R_{\rm soa}(n, \varepsilon)$ to approximate $R^*(n, \varepsilon)$, it follows from (2.9) that the minimal error probability $\mathrm{P}^*_{\mathrm{e},n}$ of any (n, M)-code is approximated by

$$\mathrm{P}^*_{\mathrm{e},n} \approx \mathrm{Q}\left(\frac{\sqrt{n}(R_n - H(P_X))}{\sqrt{V(P_X)}}\right), \tag{2.10}$$

where $R_n = \frac{\log M}{n}$ is the compression rate. The result in (2.10) implies the phenomenon of phase transition: the minimal error probability approaches one asymptotically as the blocklength increases if the asymptotic compression rate is below entropy, i.e., $\limsup_{n\to\infty} R_n < H(P_X)$, while the minimal error probability approaches zero asymptotically if $\liminf_{n\to\infty} R_n > H(P_X)$.

Upper and lower bounds on $\mathrm{P}^*_{\mathrm{e},n}$ can be derived using (2.18) and (2.44). Specifically, using (2.18) and the Berry-Esseen theorem (Theorem 1.5) yields

$$\mathrm{P}^*_{\mathrm{e},n} \le \Pr\left\{\sum_{i\in[n]} \imath(X_i|P_X) \ge nR_n\right\} \tag{2.11}$$

$$= \Pr\left\{\sum_{i\in[n]}(\imath(X_i|P_X) - H(P_X)) \geq n(R_n - H(P_X))\right\} \tag{2.12}$$

$$\leq \mathrm{Q}\left(\frac{\sqrt{n}(R_n - H(P_X))}{\sqrt{V(P_X)}}\right) + \frac{T(P_X)}{\sqrt{n(V(P_X)^3)}}. \tag{2.13}$$

Analogously, it follows from (2.44) with $\gamma = \frac{\log n}{2n}$ that

$$\mathrm{P}^*_{\mathrm{e},n} \geq \Pr\left\{\sum_{i\in[n]}\imath(X_i|P_X) \geq nR_n + \frac{1}{2}\log n\right\} - \frac{1}{\sqrt{n}} \tag{2.14}$$

$$\geq \mathrm{Q}\left(\frac{\sqrt{n}(R_n - H(P_X) + \frac{\log n}{2n})}{\sqrt{V(P_X)}}\right) - \frac{T(P_X)}{\sqrt{n(V(P_X)^3)}} - \frac{1}{\sqrt{n}}. \tag{2.15}$$

2.1.3 *Proof of Second-Order Asymptotics*

The proof of Theorem 2.1 consists of two parts: achievability and converse. In the achievability part, a code is constructed and proved to achieve an error probability of ε with the rate in (2.7). In the converse part, an impossibility result is proved for any code, showing that to achieve the given error probability ε, the compression rate can not be smaller. The proof presented below is based on [7, Chapters 2.2-2.3], using the information spectrum method [8].

2.1.3.1 Achievability

Fix any $(n, M) \in \mathbb{N}^2$. Define the following set

$$\mathcal{A}_n := \left\{x^n \in \mathcal{X}^n : \sum_{i\in[n]}\imath(x_i|P_X) < \log M\right\}. \tag{2.16}$$

The achievability part is based on the following coding scheme. Given any source sequence x^n, if $x^n \in \mathcal{A}_n$, assign a unique index for x^n; otherwise, the encoder declares an error by mapping x^n into a fixed number 0. This way, the error probability of the above code satisfies

$$\mathrm{P}_{\mathrm{e},n} = \Pr\{X^n \notin \mathcal{A}_n\} \tag{2.17}$$

$$= \Pr\left\{\sum_{i\in[n]}\imath(X_i|P_X) \geq \log M\right\}. \tag{2.18}$$

The rate R_n of the above code satisfies

$$R_n = \frac{\log(|\mathcal{A}_n| + 1)}{n} \le \frac{\log(M+1)}{n}, \tag{2.19}$$

where (2.19) holds since when $x^n \in \mathcal{A}_n$, $P_X^n(x^n) > \frac{1}{M}$ and thus

$$1 \ge \sum_{x^n \in \mathcal{A}_n} P_X^n(x^n) > \sum_{x^n \in \mathcal{A}_n} \frac{1}{M} = \frac{|\mathcal{A}_n|}{M}. \tag{2.20}$$

To apply the Berry-Esseen theorem (Theorem 1.5), define the following third absolute moment function:

$$T(P_X) := \mathbb{E}[|\imath(X|P_X) - H(P_X)|^3]. \tag{2.21}$$

Furthermore, given any $\varepsilon \in (0,1)$, let

$$\varepsilon_n = \varepsilon - \frac{T(P_X)}{\sqrt{n(V(P_X))^3}} \tag{2.22}$$

and choose M such that

$$\log M = nH(P_X) + \sqrt{nV(P_X)}\mathrm{Q}^{-1}(\varepsilon_n). \tag{2.23}$$

It follows from (2.18) that the error probability satisfies

$$\mathrm{P}_{\mathrm{e},n} \le \Pr\left\{ \sum_{i\in[n]} (\imath(X_i|P_X) - H(P_X)) \ge \sqrt{nV(P_X)}\mathrm{Q}^{-1}(\varepsilon_n) \right\} \tag{2.24}$$

$$= \Pr\left\{ \frac{1}{n} \sum_{i\in[n]} (\imath(X_i|P_X) - H(P_X)) \ge \sqrt{\frac{V(P_X)}{n}}\mathrm{Q}^{-1}(\varepsilon_n) \right\} \tag{2.25}$$

$$\le \varepsilon_n + \frac{T(P_X)}{\sqrt{n(V(P_X))^3}} \tag{2.26}$$

$$= \varepsilon. \tag{2.27}$$

Furthermore, it follows from (2.19) that

$$R_n \le \frac{\log(M+1)}{n} \tag{2.28}$$

$$= H(P_X) + \frac{\sqrt{V(P_X)}\mathrm{Q}^{-1}(\varepsilon_n)}{\sqrt{n}} + \frac{1}{n} \tag{2.29}$$

$$= H(P_X) + \frac{\sqrt{V(P_X)}\mathrm{Q}^{-1}(\varepsilon)}{\sqrt{n}} + O\Big(\frac{1}{n}\Big). \tag{2.30}$$

where (2.30) follows from the Taylor expansion of $\mathrm{Q}^{-1}(\varepsilon_n)$ around ε. Specifically, it follows from [6, Eq. (51)-(53)] that

$$\mathrm{Q}^{-1}(\varepsilon_n) = \Phi^{-1}(1-\varepsilon_n) \tag{2.31}$$

$$= \Phi^{-1}\left(1-\varepsilon+\frac{T(P_X)}{\sqrt{n(V(P_X))^3}}\right) \tag{2.32}$$

$$= \Phi^{-1}\left(\Phi(\mathrm{Q}^{-1}(\varepsilon))+\frac{T(P_X)}{\sqrt{n(V(P_X))^3}}\right) \tag{2.33}$$

$$= \mathrm{Q}^{-1}(\varepsilon) + \frac{T(P_X)}{\sqrt{n(V(P_X))^3}}(\Phi^{-1})'(\xi_n), \tag{2.34}$$

where (2.31) and (2.33) follow since $\mathrm{Q}^{-1}(a) = \Phi^{-1}(1-a)$ and (2.34) follows from Taylor expansion of $\Phi^{-1}(\cdot)$ around ε for $\xi_n \in [\Phi(\mathrm{Q}^{-1}(\varepsilon)), \Phi(\mathrm{Q}^{-1}(\varepsilon)) + \frac{T(P_X)}{\sqrt{n(V(P_X))^3}}]$ and $(\Phi^{-1})'(a) = \frac{1}{\sqrt{2\pi}}\exp(-\frac{a^2}{2})$. The achievability part is now completed.

2.1.3.2 Converse

The converse part is proved analogously, using the information spectrum method. Fix any (n, M)-code with encoder f and decoder ϕ. Define the following set of correctly decoded source sequences

$$C_n := \big\{x^n \in \mathcal{X}^n : \phi(f(x^n)) = x^n\big\}. \tag{2.35}$$

Analogously to $\mathcal{A}_n$ in (2.16), given any positive real number $\gamma \in \mathbb{R}_+$, define the set

$$\mathcal{B}_n(\gamma) := \Big\{x^n \in \mathcal{X}^n : \sum_{i\in[n]} \imath(x_i|P_X) \geq \log M + n\gamma\Big\}. \tag{2.36}$$

It follows that

$$\Pr\{X^n \in \mathcal{B}_n(\gamma)\}$$

$$= \Pr\{X^n \in (\mathcal{B}_n(\gamma)\cap C_n^{\mathrm{c}})\} + \Pr\{X^n \in (\mathcal{B}_n(\gamma)\cap C_n)\} \tag{2.37}$$

$$\leq \Pr\{X^n \in C_n^{\mathrm{c}}\} + \Pr\{X^n \in (\mathcal{B}_n(\gamma)\cap C_n)\} \tag{2.38}$$

$$\leq \mathrm{P}_{\mathrm{e},n} + \Pr\{X^n \in (\mathcal{B}_n(\gamma)\cap C_n)\}, \tag{2.39}$$

where (2.39) follows since $\Pr\{X^n \in \mathcal{C}_n^{\mathrm{c}}\} = \Pr\{\hat{X}^n \neq X^n\}$ is the error probability $\mathrm{P}_{\mathrm{e},n}$. Furthermore,

$$\Pr\{X^n \in (\mathcal{B}_n(\gamma) \cap \mathcal{C}_n)\} = \sum_{x^n \in (\mathcal{B}_n(\gamma) \cap \mathcal{C}_n)} P_X^n(x^n) \tag{2.40}$$

$$\leq \sum_{x^n \in (\mathcal{B}_n(\gamma) \cap \mathcal{C}_n)} \frac{\exp(-n\gamma)}{M} \tag{2.41}$$

$$\leq \frac{|\mathcal{C}_n| \exp(-n\gamma)}{M} \tag{2.42}$$

$$\leq \exp(-n\gamma), \tag{2.43}$$

where (2.40) follows since for any $x^n \in \mathcal{B}_n(\gamma)$, $P_X(x^n) \leq \frac{\exp(-n\gamma)}{M}$ and (2.43) follows since for any (n, M)-code, the size of $\mathcal{C}_n$ is no greater than M, which corresponds to one-to-one mapping of at most M source sequences.

Combining (2.39) and (2.43), it follows that the error probability of any (n, M)-code satisfies

$$\mathrm{P}_{\mathrm{e},n} \geq \Pr\left\{ \sum_{i \in [n]} \imath(X_i|P_X) \geq \log M + n\gamma \right\} - \exp(-n\gamma). \tag{2.44}$$

Given any $\varepsilon \in (0, 1)$, let

$$\varepsilon_n' := \varepsilon + \frac{T(P_X)}{\sqrt{n(V(P_X))^3}} + \frac{1}{\sqrt{n}}. \tag{2.45}$$

Consider any M such that

$$\log M = nH(P_X) + \sqrt{nV(P_X)}\mathrm{Q}^{-1}(\varepsilon_n') - \frac{1}{2}\log n. \tag{2.46}$$

Using (2.44) with $\gamma = \frac{\log n}{2n}$ and the Berry-Esseen theorem, the error probability of any (n, M)-code satisfies

$$\mathrm{P}_{\mathrm{e},n} \geq \Pr\left\{ \sum_{i \in [n]} (\imath(X_i|P_X) - H(P_X)) \geq \sqrt{nV(P_X)}\mathrm{Q}^{-1}(\varepsilon_n') \right\} - \frac{1}{\sqrt{n}} \tag{2.47}$$

$$\geq \varepsilon. \tag{2.48}$$

Combining (2.46) and (2.48) and applying the Taylor expansion of $\mathrm{Q}^{-1}(\varepsilon_n')$ around ε yield

$$R^*(n,\varepsilon) \geq H(P_X) + \frac{\sqrt{V(P_X)}\mathrm{Q}^{-1}(\varepsilon_n')}{\sqrt{n}} - \frac{\log n}{2n} \tag{2.49}$$

$$= H(P_X) + \frac{\sqrt{V(P_X)}\mathrm{Q}^{-1}(\varepsilon)}{\sqrt{n}} - \frac{\log n}{2n} + O\Big(\frac{1}{n}\Big). \tag{2.50}$$

The converse proof is now completed.

2.2 Lossy Source Coding

Lossless source coding is restricted to DMS. For any continuous source, including the Gaussian memoryless source (GMS), lossless source coding is impossible. This is because an infinite number of bits is required to represent any real number. To solve the above problem, Shannon pioneered the study of lossy source coding [9]. The only difference in lossy source coding beyond the lossless case is to allow imperfect recovery, where the reproduced source sequence $\hat{X}^n$ can be different from the original source sequence X^n. With this slight modification, continuous source sequences can be compressed, and DMS can be compressed more efficiently at lower rates. For both DMS and GMS, Shannon [9] derived the first-order asymptotics for the minimal compression rate as the blocklength tends to infinity and named the limit as the rate-distortion function. Several decades later, Ingber and Kochman [10] pioneered the low-latency analysis of lossy source coding by deriving second-order asymptotics. Subsequently, Kostina and Verdú [11] gave an alternative proof that holds for roughly all memoryless sources under mild conditions and provided various numerical examples to illustrate the tightness of second-order asymptotics. In contrast to lossless source coding, the third-order asymptotics is generally unknown for the lossy counterpart except [11, Theorem 27], where a DMS with uniform distribution is considered.

2.2.1 Problem Formulation

Let $(\mathcal{X}, \hat{\mathcal{X}})$ be a pair of alphabets. Fix any distribution $P_X \in \mathcal{P}(\mathcal{X})$ and two integers $(n, M) \in \mathbb{N}^2$. Assume that the source sequence X^n is generated i.i.d. from P_X. Similarly to the lossless case illustrated in Fig. 2.1, in lossy source coding, the source sequence X^n is compressed into an index $S \in \mathcal{M} = [M]$ and recovered as $\hat{X}^n$ by a decoder using the compressed index. The only difference is that one does not require perfect recovery with $\hat{X}^n = X^n$.

To quantity the imperfectness between the original and reproduced source sequences $(x^n, \hat{x}^n)$, a distortion measure $d : \mathcal{X} \times \hat{\mathcal{X}} \to \mathbb{R}_+$ is used in an additive manner, yielding sequence distortion $d(x^n, \hat{x}^n) = \frac{1}{n}\sum_{i \in [n]} d(x_i, \hat{x}_i)$. The Hamming distortion measure is usually used for DMS with finite alphabets $(\mathcal{X}, \hat{\mathcal{X}})$.

Definition 2.2 Given any $(x, \hat{x}) \in \mathcal{X} \times \hat{\mathcal{X}}$, the Hamming distortion measure satisfies

$$d(x, \hat{x}) = \mathbb{1}\{x \neq \hat{x}\}. \tag{2.51}$$

The quadratic distortion measure is usually used for continuous memoryless sources, such as GMS.

Definition 2.3 Given any $(x, \hat{x}) \in \mathcal{X} \times \hat{\mathcal{X}}$, the quadratic distortion measure satisfies

$$d(x, \hat{x}) = (x - \hat{x})^2. \tag{2.52}$$

An (n, M)-code for lossy source coding is defined exactly the same as in Definition 2.1 with an encoder $f : \mathcal{X}^n \to \mathcal{M}$ and a decoder $\phi : \mathcal{M} \to \hat{\mathcal{X}}^n$. To evaluate the performance of a code for lossy source coding, two criteria are usually used: expected distortion and excess-distortion probability. Fix any positive real number $D \in \mathbb{R}_+$ as the tolerable distortion level. Under the expected distortion criterion, one calculates the average distortion and compares it to the target distortion level D

$$\mathbb{E}[d(X^n, \hat{X}^n)] \leq D, \tag{2.53}$$

while under the excess-distortion probability criterion, one calculates the following probability of incurring a distortion level greater than the target distortion level D:

$$\mathrm{P}_{\mathrm{e},n}(D) := \Pr\{d(\hat{X}^n, X^n) > D\}. \tag{2.54}$$

Recall that $R_n = \frac{\log M}{n}$ is the compression rate. Analogously to (2.2), the theoretical benchmark for lossy source coding is defined as follows under both criteria:

$$R^*(n, D) := \inf\left\{R_n : \exists \text{ an } (n, M)\text{–code s.t. } \mathbb{E}[d(X^n, \hat{X}^n)] \leq D\right\}, \tag{2.55}$$

$$R^*(n, D, \varepsilon) := \inf\left\{R_n : \exists \text{ an } (n, M)\text{–code s.t. } \mathrm{P}_{\mathrm{e},n}(D) \leq \varepsilon\right\}. \tag{2.56}$$

The study of lossy source coding characterizes $R^*(D)$ under the expected distortion criterion and $R^*(n, D, \varepsilon)$ under the excess-distortion probability criterion. As illustrated in [7, Chapter 3.1], both criteria lead to the same first-order asymptotics as the blocklength tends to infinity when a bounded distortion measure is used, i.e., $\max_{(x,\hat{x}) \in \mathcal{X} \times \hat{\mathcal{X}}} d(x, \hat{x}) < \infty$. The same holds true for GMS under the quadratic distortion measure although the quadratic distortion measure is unbounded. Thus, for most cases,

$$\lim_{n\to\infty} R^*(n, D) = \lim_{n\to\infty} R^*(n, D, \varepsilon). \tag{2.57}$$

The expected distortion criterion was proposed by Shannon and used in a sequence of follow-up studies in first-order asymptotics. In the long quest for second-order asymptotics of lossy source coding, one finds that the excess-distortion probability criterion is more appropriate since it resembles the error probability in lossless source coding and allows one to apply similar proof techniques, including the Berry-Esseen theorem. Till now, second-order asymptotics is only available for $R^*(n, D, \varepsilon)$:

$$R^*(n, D, \varepsilon) = a + \frac{b}{\sqrt{n}} + O\Big(\frac{\log n}{n}\Big). \tag{2.58}$$

Shannon [9] bounded a explicitly and named a the rate-distortion function. Several decades later, in 2010s, the explicit formula for b was characterized first by Ingber and Kochman [10] for DMS and GMS. Subsequently, Kostina and Verdú [10] generalized the results to almost all memoryless sources with alternative proofs and provided abundant numerical examples. The third-order term $O\big(\frac{\log n}{n}\big)$ is left open.

2.2.2 *Distortion-Tilted Information Density*

The distortion-tilted information density takes the same role for lossy source coding as the entropy density for lossless source coding. In particular, the first and second-order asymptotics are functions of the first and second-order statistics of the distortion-tilted information density random variable. In this section, the definition, properties, and example of this information density are presented.

Fix any distribution $P_X \in \mathcal{P}(\mathcal{X})$ and distortion level $D \in \mathbb{R}_+$. The rate-distortion function $R(P_X, D)$ is defined as

$$R(P_X, D) := \min_{P_{\hat{X}|X} \in \mathcal{P}(\hat{\mathcal{X}}|\mathcal{X}):\ \mathbb{E}[d(X,\hat{X})] \le D} I(P_X, P_{X|\hat{X}}). \tag{2.59}$$

Note that $R(P_X, D)$ decreases in D. When $D \ge \overline{D} := \min_{\hat{x}\in\hat{\mathcal{X}}} \mathbb{E}[d(X, \hat{x})]$, $R(P_X, D) = 0$. On the other hand, when $0 < D < \underline{D} := \min_{P_{\hat{X}|X}} \mathbb{E}[d(X, \hat{X})]$, $R(P_X, D) = \infty$. Under the Hamming distortion measure, $\underline{D} = 0$ and

$$R(P_X, D) \le R(P_X, 0) = H(P_X). \tag{2.60}$$

Thus, lossy source coding allows lower compression rates. To some extent, lossless source coding is a special case of lossy source coding with zero distortion level under the Hamming distortion measure.

The rate-distortion function $R(P_X, D)$ is a convex optimization problem since the constraint is linear in $P_{\hat{X}|X}$ and the objective function $I(P_X, P_{X|\hat{X}})$ is convex in $P_{\hat{X}|X}$ [12, Theorem 2.7.4]. Thus, for any $D \in [\underline{D}, \overline{D}]$, there exists an optimizer $P^*_{\hat{X}|X}$ that achieves $R(P_X, D)$. Let $P^*_{\hat{X}}$ be induced by P_X and $P^*_{\hat{X}|X}$. Furthermore, the optimal Lagrange multiplier λ^* for the dual problem of $R(P_X, D)$ satisfies

$$\lambda^* = -\frac{\partial R(P_X, D')}{\partial D'}\bigg|_{D'=D}. \tag{2.61}$$

One can now define the distortion-tilted information density [10, Proposition 7] (see also [11, Definition 6]), which plays the same role as the entropy density for lossless source coding.

Definition 2.4 Given $x \in \mathcal{X}$, the *distortion-tilted information density* is defined as

$$\jmath(x|D, P_X) := -\log \mathbb{E}_{P^*_{\hat{X}}}[\exp(\lambda^* D - \lambda^* d(x, \hat{X}))]. \tag{2.62}$$

Definition 2.4 appeared implicitly in early studies [13–15].

Let $P^*_{X|\hat{X}}$ be induced by the the optimal test channel $P^*_{\hat{X}|X}$ and the source distribution P_X. For any $(x, \hat{x}) \in \mathcal{X} \times \hat{\mathcal{X}}$, define the following *mutual information density*

$$\imath(x; \hat{x}) := \log \frac{P^*_{X|\hat{X}}(x|\hat{x})}{P_X(x)}. \tag{2.63}$$

Csiszár [16] proved the properties of the distortion-tilted information density.

Lemma 2.1 (Properties of Distortion-Tilted Information Density)

1. *Given any $\hat{x} \in \operatorname{supp}(P^*_{\hat{X}})$, the distortion-tilted information density has the following alternative form:*

$$\jmath(x|D, P_X) = \imath(x; \hat{x}) + \lambda^* d(x, \hat{x}) - \lambda^* D. \tag{2.64}$$

2. *The expected value of the distortion-titled information density random variable is the rate-distortion function, i.e.,*

$$R(P_X, D) = \mathbb{E}_{P_X}[\jmath(X|D, P_X)]. \tag{2.65}$$

3. *For any $\hat{x} \in \hat{\mathcal{X}}$,*

$$\mathbb{E}[\exp(\lambda^* D - \lambda^* d(X, \hat{x}) + \jmath(X|D, P_X))] \le 1. \tag{2.66}$$

Claim 1 helps calculate the distortion-tilted information density. Claim 2 parallels the corresponding result for lossless source coding, where entropy is the expected value of the entropy density random variable. As will be presented soon, the second-order term of lossy source coding depends on variance of the distortion-tilted information density. Claim 3 is used to derive a finite blocklength converse result that holds for any memoryless source.

In the following, the distortion-tilted information density is calculated for GMS under the quadratic distortion measure, and used to verify the properties in Lemma 2.1. Fix any $\sigma \in \mathbb{R}_+$. Let $P_X = \mathcal{N}(\cdot; 0, \sigma^2)$ denote the pdf of a Gaussian random variable with zero mean and variance σ^2. Fix any $0 < D < \sigma^2$. The rate-distortion function of GMS satisfies

$$R(P_X, D) = \frac{1}{2}\log\frac{\sigma^2}{D}. \tag{2.67}$$

The optimal test channel $P^*_{\hat{X}|X}$ satisfies that $(X - \hat{X}) \sim \mathcal{N}(\cdot; 0, D)$ and the induced marginal distribution $P^*_{\hat{X}} = \mathcal{N}(\cdot; 0, \sigma^2 - D)$. It follows from (2.61) that

$$\lambda^* = -\frac{\partial R(P_X, D')}{\partial D'}\bigg|_{D'=D} \tag{2.68}$$

$$= -\frac{\partial(\frac{1}{2}\log\sigma^2 - \frac{1}{2}\log D')}{\partial D'}\bigg|_{D'=D} \tag{2.69}$$

$$= \frac{1}{2D}. \tag{2.70}$$

Thus, for any $(x, \hat{x}) \in \mathbb{R}^2$,

$$\exp(\lambda^* D - \lambda^* d(x, \hat{x})) = \exp\left(\frac{1}{2} - \frac{(x - \hat{x})^2}{2D}\right) \tag{2.71}$$

$$= \exp\left(\frac{1}{2}\right)\exp\left(-\frac{(x - \hat{x})^2}{2D}\right). \tag{2.72}$$

Using $P^*_{\hat{X}} = \mathcal{N}(\cdot; 0, \sigma^2 - D)$ yields

$$\mathbb{E}_{P^*_{\hat{X}}}[\exp(\lambda^* D - \lambda^* d(x, \hat{X}))]$$

$$= \exp\left(\frac{1}{2}\right)\int_{-\infty}^{\infty}\frac{1}{\sqrt{2\pi(\sigma^2 - D)}}\exp\left(-\frac{(\hat{x} - x)^2}{2D} - \frac{\hat{x}^2}{2(\sigma^2 - D)}\right)\mathrm{d}\hat{x} \tag{2.73}$$

$$= \exp\left(\frac{1}{2}\left(1-\frac{x^2}{\sigma^2}\right)\right)\sqrt{D/\sigma^2}\int_{-\infty}^{\infty}\mathcal{N}\left(\hat{x};\frac{\sigma^2-D}{\sigma^2}x,\frac{(\sigma^2-D)D}{\sigma^2}\right)\mathrm{d}\hat{x} \tag{2.74}$$

$$= \exp\left(\frac{1}{2}\left(1-\frac{x^2}{\sigma^2}\right)\right)\sqrt{D/\sigma^2}, \tag{2.75}$$

where (2.75) follows since the integration in (2.74) equals one, which integrates the pdf of a Gaussian random variable with mean $\frac{\sigma^2-D}{\sigma^2}x$ and variance $\frac{(\sigma^2-D)D}{\sigma^2}$ over the whole range, and (2.74) follows from the following algebra:

$$-\frac{(\hat{x}-x)^2}{2D}-\frac{\hat{x}^2}{2(\sigma^2-D)}$$

$$= \frac{1}{-2(\sigma^2-D)D}\left((\sigma^2-D)(\hat{x}-x)^2+D\hat{x}^2\right) \tag{2.76}$$

$$= \frac{1}{-2(\sigma^2-D)D}\left(\sigma^2\hat{x}^2-2(\sigma^2-D)x\hat{x}+(\sigma^2-D)x^2\right) \tag{2.77}$$

$$= \frac{1}{-2(\sigma^2-D)D}\left(\sigma^2\left(\hat{x}-\frac{\sigma^2-D}{\sigma^2}x\right)^2-\frac{(\sigma^2-D)^2}{\sigma^2}x^2+(\sigma^2-D)x^2\right) \tag{2.78}$$

$$= \frac{1}{-2(\sigma^2-D)D}\left(\sigma^2\left(\hat{x}-\frac{\sigma^2-D}{\sigma^2}x\right)^2+\frac{D}{\sigma^2}(\sigma^2-D)x^2\right) \tag{2.79}$$

$$= \frac{\sigma^2}{-2(\sigma^2-D)D}\left(\hat{x}-\frac{\sigma^2-D}{\sigma^2}x\right)^2-\frac{x^2}{2\sigma^2}. \tag{2.80}$$

It follows from Definition 2.4 and (2.75) that the distortion-tilted information density for a GMS under the quadratic distortion measure satisfies that for any $x \in \mathbb{R}$,

$$\jmath(x|D,P_X) = -\log\mathbb{E}_{P^*_{\hat{X}}}[\exp(\lambda^*D-\lambda^*d(x,\hat{X}))] \tag{2.81}$$

$$= \frac{x^2}{2\sigma^2}+\frac{1}{2}\log\frac{\sigma^2}{D}-\frac{1}{2}. \tag{2.82}$$

Lemma 2.1 is illustrated as follows. For GMS, the mutual information density is

$$\imath(x;\hat{x}) = \log\frac{\mathcal{N}(x-\hat{x};0,D)}{\mathcal{N}(x;0,\sigma^2)} \tag{2.83}$$

$$= -\frac{(\hat{x}-x)^2}{2D}-\frac{1}{2}\log(2\pi D)+\frac{x^2}{2\sigma^2}+\frac{1}{2}\log(2\pi\sigma^2) \tag{2.84}$$

$$= \frac{x^2}{2\sigma^2}-\frac{(\hat{x}-x)^2}{2D}+\frac{1}{2}\log\frac{\sigma^2}{D}. \tag{2.85}$$

It follows that

$$\begin{aligned}
&\imath(x;\hat{x}) + \lambda^* d(x,\hat{x}) - \lambda^* D \\
&= \frac{x^2}{2\sigma^2} - \frac{(\hat{x}-x)^2}{2D} + \frac{1}{2}\log\frac{\sigma^2}{D} + \frac{(\hat{x}-x)^2}{2D} - \frac{1}{2} &(2.86)\\
&= \frac{1}{2}\frac{x^2}{\sigma^2} + \frac{1}{2}\log\frac{\sigma^2}{D} - \frac{1}{2} &(2.87)\\
&= \jmath(x|D, P_X). &(2.88)
\end{aligned}$$

Claim 1 of Lemma 2.1 is verified.

Taking expectation over both sides of (2.82) and noting that $\mathbb{E}[X^2] = \sigma^2$ yield

$$\begin{aligned}
\mathbb{E}[\jmath(X|D, P_X] &= \frac{1}{2}\log\frac{\sigma^2}{D} + \frac{1}{2\sigma^2}\mathbb{E}[X^2] - \frac{1}{2} &(2.89)\\
&= \frac{1}{2}\log\frac{\sigma^2}{D}. &(2.90)
\end{aligned}$$

Claim 2 of Lemma 2.1 is verified.

For GMS, $\text{supp}(P^*_{\hat{X}}) = \hat{\mathcal{X}} = \mathbb{R}$. Thus, using Claim 1 and (2.63) lead to

$$\begin{aligned}
\mathbb{E}[\exp(\lambda^* D - \lambda^* d(X,\hat{x}) + \jmath(X|D, P_X))] &= \mathbb{E}[\exp(\imath(X;\hat{x}))] &(2.91)\\
&= \mathbb{E}\left[\frac{P_{X|\hat{X}}(X|\hat{x})}{P_X(X)}\right] &(2.92)\\
&= \int_{-\infty}^{\infty} P_{X|\hat{X}}(x|\hat{x})\mathrm{d}x &(2.93)\\
&= 1. &(2.94)
\end{aligned}$$

Claim 3 of Lemma 2.1 is verified.

2.2.3 *Second-Order Asymptotics*

Define the following the *distortion-dispersion function*:

$$V(P_X, D) := \text{Var}_{P_X}[\jmath(X|D, P_X)], \tag{2.95}$$

which is the variance of the distortion-tilted information density.

Assume that (1) the distortion level $D \in (\underline{D}, \overline{D})$ such that the rate-distortion function $R(P_X, D)$ is positive and finite; (2) the distortion-dispersion function $V(P_X, D)$ is positive and finite; and (3) the expected value $\mathbb{E}_{P_X \times P^*_{\hat{X}}}[d(X, \hat{X})^9]$ is finite.

Theorem 2.2 *For any $\varepsilon \in (0,1)$, the minimal compression rate $R^*(n, D, \varepsilon)$ satisfies*

$$R^*(n, D, \varepsilon) = R(P_X, D) + \frac{\sqrt{V(P_X, D)}\mathrm{Q}^{-1}(\varepsilon)}{\sqrt{n}} + O\Big(\frac{\log n}{n}\Big). \tag{2.96}$$

Theorem 2.2 generalizes Theorem 2.1 to the lossy case. Under the Hamming distortion measure with $D = 0$, Theorem 2.2 recovers Theorem 2.1 up to second-order asymptotics. The remarks for Theorem 2.1 also hold for the lossy case including strong converse and phase transition.

Theorem 2.2 holds for any memoryless sources and distortion levels, as long as the assumptions are satisfied. Any DMS under the bounded distortion measure and any GMS under the quadratic distortion measure satisfy the assumptions. For GMS $P_X = \mathcal{N}(\cdot; 0, \sigma^2)$ under quadratic distortion, it follows from (2.82) that

$$V(P_X, D) = \mathrm{Var}\Big[\frac{X^2}{2\sigma^2}\Big] \tag{2.97}$$

$$= \frac{\mathbb{E}[X^4] - (\mathbb{E}[X^2])^2}{4\sigma^4} \tag{2.98}$$

$$= \frac{3\sigma^4 - \sigma^4}{4\sigma^4} \tag{2.99}$$

$$= \frac{1}{2}. \tag{2.100}$$

Thus, for any $D \in (0, \sigma^2)$,

$$R^*(n, D, \varepsilon) = \frac{1}{2}\log\frac{\sigma^2}{D} + \sqrt{\frac{1}{2n}}\mathrm{Q}^{-1}(\varepsilon) + O\Big(\frac{\log n}{n}\Big). \tag{2.101}$$

2.2.4 Proof of Second-Order Asymptotics

There are at least ways to prove Theorem 2.2. Ingber and Kochman [10, Theorem 1] used the method of types [17] for DMS under bounded distortion measures and used generalized power types [18, 19] for GMS under quadratic distortion. For any memoryless sources, Kostina and Verdú [11, Theorem 12] provided a unified proof with finite blocklength upper and lower bounds. Streamlined proof steps at the textbook level of both ways were summarized in [7, Chapters 3.4-3.5] (see also [20, Chapter 3.5] for lesser detailed proofs). However, specialized proofs for GMS under quadratic distortion were not included in previous monographs [7, 20], which will be presented below.

2.2.4.1 Achievability

The idea of the achievability part is due to Ingber and Kochman [10]. A key result is the *type covering lemma* for GMS [21], and the proof below is specialized from [22, Theorem 20], which was for the multiterminal lossy source coding problem of successive refinement.

Let $\xi \in \mathbb{R}_+$ be a positive real number to be specified later. Define a set

$$\mathcal{U}^{\xi} := \left\{ x^n \in \mathbb{R}^n : e^{-2\xi} < \frac{\|x^n\|^2}{n\sigma^2} < e^{2\xi} \right\}. \tag{2.102}$$

Similarly to [23, Eq. (35)], it follows from Cramér's theorem [24, Theorem 2.1.24] that the probability for a GMS not belong to the set is upper bounded as

$$\Pr\{X^n \notin \mathcal{U}^{\xi}\} \le 4\exp\left(-ne(\xi)\right), \tag{2.103}$$

where the exponent function is

$$e(\xi) := \frac{\exp(2\xi) - 1 - 2\xi}{2}. \tag{2.104}$$

Let $\delta \in \mathbb{R}_+$ be specified and define the integer

$$k := \left\lceil \frac{e^{2\xi} - e^{-2\xi}}{\delta} \right\rceil + 1. \tag{2.105}$$

Define the following constant and power type classes for each $i \in [k]$:

$$\Lambda(i) := \sigma^2 e^{-2\xi} + (i-1)\delta\sigma^2 \tag{2.106}$$

$$\mathcal{U}_i := \left\{ x^n : \Lambda(i-1) \le \frac{\|x^n\|^2}{n} \le \Lambda(i) \right\}, \tag{2.107}$$

where $\mathcal{U}_i$ collects GMS sequences with normalized power in the range of $(\Lambda(i-1), \Lambda(i)]$. It follows from (2.102) and (2.107) that $\mathcal{U}^{\xi} \subset \bigcup_{i=1}^{k} \mathcal{U}_i$. For any $i \in [k]$ and any sequence $x^n \in \mathcal{U}_i$, the power type of x^n is defined as i. On other other hand, if $x^n \notin \mathcal{U}^{\xi}$, the power type of x^n is defined as 0.

Lemma 2.2 *Fix any $i \in [k]$. Given $x^n \in \mathcal{U}_i$, there exists a set $\mathcal{B}_i \subset \mathbb{R}^n$ such that*

$$\mathcal{U}_i \subset \bigcup_{\hat{x}^n \in \mathcal{B}_i} \left\{ x^n : \|x^n - \hat{x}^n\|^2 \le D_1 \right\}. \tag{2.108}$$

$$\log|\mathcal{B}_i| \le \frac{n}{2}\log\frac{\Lambda(i)}{D_1} + \frac{5}{2}\log n + \log 6. \tag{2.109}$$

Lemma 2.2 follows from [21, Theorem 1.2] concerning sphere covering. Specifically, to make sure that all GMS sequences in $\mathcal{U}_i$ can be covered with a codeword within distortion level D, $6n^{5/2}(\Lambda(i)/D_1)^{n/2}$ codewords are required, which yields the desired bound in (2.109). Lemma 2.2 implies the existence of a deterministic $(n, |\mathcal{B}|)$-code such that all GMS sequences in $\mathcal{U}_i$ incurs a distortion less than D.

Using the power types and the type covering lemma, an (n, M)-code for lossy source coding of GMS can be constructed as follows. For each $i \in [k]$, let $\mathcal{B}_i$ in Lemma 2.2 be the set of codewords for power type class i. If $|\mathcal{B}_i| > M$, the codewords for power type i includes only the first M codewords of $\mathcal{B}_i$. Given a source sequence X^n, if $X^n \notin \mathcal{U}^\xi$, the encoder f sends the power type 0, indicating that an excess-distortion error occurs. Otherwise, the encoder f first sends the power type of X^n using $\log k$ nats and then sends the index of the codeword in $\mathcal{B}_i$ that has minimum quadratic loss with respect to X^n. It follows from Lemma 2.2 that if $\log M \geq \frac{n}{2}\log\frac{\Lambda(i)}{D_1} + \frac{5}{2}\log n + \log k + \log 6$, no excess-distortion event occurs for GMS in the power type class $\mathcal{U}_i$. Otherwise, an excess-distortion event occurs. Thus, the excess-distortion probability of the above (n, M)-code satisfies

$$\begin{aligned} \mathrm{P}_{\mathrm{e},n}(D) &\leq \Pr\{X^n \notin \mathcal{U}^\xi\} \\ &\quad + \sum_{i\in[K]} \Pr\{X^n \in \mathcal{U}_i\}\mathbb{1}\bigg(\log M < \frac{n}{2}\log\frac{\Lambda(i)}{D_1} + \frac{5}{2}\log n + \log k + \log 6\bigg). \end{aligned} \tag{2.110}$$

The first term in (2.110) was bounded in (2.103). Recall that $R_n = \frac{\log M}{n}$ is the compression rate. For simplicity, let

$$\eta_n := R_n - \frac{5\log n + 2\log k + 2\log 6}{2n}. \tag{2.111}$$

The second term in (2.110) is bounded as follows:

$$\begin{aligned} &\sum_{i\in[K]} \Pr\{X^n \in \mathcal{U}_i\}\mathbb{1}\bigg(\eta_n \leq \frac{1}{2}\log\frac{\Lambda(i)}{D}\bigg) \\ &= \sum_{i\in[K]} \Pr\{X^n \in \mathcal{U}_i\}\mathbb{1}\big(\Lambda_i \geq D\exp(2\eta_n)\big) &(2.112)\\ &\leq \sum_{i\in[K]} \Pr\{X^n \in \mathcal{U}_i\}\mathbb{1}\Big(\|X^n\|^2 + n\delta\sigma^2 \geq nD\exp(2\eta_n)\Big) &(2.113)\\ &= \Pr\Big\{\|X^n\|^2 + n\delta\sigma^2 \geq nD\exp(2\eta_n)\Big\} &(2.114)\\ &= \Pr\bigg\{\frac{1}{n}\sum_{i\in[n]}\frac{X_i^2}{\sigma^2} \geq \frac{D\exp(2\eta_n)}{\sigma^2} - \delta\bigg\}, &(2.115) \end{aligned}$$

where (2.113) follows since (2.107) implies that if $X^n \in \mathcal{U}_i$, $\|X^n\|^2 + n\delta\sigma^2 \geq n\Lambda(i)$.

Note that $Y_i := \frac{X_i^2}{\sigma^2}$ is χ_1^2-distributed for each $i \in [n]$. Combining (2.103), (2.110), and (2.115) yields the following finite blocklength upper bound for the excess-distortion probability:

$$\mathrm{P}_{\mathrm{e},n}(D) \leq 4\exp\left(-ne(\xi)\right) + \Pr\left\{\frac{1}{n}\sum_{i\in[n]} Y_i \geq \frac{D\exp(2\eta_n)}{\sigma^2} - \delta\right\}. \tag{2.116}$$

The second-order asymptotics is proved as follows. Fix any $n \in \mathbb{N}$. Set $\xi = n^{-1/3}$ and $\delta = 1/n$. Using (2.104) and the Taylor expansion of $\exp(2\xi)$,the first term on the right-hand-side of (2.116) satisfies that for n sufficiently large,

$$4\exp(-ne(\xi)) = 4\exp(-n(2\xi^2 + o(\xi^2))) \tag{2.117}$$

$$= 4\exp\left(-2n^{1/3} + o(n^{1/3})\right) \tag{2.118}$$

$$\leq \frac{1}{n} \to 0. \tag{2.119}$$

Furthermore, it follows from the definition of k in (2.105) with Taylor expansion that for n sufficiently large,

$$k = \left\lceil \frac{1 + 2\xi - (1 - 2\xi) + o(\xi)}{\delta} \right\rceil + 1 \tag{2.120}$$

$$= \lceil 4n^{2/3} + O(n^{2/3}) \rceil + 1 \tag{2.121}$$

$$\leq n. \tag{2.122}$$

Given any $\varepsilon \in (0, 1)$, let

$$\varepsilon_n^{\mathrm{GMS}} := \varepsilon - 4\exp\left(-ne(\xi)\right) - \frac{8}{\sqrt{2n}}. \tag{2.123}$$

Choose M such that

$$\log M = \frac{n}{2}\log\frac{\sigma^2}{D} + \sqrt{\frac{n}{2}}\mathrm{Q}^{-1}(\varepsilon_n^{\mathrm{GMS}}) + \frac{7}{2}\log n + \log 6 + 1. \tag{2.124}$$

It follows from (2.111) and (2.122) that

$$\eta_n \geq \frac{1}{2}\log\frac{\sigma^2}{D} + \sqrt{\frac{1}{2n}}\mathrm{Q}^{-1}(\varepsilon_n^{\mathrm{GMS}}) + \frac{1}{n}, \tag{2.125}$$

and thus

$$\Pr\left\{\frac{1}{n}\sum_{i\in[n]} Y_i \geq \frac{D\exp(2\eta_n)}{\sigma^2} - \frac{1}{n}\right\}$$

$$\leq \Pr\left\{\frac{1}{n}\sum_{i\in[n]} Y_i \geq \exp\left(2\sqrt{\frac{1}{2n}}\mathrm{Q}^{-1}(\varepsilon_n^{\mathrm{GMS}}) + \frac{1}{n}\right) - \frac{1}{n}\right\} \tag{2.126}$$

$$\leq \Pr\left\{\frac{1}{n}\sum_{i\in[n]} Y_i \geq 1 + 2\sqrt{\frac{1}{2n}}\mathrm{Q}^{-1}(\varepsilon_n^{\mathrm{GMS}})\right\} \tag{2.127}$$

$$\leq \Pr\left\{\frac{1}{n}\sum_{i\in[n]} (Y_i - 1) \geq \sqrt{\frac{4}{2n}}\mathrm{Q}^{-1}(\varepsilon_n^{\mathrm{GMS}})\right\} \tag{2.128}$$

$$\leq \varepsilon_n^{\mathrm{GMS}} + \frac{8}{\sqrt{2n}}, \tag{2.129}$$

where (2.129) follows from the Berry-Esseen theorem and the fact that for each $i \in [n]$, $Y_i = \frac{X_i^2}{\sigma^2}$ is non-negative and χ_1^2-distributed, which implies that $\mathbb{E}[Y_i] = 1$, $\mathrm{Var}[Y_i] = 2$ and

$$\mathbb{E}[|Y_i - \mathbb{E}[Y_i]|^3] = \mathbb{E}[|Y_i - 1|^3] \tag{2.130}$$

$$= \mathbb{E}[(1 - Y_i)^3 \mathbb{1}(0 \leq Y_i < 1)] + \mathbb{E}[(Y_i - 1)^3 \mathbb{1}(Y_i \geq 1)] \tag{2.131}$$

$$\leq 1 + \mathbb{E}[Y_i^3] \tag{2.132}$$

$$= 1 + 15 = 16. \tag{2.133}$$

The achievability proof for GMS under quadratic distortion is completed by combining (2.116), (2.124), and (2.129) and using the Taylor expansion of $\mathrm{Q}^{-1}(\cdot)$.

2.2.4.2 Converse

The converse proof is due to Kostina and Verdú [11], which holds for arbitrary memoryless sources and is based on the following finite blocklength lower bound to the excess-distortion probability.

Lemma 2.3 *Given any $\gamma \in \mathbb{R}_+$, any (n, M)-code satisfies that*

$$\mathrm{P}_{\mathrm{e},n}(D) \geq \Pr\left\{\sum_{i\in[n]} \jmath(X_i|D, P_X) \geq \log M + n\gamma\right\} - \exp(-n\gamma). \tag{2.134}$$

Lemma 2.3 generalizes the finite blocklength converse bound in (2.44) for the lossless source coding to the lossy setting, where the entropy density is replaced by the distortion-tilted information density. Lemma 2.3 was proved by Kostina and Verdú [11, Theorem 7] (see also [7, Theorem 3.4]). The key idea is similar to the lossless case, and Claim 3 of Lemma 2.1 is used to yield the $\exp(-n\gamma)$ term.

The rest of the proof parallels those for the lossless case. Choose $\gamma = \frac{\log n}{2}$. It follows from Lemma 2.3 that the excess-distortion probability of any (n, M)-code satisfies

$$\mathrm{P}_{\mathrm{e},n}(D) \geq \Pr_{P_X^n}\left\{\sum_{i\in[n]} \jmath(X_i|P_X, D) \geq \log M + \frac{1}{2}\log n\right\} - \frac{1}{\sqrt{n}}. \tag{2.135}$$

Let $T(P_X, D)$ be the third absolute moment of the distortion-tilted information density, i.e.,

$$T(P_X, D) := \mathbb{E}\Big[\big|\jmath(X|D, P_X) - R(P_X, D)\big|^3\Big]. \tag{2.136}$$

For GMS, it follows from (2.82) and (2.133) that

$$R(P_X, D) = \frac{1}{2}\log\frac{\sigma^2}{D}, \tag{2.137}$$

$$V(P_X, D) = \frac{1}{2}, \tag{2.138}$$

$$T(P_X, D) = \mathbb{E}\left[\left|\frac{X^2/\sigma^2 - 1}{2}\right|^3\right] \leq 2. \tag{2.139}$$

Given any $\varepsilon \in (0, 1)$, let

$$\varepsilon_n^{\mathrm{rd}} := \varepsilon + \frac{4\sqrt{2}+1}{\sqrt{n}}. \tag{2.140}$$

Consider any M such that

$$\log M := \frac{n}{2}\log\frac{\sigma^2}{D} + \sqrt{\frac{n}{2}}\mathrm{Q}^{-1}(\varepsilon_n^{\mathrm{rd}}) - \frac{1}{2}\log n. \tag{2.141}$$

Applying the Berry-Esseen theorem to the first term in (2.135) yields

$$\mathrm{P}_{\mathrm{e},n}(D) \geq \varepsilon. \tag{2.142}$$

The converse proof is completed by using (2.141) and the Taylor expansion of $\mathrm{Q}^{-1}(\cdot)$.

2.3 Channel Coding

Channel coding aims to combat noise over the channel so that messages can be reliably transmitted. Shannon's well-known channel coding theorem [1] states that for any memoryless channel, the capacity is the maximal message rate for data transmission to ensure that the error probability vanishes to zero as the blocklength tends to infinity. Subsequently, Strassen [3] pioneered the study of low-latency channel coding by deriving second-order asymptotics, which was rediscovered by Hayashi [25] and by Polyanskiy et al. [26]. Further refined third-order asymptotics was derived by Tomamichel and Tan [27, 28].

2.3.1 Problem Formulation

Let $(\mathcal{X}, \mathcal{Y})$ be two alphabets, $P_X \in \mathcal{P}(\mathcal{X})$ be a distribution and $P_{Y|X} \in \mathcal{P}(\mathcal{Y}|\mathcal{X})$ be a conditional distribution. Fix two integers $(n, M) \in \mathbb{N}^2$ and recall that $\mathcal{M} = [M] = \{1, \ldots, M\}$ denotes the set of all integers from 1 to M. Assume that W is uniformly distributed over $\mathcal{M}$. As shown in Fig. 2.3, the task of channel coding is to transmit a message W using channel inputs X^n over a memoryless channel $P_{Y|X}^n$ so that the receiver side can recover the message W using the noisy channel output Y^n. Note that by memoryless, it means that for each $i \in [n]$, Y_i is the output of passing X_i over the same channel $P_{Y|X}$.

Three types of channels are commonly used.

Definition 2.5 A channel $P_{Y|X}$ is said to be a Binary Symmetric Channel (BSC) with parameter $\zeta \in (0, 1)$ if $\mathcal{X} = \mathcal{Y} = \{0, 1\}$ and for any $(x, y) \in \mathcal{X} \times \mathcal{Y}$,

$$P_{Y|X}(y|x) = \zeta \mathbb{1}(y \neq x) + (1 - \zeta)\mathbb{1}(y = x). \tag{2.143}$$

A BSC with parameter ζ flips each input symbol with probability ζ.

Definition 2.6 A channel $P_{Y|X}$ is said to be a Binary Erasure Channel (BEC) with parameter $\tau \in (0, 1)$ if $\mathcal{X} = \{0, 1\}$, $\mathcal{Y} = \{0, 1, \mathrm{e}\}$ and

$$P_{Y|X}(y|x) = (1 - \tau)^{\mathbb{1}(y=x)} \tau^{\mathbb{1}(y=\mathrm{e})}. \tag{2.144}$$

A BEC with parameter τ erases each transmitted randomly with probability τ.

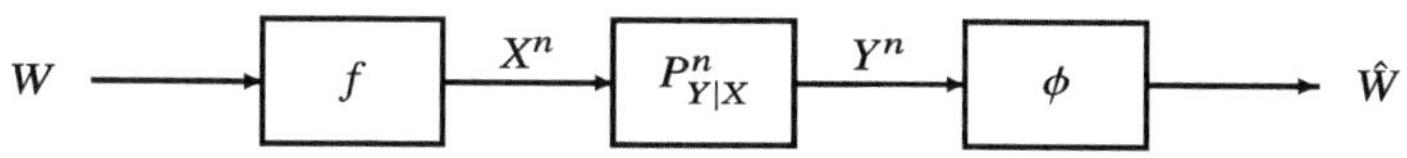

Fig. 2.3 System model for channel coding

Definition 2.7 A channel $P_{Y|X}$ is said to be an AWGN channel with parameter $\sigma \in \mathbb{R}_+$ if $\mathcal{X} = \mathcal{Y} = \mathbb{R}$ and for any $(x, y) \in \mathcal{X} \times \mathcal{Y}$,

$$P_{Y|X}(y|x) = \frac{1}{\sqrt{2\pi\sigma^2}} \exp\left(-\frac{(y-x)^2}{2\sigma^2}\right). \tag{2.145}$$

An AWGN channel distorts each transmit symbol with Gaussian noise. For AWGN channels, a power constraint is imposed on channel inputs X^n. Otherwise the transmission rate can be unbounded with an infinite power. Given a positive real number $P \in \mathbb{R}_+$, the average power constraint requires that $\mathbb{E}[\|X^n\|^2] \leq nP$, while the peak power constraint requires that $\|X^n\|^2 \leq nP$ with probability one. If $\mathcal{X} = \{-1, 1\}$, for any $P \in \mathbb{R}_+$, the above channel is known as the binary-input AWGN channel with x replaced by $\sqrt{P}x$. Binary-input AWGN channels model many practical communication scenarios and P is again the power constraint.

A code specifies how the channel input is chosen and how the estimate is generated.

Definition 2.8 An (n, M)-code for channel coding consists of an encoder $f : \mathcal{M} \to \mathcal{X}^n$ and a decoder $\phi : \mathcal{Y}^n \to \mathcal{M}$.

To evaluate the performance any (n, M)-code, the message rate $R_n := \frac{\log M}{n}$ and the error probability $\mathrm{P}_{\mathrm{e},n} := \Pr\{\hat{W} \neq W\}$ are considered, where W is the transmitted message and $\hat{W}$ is the message estimated by decoder ϕ. There is a tradeoff among the blocklength n, the message size M, and the error probability. Thus, the study of channel coding is to characterize the tradeoff between the compression rate R_n and the error probability $\mathrm{P}_{\mathrm{e},n}$ given any blocklength $n \in \mathbb{N}$. At finite blocklength, the theoretical benchmark is the maximal transmission rate subject to an error probability of at most ε, i.e.,

$$R_{\mathrm{c}}^*(n, \varepsilon) := \sup\left\{R_n : \exists \text{ an } (n, M)\text{−code s.t. } \mathrm{P}_{\mathrm{e},n} \leq \varepsilon\right\}. \tag{2.146}$$

It follows from [1, 3, 25–28] that for almost any discrete memoryless channel (DMC) and AWGN channels

$$R_{\mathrm{c}}^*(n, \varepsilon) = a + \frac{b}{\sqrt{n}} + \frac{c \log n}{n} + O\left(\frac{1}{n}\right), \tag{2.147}$$

where $(a, b, c) \in \mathbb{R}^3$ are the coefficients for the first, second and third dominant terms. Shannon [1] bounded the first-order term $\lim_{\varepsilon \to 0} \lim_{n \to \infty} R^*(n, \varepsilon)$ with vanishing error probability explicitly and show that $a = C(P_{Y|X})$ when $\varepsilon \to 0$, where $C(P_{Y|X})$ denotes the channel capacity to be defined in (2.149). Strassen [3] and Hayashi [25] bounded the first two terms for any $\varepsilon \in (0, 1)$, showing that $a = C(P_X)$ regardless of ε and characterized b as a function of the channel variation and the error probability. Finally, Polyanskiy et al. [26] derived finite blocklength upper and lower bounds, characterized the first two terms for most channels and obtained the third-order terms for several special cases including the BSC. Finally, for most DMC and AWGN channels, the third-order terms were derived by Tomamichel and Tan [27, 28], who showed that $c = \frac{1}{2}$ under certain conditions.

2.3.2 *Second-Order Asymptotics*

2.3.2.1 DMC

For a DMC, both the input $\mathcal{X}$ and the output alphabet $\mathcal{Y}$ are finite. Given any input distribution $P_X \in \mathcal{P}(\mathcal{X})$ and channel $P_{Y|X} \in \mathcal{P}(\mathcal{Y}|\mathcal{X})$, P_Y be the induced marginal distribution on $\mathcal{Y}$. For any $(x, y) \in \mathcal{X} \times \mathcal{Y}$, define the following mutual information density

$$\imath_{\mathrm{c}}(x; y) := \log \frac{P_{Y|X}(y|x)}{P_Y(y)}. \tag{2.148}$$

The capacity for a channel $P_{Y|X}$ is defined as

$$C(P_{Y|X}) := \max_{P_X \in \mathcal{P}(\mathcal{X})} \mathbb{E}[\imath_{\mathrm{c}}(X; Y)]. \tag{2.149}$$

Let the capacity-achieving set $\mathcal{P}_{\mathrm{ca}}$ be the set of distributions P_X^* achieving capacity in (2.149). For any $\varepsilon \in [0, 1)$, define the following *channel dispersion function*

$$V(P_{Y|X}, \varepsilon) := \begin{cases} V_- := \min_{P_X \in \mathcal{P}_{\mathrm{ca}}} \mathrm{Var}[\imath_{\mathrm{c}}(X; Y)] & \text{if } \varepsilon < 0.5, \\ V_+ := \max_{P_X \in \mathcal{P}_{\mathrm{ca}}} \mathrm{Var}[\imath_{\mathrm{c}}(X; Y)] & \text{if } \varepsilon \geq 0.5. \end{cases} \tag{2.150}$$

The case of $\varepsilon < 0.5$ is of interest since a small error probability is desired.

Assume that the channel dispersion $V(P_{Y|X}, \varepsilon)$ is finite.

Theorem 2.3 (Second-Order Asymptotics for DMC) *For any $\varepsilon \in (0, 1)$, the maximal transmission rate $R_{\mathrm{c}}^*(n, \varepsilon)$ satisfies*

$$R_{\mathrm{c}}^*(n, \varepsilon) = C(P_{Y|X}) - \frac{\sqrt{V(P_{Y|X}, \varepsilon)}\mathrm{Q}^{-1}(\varepsilon)}{\sqrt{n}} + o\Big(\frac{1}{\sqrt{n}}\Big). \tag{2.151}$$

An implicit assumption is needed for Theorem 2.3, which requires the third absolute moment of the mutual information density is finite for capacity achieving input distributions. This assumption is satisfied by most memoryless channels with finite input or output alphabet, including DMC. It was shown by Tomamichel and Tan [27] that $o\Big(\frac{1}{\sqrt{n}}\Big) = \frac{\log n}{2n}$ for a large set of DMS including the BSC. However, the above result does not hold for BEC [26, Theorem 52], for which the third-order term satisfies $o\left(\frac{1}{\sqrt{n}}\right) = O\left(\frac{1}{n}\right)$ [26, Theorem 53].

Recall that $H_{\mathrm{b}}(\cdot)$ is the binary entropy function. For a BSC with parameter $\xi \in (0, 1)$, the capacity [12, Eq. (7.7)] is

$$C(P_{Y|X}) = \log 2 - H_{\mathrm{b}}(\xi). \tag{2.152}$$

The capacity achieving input distribution is unique and given by $P_X^* = [0.5, 0.5]$. The induced channel output distribution is $P_Y^* = [0.5, 0.5]$. For any $(x, y) \in \{0, 1\}^2$, it follows from (2.148) that the mutual information density is

$$\imath_{\mathrm{c}}(x; y) = \mathbb{1}(y \neq x) \log(\zeta) + \mathbb{1}(y = x) \log(1 - \zeta) + \log 2. \tag{2.153}$$

It can be verified that

$$\mathbb{E}_{P_X^* \times P_{Y|X}}[\imath_{\mathrm{c}}(X; Y)] = (1 - \xi)(\log(1 - \xi) + \log 2) + \xi(\log \xi + \log 2) \tag{2.154}$$

$$= \log 2 - H_{\mathrm{b}}(\xi). \tag{2.155}$$

For any $\varepsilon \in (0, 1)$ such that $\varepsilon \neq 0.5$, the channel dispersion function is

$$V(P_{Y|X}, \varepsilon) = \mathrm{Var}_{P_X^* \times P_{Y|X}}[\imath_{\mathrm{c}}(X; Y)] \tag{2.156}$$

$$= (1 - \xi)\xi^2(\log(1 - \xi) - \log(\xi))^2 + \xi(1 - \xi)^2(\log(\xi) - \log(1 - \xi))^2 \tag{2.157}$$

$$= \xi(1 - \xi)(\log(1 - \xi) - \log(\xi))^2. \tag{2.158}$$

In Fig. 2.4, the first-, second-, and third-order asymptotics are plotted for BSC with parameter $\xi = 0.1$ when error probability is $\varepsilon = 0.001$. It is observed that Shannon's first-order asymptotic result deviates from refined second- and third-order asymptotics finite blocklength.

For a BEC channel with parameter $\tau \in (0, 1)$, the capacity [12, Eq. (7.16)] is

$$C(P_{Y|X}) = (1 - \tau) \log 2. \tag{2.159}$$

The capacity achieving input distribution is unique and given $P_X^* = \frac{1}{2}$. The induced optimal output distribution satisfies $P_Y^*(0) = P_Y^*(1) = \frac{1-\tau}{2}$ and $P_Y^*(\mathrm{e}) = \tau$. For any $(x, y) \in \{0, 1\} \times \{0, 1, \mathrm{e}\}$, it follows from (2.148) that the mutual information density is

$$\imath_{\mathrm{c}}(x; y) = \mathbb{1}(y = x) \log(1 - \tau) - \mathbb{1}(y \neq \mathrm{e}) \log\left(\frac{1 - \tau}{2}\right). \tag{2.160}$$

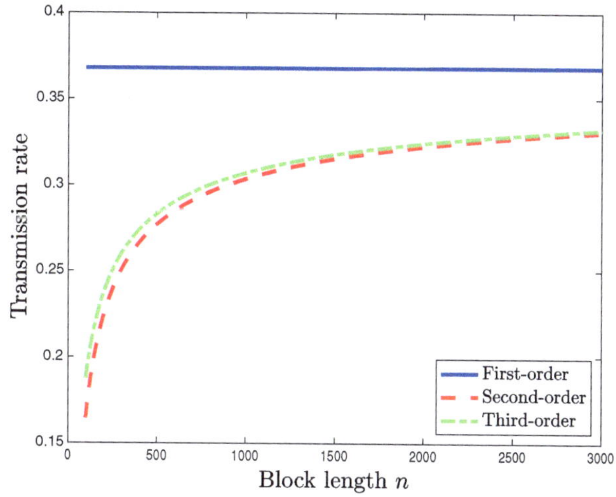

Fig. 2.4 Plot of first-, second-, and third-order asymptotics for BSC with parameter $\xi = 0.1$ when error probability is $\varepsilon = 0.001$

It can be verified that

$$\mathbb{E}[\imath_c(X;Y)] = (1-\tau)\log(1-\tau) - (1-\tau)\log\left(\frac{1-\tau}{2}\right) \tag{2.161}$$

$$= (1-\tau)\log 2. \tag{2.162}$$

For any $\varepsilon \in (0,1)$, the channel dispersion is

$$V(P_{Y|X},\varepsilon) = \mathrm{Var}[\imath_c(X;Y)] \tag{2.163}$$

$$= \mathbb{E}[(\imath_c(X;Y) - (1-\tau)\log 2)^2] \tag{2.164}$$

$$= \Pr\{X=Y\}\tau^2(\log 2)^2 + \Pr\{Y=\mathrm{e}\}((1-\tau)\log 2)^2 \tag{2.165}$$

$$= (1-\tau)\tau^2(\log 2)^2 + \tau(1-\tau)^2(\log 2)^2 \tag{2.166}$$

$$= \tau(1-\tau)\log 2. \tag{2.167}$$

Theorem 2.3 also holds for binary-input AWGN channels. Fix $\sigma = 1$. Recall that $\mathcal{N}(y;0,\sigma^2)$ denotes the pdf of the Gaussian distribution with mean 0 and variance σ^2 with y being the realization variable. Let $\mathcal{N}^{\mathrm{mix}}(y;P,\sigma^2)$ be the pdf of the random variable Y such that $Y \sim \mathcal{N}(\cdot;P,\sigma^2)$ with probability 0.5 and $Y \sim \mathcal{N}(\cdot;-P,\sigma^2)$ with probability 0.5. For any binary-input AWGN channel, the capacity achieving distribution is $P_X^* = [0.5, 0.5]$. Thus, given $(x,y) \in \{-1,1\}\times\mathbb{R}$, the information density for a binary-input AWGN channel is

$$\imath_c(x;y) = \log \mathcal{N}(y-x;0,\sigma^2) - \log(\mathcal{N}^{\mathrm{mix}}(y;P,\sigma^2)). \tag{2.168}$$

The capacity of binary-input AWGN channels does not have explicit formulas, which can be calculated using integrals numerically as

$$C(P_{Y|X}) = \mathbb{E}[\imath_c(X;Y)] \tag{2.169}$$

$$= \mathbb{E}[\log \mathcal{N}(Y-X;0,\sigma^2)] - \mathbb{E}[\log \mathcal{N}^{\text{mix}}(Y;P,\sigma^2)] \tag{2.170}$$

$$= -\frac{1}{2}\log(2\pi\sigma^2) - \frac{\mathbb{E}[(Y-X)^2]}{2\sigma^2} + \log 2 + \frac{1}{2}\log(2\pi\sigma^2)$$
$$- \mathbb{E}\Big[\exp\Big(-\frac{(Y-P)^2}{2\sigma^2}\Big) + \exp\Big(-\frac{(Y-\sqrt{P})^2}{2\sigma^2}\Big)\Big] \tag{2.171}$$

$$= \log 2 - \frac{1}{2} - \mathbb{E}\Big[\exp\Big(-\frac{(Y-\sqrt{P})^2}{2\sigma^2}\Big) + \exp\Big(-\frac{(Y+P)^2}{2\sigma^2}\Big)\Big]. \tag{2.172}$$

Since the capacity achieving input distribution is unique, the channel dispersion function $V(P_{Y|X},\varepsilon)$ remains the same for any $\varepsilon \in (0,1)$, which can be calculated numerically using

$$V(P_{Y|X},\varepsilon) := \text{Var}[\imath_c(X;Y)] \tag{2.173}$$

$$= \mathbb{E}\Big[\big(\log \mathcal{N}(Y-X;0,\sigma^2)\big)^2\Big] + \mathbb{E}\Big[\big(\log \mathcal{N}^{\text{mix}}(Y;P,\sigma^2)\big)^2\Big]$$
$$- 2\mathbb{E}\big[\log \mathcal{N}(Y-X;0,\sigma^2)\log \mathcal{N}^{\text{mix}}(Y;P,\sigma^2)\big]$$
$$- (C(P_{Y|X}))^2. \tag{2.174}$$

In Fig. 2.5, the first- and second-order asymptotics are plotted for a binary-input AWGN channel with $P=5$ when error probability is $\varepsilon = 0.01$.

2.3.2.2 AWGN Channels

Theorem 2.3 does not hold for AWGN channels with the continuous input alphabet. For simplicity, set $\sigma = 1$ so that the signal to noise ratio (SNR) $\frac{P}{\sigma^2} = P$. Furthermore, define the following functions

$$C(P) := \log(1+P), \tag{2.175}$$

$$V(P) := \frac{P(P+2)}{2(P+1)^2} \tag{2.176}$$

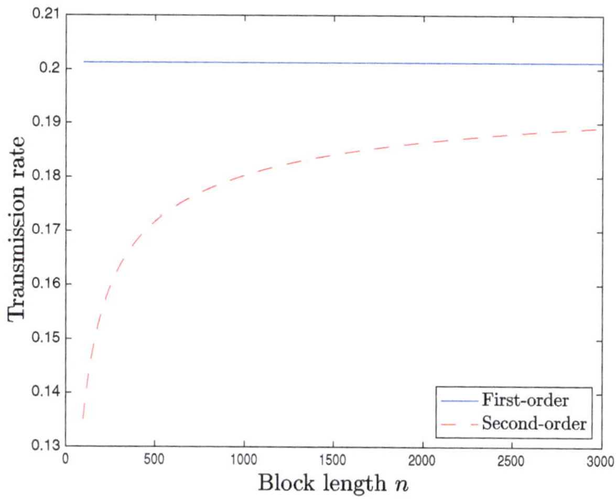

Fig. 2.5 Plot of first- and second-order asymptotics for a binary-input AWGN channel with $P = 5$ when error probability is $\varepsilon = 0.01$

Theorem 2.4 (Second-Order Asymptotics for AWGN Channels) *For any* $\varepsilon \in (0, 1)$, *the maximal transmission rate* $R_c^*(n, \varepsilon)$ *for AWGN channels satisfies*

$$R_c^*(n, \varepsilon) = C(P) - \frac{\sqrt{V(P)}\mathrm{Q}^{-1}(\varepsilon)}{\sqrt{n}} + \frac{\log n}{2n} + O\Big(\frac{1}{n}\Big). \tag{2.177}$$

The first-order asymptotics was shown by Shannon [1]. The second-order asymptotics was firstly proved by Hayashi [25, Theorem 5], who bounded the remainder term as $O(\frac{1}{\sqrt{n}})$. Subsequently, Polyanskiy, Poor, and Verdú [26, Theorem 54] showed that the third-order term is bounded from below by $O(\frac{1}{n})$ and from above by $O(\frac{\log n}{n})$. The exact third-order term was proved by Tan and Tomamichel [28]. Bounds on the remainder fourth-order term was recently obtained by Yavas, Kostina and Effros [29, Theorem 5]. In Fig. 2.6, the first-, second-, third-order asymptotics and the fourth-order upper bound are plotted for an AWGN channel with $P = 3$ when error probability is $\varepsilon = 0.1$.

For AWGN channels, the first-order asymptotic result of Shannon [1] is achieved by Gaussian codewords generated i.i.d. from the Gaussian distribution with mean zero and variance P under average power constraint. However, for second-order asymptotics, under both the average and maximal power constraint, the optimal performance is achieved by spherical codewords generated uniformly over the sphere of radius $\sqrt{nP}$ so that each codeword satisfies the maximal power constraint with equality. If one mistakenly applies the i.i.d. Gaussian codewords under the average power constraint, the achieved second-order term is suboptimal with a

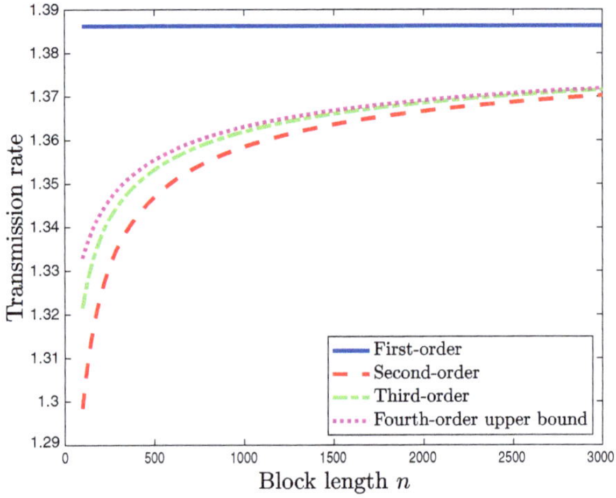

Fig. 2.6 Plot of first-, second-, third-order asymptotics and the fourth-order upper bound for an AWGN channel with $P = 3$ when error probability is $\varepsilon = 0.1$

larger channel dispersion (cf. (2.185)). Specifically, this corresponds to applying Theorem 2.3 to AWGN channels. In terms of the capacity, the optimizer $P_X \in \mathcal{P}(\mathcal{X})$ is restricted to input distributions P_X such that $\mathbb{E}[X^2] \leq P$. It follows from [12, Eq. (9.17)] that

$$C(P_{Y|X}) = \frac{1}{2}\log\left(1 + \frac{P}{\sigma^2}\right). \tag{2.178}$$

The capacity achieving input distribution is $P_X^* = \mathcal{N}(\cdot; 0, P)$. Thus, the optimal output distribution is $P_Y^* = \mathcal{N}(\cdot; 0, \sigma^2 + P)$. For any $(x, y) \in \mathbb{R}^2$, it follows from (2.148) that the mutual information density is

$$\imath(x; y) = \log\frac{\mathcal{N}(y - x; 0, \sigma^2)}{\mathcal{N}(y; 0, P + \sigma^2)} \tag{2.179}$$

$$= -\frac{1}{2}\log(2\pi\sigma^2) - \frac{(y - x)^2}{2\sigma^2} + \frac{1}{2}\log(2\pi(\sigma^2 + P)) + \frac{y^2}{2(\sigma^2 + P)} \tag{2.180}$$

$$= \frac{y^2}{2(\sigma^2 + P)} - \frac{(y - x)^2}{2\sigma^2} + \frac{1}{2}\log\left(1 + \frac{P}{\sigma^2}\right). \tag{2.181}$$

Note that

$$\mathbb{E}[\imath(X; Y)] = \frac{\mathbb{E}[Y^2]}{2(\sigma^2 + P)} - \frac{\mathbb{E}[(Y - X)^2]}{2\sigma^2} + \frac{1}{2}\log\left(1 + \frac{P}{\sigma^2}\right) \tag{2.182}$$

$$= \frac{\sigma^2 + P}{2(\sigma^2 + P)} - \frac{\sigma^2}{2\sigma^2} + \frac{1}{2}\log\left(1 + \frac{P}{\sigma^2}\right) \tag{2.183}$$

$$= \frac{1}{2}\log\left(1+\frac{P}{\sigma^2}\right). \tag{2.184}$$

Furthermore, for any $\varepsilon \in (0,1)$, it follows from [30, Eq. (88)-(92)] that the channel dispersion function is

$$V(P_{Y|X},\varepsilon) = \mathrm{Var}[\imath(X;Y)] = V(P) + \frac{1}{2}\left(\frac{P}{P+1}\right)^2. \tag{2.185}$$

Thus, for any $\varepsilon \in (0, 0.5)$, the second-order asymptotics with dispersion $V(P_{Y|X},\varepsilon)$ is strictly smaller than the optimal result in Theorem 2.4 since $\mathrm{Q}^{-1}(\varepsilon)$ is positive.

The above superiority of spherical codewords over i.i.d. Gaussian codewords for second-order asymptotics of AWGN channels generalized to other additive noise channels [30, Theorem 1].

2.3.3 Proof of Second-Order Asymptotics

Detailed proof steps for second-order asymptotics of both DMC and AWGN channels are well summarized at the textbook level [20, Chapters 4]. However, the above proof is highly involved since it characterizes the third-order term exactly. This section presents a simpler achievability proof with exact second-order asymptotics using the random coding union bound [26, Theorem 16].

Recall the definition of the information density $\imath_c(x;y)$ in (2.148). For any $(x^n, y^n) \in \mathcal{X}^n \times \mathcal{Y}^n$, define the accumulated mutual information density as

$$\imath_c(x^n;y^n) := \sum_{i\in[n]} \imath_c(x_i;y_i). \tag{2.186}$$

The random coding union bound [26, Theorem 16] yields the following finite blocklength upper bound.

Lemma 2.4 *Given any $(n, M) \in \mathbb{N}^2$ and any $P_X \in \mathcal{P}(\mathcal{X})$, there exists an (n, M)-code for channel coding over an binary-input AWGN channel such that*

$$\varepsilon \le \mathbb{E}\big[\min\big\{1, M\Pr\big\{\imath_c(\bar{X}^n;Y^n) \ge \imath_c(X^n;Y^n)|X^n,Y^n\big\}\big\}\big]. \tag{2.187}$$

where $(X^n, \bar{X}^n, Y^n)$ are jointly distributed as $P_X^n(X^n)P_X^n(\bar{X}^n)(P_{Y|X})^n(Y^n|X^n)$.

Proof Let $\mathbf{X}^n = (X^n(1),\ldots,X^n(M))$ be M independent codewords, each generated i.i.d. from P_X. Consider the following (n, M)-code. Given each message $W \in [M]$, the encoder f transmits the codeword $X^n(W)$ as the channel input. Given the noisy channel output Y^n, the decoder applies the maximal mutual information density decoder and estimates the message as $\hat{W}$, i.e.,

$$\hat{W} = \arg\max_{m\in[M]} \imath_{\mathrm{c}}(X^n(m); Y^n) \tag{2.188}$$

For the above code, an error event occurs if there exists a message $m \neq W$ such that $\imath_{\mathrm{c}}(X^n(m); Y^n) \leq \imath_{\mathrm{c}}(X^n(W); Y^n)$. It follows that

$$\mathrm{P}_{\mathrm{e},n} \leq \Pr\left\{\exists\, m \in [M]:\ m \neq W,\ \imath_{\mathrm{c}}(X^n(\bar{W}); Y^n) \leq \imath_{\mathrm{c}}(X^n(W); Y^n)\right\} \tag{2.189}$$

$$= \frac{1}{M}\sum_{w\in[M]} \Pr\left\{\exists\, m \in [M]:\ m \neq w,\ \imath_{\mathrm{c}}(X^n(m); Y^n) \leq \imath_{\mathrm{c}}(X^n(w); Y^n)\right\}. \tag{2.190}$$

where (2.190) follows since the message W is uniformly distributed. For each $w \in [M]$, define the error event $\mathcal{E}(w) := \{\exists\, m \in [M]:\ m \neq w,\ \imath_{\mathrm{c}}(X^n(m); Y^n) \leq \imath_{\mathrm{c}}(X^n(w); Y^n)\}$. The probability term in (2.190) can be further upper bounded as follows:

$$\Pr\left\{\mathcal{E}(w)\right\} = \mathbb{E}\Big[\Pr\{\mathcal{E}(w)\big|X^n(w), Y^n\}\Big] \tag{2.191}$$

$$= \mathbb{E}\Big[\min\left\{\Pr\{\mathcal{E}(w)|X^n(w), Y^n\}, 1\right\}\Big]. \tag{2.192}$$

where (2.191) follows from the law of total expectation that $\mathbb{E}[f(X, Y)] = \mathbb{E}_Y[\mathbb{E}_X[f(X, Y)|Y]]$, and (2.192) follows since any probability term is upper bounded by one. Furthermore, given any (x^n, y^n), it follows that

$$\Pr\{\mathcal{E}(w)|X^n(w) = x^n, Y^n = y^n\}$$

$$\leq \sum_{m\in[M]:\ m\neq w} \Pr\{\imath_{\mathrm{c}}(X^n(m); Y^n) \leq \imath_{\mathrm{c}}(x^n; y^n)\} \tag{2.193}$$

$$\leq \sum_{m\in[M]:\ m\neq w} \Pr\{\imath_{\mathrm{c}}(\bar{X}^n; Y^n) \leq \imath_{\mathrm{c}}(x^n; y^n)\} \tag{2.194}$$

where (2.194) follows since each codeword is independent of each other and generated i.i.d. from P_X.

Combining (2.192) and (2.194) and using again the i.i.d. nature of codeword lead to

$$\Pr\left\{\mathcal{E}(w)\right\} \leq \mathbb{E}\Big[\min\left\{M\Pr\{\imath_{\mathrm{c}}(\bar{X}^n; Y^n) \leq \imath_{\mathrm{c}}(X^n(w); Y^n)|X^n(w), Y^n\}, 1\right\}\Big] \tag{2.195}$$

$$= \mathbb{E}\Big[\min\left\{M\Pr\{\imath_{\mathrm{c}}(\bar{X}^n; Y^n) \leq \imath_{\mathrm{c}}(X^n; Y^n)|X^n, Y^n\}, 1\right\}\Big] \tag{2.196}$$

The proof of Lemma 2.4 is completed by combining (2.190) and (2.196).

The expectation term in Lemma 2.4 is upper bounded as follows:

$$\begin{aligned}
&\mathbb{E}\left[\min\left\{1, M \Pr\left\{\imath_c(\bar{X}^n; Y^n) \geq \imath_c(X^n; Y^n) | X^n, Y^n\right\}\right\}\right] \\
&\leq \Pr\left\{M \Pr\left\{\imath_c(\bar{X}^n; Y^n) \geq \imath_c(X^n; Y^n) | X^n, Y^n\right\} > \frac{1}{\sqrt{n}}\right\} + \frac{1}{\sqrt{n}} &(2.197)\\
&\leq \Pr\left\{M \exp(-\imath_c(X^n; Y^n)) \geq \frac{1}{\sqrt{n}}\right\} + \frac{1}{\sqrt{n}} &(2.198)\\
&= \Pr\left\{\log M - \imath_c(X^n; Y^n) \geq -\frac{1}{2}\log n\right\} + \frac{1}{\sqrt{n}} &(2.199)\\
&= \Pr\left\{\sum_{i\in[n]} \imath_c(X_i; Y_i) \leq \log M + \frac{1}{2}\log n\right\} + \frac{1}{\sqrt{n}}, &(2.200)
\end{aligned}$$

where (2.197) follows from [30, Eq. (37)], which states that $\mathbb{E}[\min\{1, J\}] \leq \Pr\{J > \frac{1}{\sqrt{n}}\} + \frac{1}{\sqrt{n}}$ for any $n \in \mathbb{N}$, (2.198) follows from the change of measure technique, which states that given (X^n, Y^n),

$$\begin{aligned}
\Pr\{\imath(\bar{X}^n; y^n) \geq \imath_c(X^n; Y^n)\} &= \sum_{\bar{x}^n} P_X^n(\bar{x}^n)\mathbb{1}(\imath(\bar{x}^n; Y^n) \geq \imath_c(X^n; Y^n)) &(2.201)\\
&\leq \sum_{\bar{x}^n} P_{X|Y}^n(\bar{x}^n|Y^n)\exp(-\imath_c(X^n; Y^n)) &(2.202)\\
&= \exp(-\imath(X^n; Y^n)), &(2.203)
\end{aligned}$$

where (2.202) follows from the fact that $\imath(x^n; y^n) \geq \imath_c(X^n; y^n)$ and the definition of the accumulated mutual information density in (2.186), which leads to the desired upper bound for $P_X^n(\tilde{x}^n)$ in (2.202).

Let

$$T_+ := \max_{P_X \in \mathcal{P}_{\mathrm{ca}}} \mathbb{E}[|\imath_c(X; Y) - C(P_{Y|X})|^3] \tag{2.204}$$

denote the maximal third absolute moment of the mutual information density for capacity achieving input distributions. Note that T_+ is bounded for any channel with discrete input alphabet $\mathcal{X}$ [26, Lemma 46]. Given any $\varepsilon \in (0, 1)$, let

$$\varepsilon_n^{\mathrm{c}} := \varepsilon - \frac{1}{\sqrt{n}} - \frac{T(P_{Y|X})}{\sqrt{nV(P_{Y|X}, \varepsilon)^3}}. \tag{2.205}$$

Choose M such that

$$\log M = nC(P_{Y|X}) - \sqrt{nV(P_{Y|X}(\varepsilon))}\mathrm{Q}^{-1}(\varepsilon_n^{\mathrm{c}}) - \frac{1}{2}\log n. \tag{2.206}$$

Combining Lemma 2.4 and (2.200), and applying the Berry-Esseen theorem yield

$$\mathrm{P}_{\mathrm{e},n} \leq \varepsilon. \tag{2.207}$$

The achievability proof of Theorem 2.3 is completed by using the Taylor expansion of $\mathrm{Q}^{-1}(\cdot)$, the choice of M in (2.206) and the definition of rate $R_n = \frac{\log M}{n}$.

2.4 Joint Source-Channel Coding

The goal of communication is to transmit a source over a noisy channel, which is naturally a joint source-channel coding problem. Asymptotically as the blocklength tends to infinity, Shannon [1, 9] showed that it is optimal to apply source coding and channel coding separately in a sequential manner, so that the redundancy of a source is first removed and the compressed version is transmitted over the noisy channel. Note that in source coding, blocklength refers to the length of the source sequence while in channel coding, blocklength refers to the number of channel uses. Under the low-latency constraint, Shannon's separation theorem fails: joint source-channel coding achieves strictly better performance than separate source-channel coding [31, 32].

2.4.1 Problem Formulation

Fix four alphabets $\mathcal{S}, \mathcal{X}, \mathcal{Y}, \hat{\mathcal{S}}$ and two integers $(k, n) \in \mathbb{N}^2$. Let $P_S \in \mathcal{P}(\mathcal{S})$ denote the source distribution and $P_{Y|X} \in \mathcal{P}(\mathcal{Y}|\mathcal{X})$ denote the channel transition matrix. Assume that the source sequence S^k is generated i.i.d. from P_S. As shown in Fig. 2.7, a source sequence S^k is encoded into a channel input X^n, which is transmitted over the memoryless channel $P_{Y|X}^n$ and yields noisy channel output Y^n. Subsequently, a decoder produces source estimate $\hat{S}^k$. Similarly to lossy source coding, the estimate in JSCC is allowed to be imperfect and the distortion is

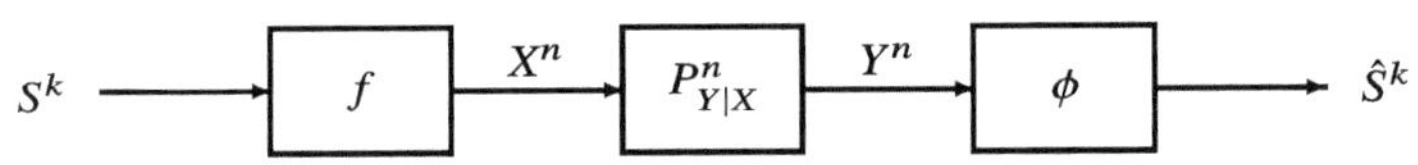

Fig. 2.7 System model for joint source-channel coding

measured via $d(S^k, \hat{S}^k) := \frac{1}{k}\sum_{i\in[k]} d(S_i, \hat{S}_i)$, where the distortion function $d(\cdot)$ was defined in Sect. 2.2.1

A code specifies how encoding and decoding is done.

Definition 2.9 An (k, n)-code for JSCC consists of an encoder $f : \mathcal{S}^k \to \mathcal{X}^n$ and a decoder $\phi : \mathcal{Y}^n \to \hat{\mathcal{S}}^k$.

Thus, $\hat{S}^k = \phi(f(S^k))$. For JSCC, the rate is defined as $R = \frac{k}{n}$, which measures the number of source symbols that can be transmitted per channel use. The reliability is measured via the following excess-distortion probability with respect to a distortion level $D \in \mathbb{R}_+$, analogously to lossy source coding:

$$\mathrm{P_e}(n, k, D) := \Pr\{d(\hat{S}^k, S^k) > D\}. \tag{2.208}$$

However, different from (2.54) for lossy source coding, the excess-distortion probability here is calculated over the channel distribution in addition to the source distribution.

Under the low-latency constraint, given any $\varepsilon \in (0, 1)$, the theoretical benchmark is defined as follows:

$$R^*_{\text{jscc}}(n, D, \varepsilon) := \sup\left\{\frac{k}{n} : \exists \text{ an } (k, n)\text{-code for JSCC s.t. } \mathrm{P_e}(n, k, D) \le \varepsilon\right\}. \tag{2.209}$$

Note that $R^*_{\text{jscc}}(n, D)$ is the maximal number of source symbols per channel use that can be transmitted with blocklength n when the excess-distortion probability with respect to D is no greater than ε.

Fix an integer M and an alphabet $\mathcal{M} = [M]$. In contrast to JSCC, as shown in Fig. 2.8, SSCC separates the communication process into source coding with encoder f_s and decoder ϕ_s, and channel coding with encoder f_c and decoder ϕ_c. The source sequence is first compressed by the source encoder f_s into an index $W \in \mathcal{M}$, which is further encoded into channel input X^n by the channel encoder f_c. Subsequently, the memoryless noisy channel yields the noisy output Y^n, which is further decoded into an index estimate $\hat{W}$ by ϕ_c and finally processed by ϕ_s to generate source estimate $\hat{S}^k$. A code for SSCC is defined analogously.

Definition 2.10 An (k, n, M)-code for SSCC consists of an (k, M)-code for source coding per Definition 2.1 and an (n, M)-code for channel coding per Definition 2.8.

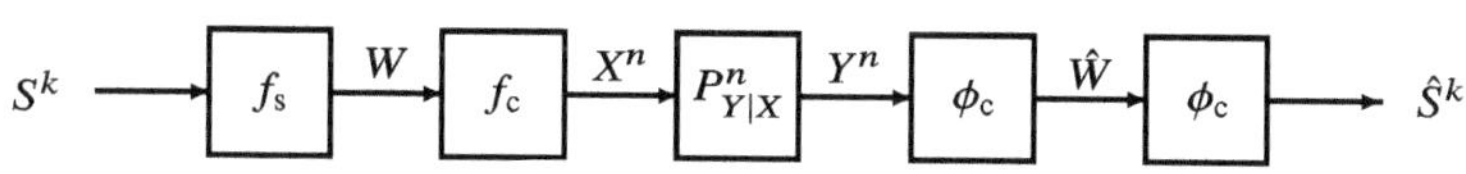

Fig. 2.8 System model for separate source-channel coding

Analogously to (2.209), the low-latency theoretical benchmark for SSCC is defined as follows:

$$R^*_{\text{sscc}}(n, D, \varepsilon) := \sup\left\{\frac{k}{n} : \exists \text{ an } (k, n, M)\text{-code for SSCC s.t. } \mathrm{P_e}(n, k, D) \leq \varepsilon\right\}. \tag{2.210}$$

Recall the definitions of the rate-distortion function $R(\cdot)$ in (2.59) and the capacity $C(P_{Y|X})$ in (2.149). Shannon [9] showed that as $n \to \infty$, both JSCC and SSCC achieve the same first-order asymptotics.

$$\lim_{n\to\infty} R^*_{\text{jscc}}(n, D, \varepsilon) = \lim_{n\to\infty} R^*_{\text{sscc}}(n, D, \varepsilon) = \frac{C(P_{Y|X})}{R(P_S, D)} =: \rho(P_S, P_{Y|X}, D). \tag{2.211}$$

This result guides the design of SSCC in practical communication systems for decades. The low-latency study of JSCC was pioneered by Wang, Ingber, and Kochman [31], who derived second-order asymptotics for transmitting a DMS over a DMC and demonstrated the inferiority of SSCC. Subsequently, the results in [31] were generalized by Kostina and Verdú [32] to transmit an arbitrary memoryless source over an arbitrary memoryless channels under mild conditions. Furthermore, using [11, 26], Kostina and Verdú derived finite blocklength upper and lower bounds for the excess-distortion probability of optimal codes. It follows from [9, 31, 32] that for any $\varepsilon \in (0, 1)$

$$R^*_{\text{jscc}}(n, D, \varepsilon) = a + \frac{b}{\sqrt{n}} + \frac{c \log n}{n}. \tag{2.212}$$

The first-order term $a = \rho(P_S, P_{Y|X}, D)$ strengthens Shannon's result with strong converse. The second-order term b is a function of the source dispersion $V(P_S, D)$ in (2.95), channel dispersion $V(P_{Y|X}, \varepsilon)$ in (2.150) and excess-distortion probability ε. The third-order term c is uncharacterized but upper bounded by $\frac{1}{2}$ for several cases including transmitting a GMS over an AWGN channel and the lossless JSCC problem.

2.4.2 *Second-Order Asymptotics*

For any $\varepsilon \in (0, 1)$, define the dispersion for JSCC as

$$V_{\text{jscc}}(P_S, P_{Y|X}, D, \varepsilon) := \frac{\rho(P_S, P_{Y|X}, D)V(P_S, D) + V(P_{Y|X}, \varepsilon)}{(R(P_S, D))^2}. \tag{2.213}$$

Theorem 2.5 *Given any* $\varepsilon \in (0,1)$, *the maximal transmission rate for JSCC satisfies*

$$R^*_{\mathrm{jscc}}(n,D,\varepsilon)=\rho(P_S,P_{Y|X},D)-\sqrt{\frac{\mathrm{V}_{\mathrm{jscc}}(P_S,P_{Y|X},D,\varepsilon)}{n}}\mathrm{Q}^{-1}(\varepsilon)+O\Big(\frac{\log n}{n}\Big). \tag{2.214}$$

The proof of Theorem 2.5 generalizes the proofs of second-order asymptotics for lossy source coding and channel coding judiciously. In particular, a valid achievability scheme is to connect source coding and channel coding via the idea of unequal error protection [31]. Specifically, the source sequence can be classified into various types or power types, and for each type, a SSCC code can be used with a particular chosen parameter M. Similar idea was also used by Kostina and Verdú in their non-asymptotic achievability bound [32].

Fix any $\varepsilon \in (0,1)$. Theorem 2.5 implies that the second-order term b for JSCC is

$$L_{\mathrm{jscc}}(P_S,P_{Y|X},D,\varepsilon)=-\sqrt{\mathrm{V}_{\mathrm{jscc}}(P_S,P_{Y|X},D,\varepsilon)}\mathrm{Q}^{-1}(\varepsilon). \tag{2.215}$$

Correspondingly, the second-order term b for SSCC [32, Eq. (126)] equals

$$L_{\mathrm{sscc}}(P_S,P_{Y|X},D,\varepsilon)=\max_{\substack{(\varepsilon_1,\varepsilon_2)\in(0,1)^2:\\ \varepsilon_1+\varepsilon_2\le\varepsilon}}\Bigg(\frac{-\sqrt{\rho(P_S,P_{Y|X},D)V(P_S,D)}\mathrm{Q}^{-1}(\varepsilon_1)}{R(P_S,D)} - \frac{\sqrt{\mathrm{V}(P_{Y|X},\varepsilon_2)}\mathrm{Q}^{-1}(\varepsilon_2)}{R(P_S,D)}\Bigg). \tag{2.216}$$

Note that $L_{\mathrm{sscc}}(P_S,P_{Y|X},D,\varepsilon)=L_{\mathrm{jscc}}(P_S,P_{Y|X},D,\varepsilon)$ when either the source dispersion $R(P_S,D)=0$ or the channel dispersion $\mathrm{V}(P_{Y|X},\varepsilon)=0$. Otherwise, $L_{\mathrm{sscc}}(P_S,P_{Y|X},D,\varepsilon)<L_{\mathrm{jscc}}(P_S,P_{Y|X},D,\varepsilon)$. Thus, generally, JSCC achieves better second-order asymptotics. In Fig. 2.9, the first- and second-order asymptotics for JSCC and SSCC are plotted for transmitting a GMS over an AWGN channel, where $P_S=\mathcal{N}(\cdot;0,1)$, transmission power $P=3$, noise power $\sigma^2=1$ and $\varepsilon=0.01$.

Intuitively, the performance loss of SSCC results from the separation operation. In SSCC, an excess-distortion event occurs when either the source coding rate $R_n^{\mathrm{s}}=\frac{\log M}{k}$ is too small with respect to a function $f(S^k,D)$ of the source and distortion level, or the channel coding rate $R_n^{\mathrm{c}}=\frac{\log M}{n}$ is too large with respect to another function $g(X^n,Y^n)$ of the channel input and output. However, in JSCC, an error occurs if $\frac{R_n^{\mathrm{c}}}{R_n^{\mathrm{s}}}=\frac{k}{n}$ is too large with respect to $\frac{g(X^n,Y^n)}{f(S^k,D)}$. Since

$$\mathbb{1}\big(R_n^{\mathrm{s}}\ge f(S^k,D)\text{ and }R_n^{\mathrm{c}}\le g(X^n,Y^n)\big)\le \mathbb{1}\Bigg(\frac{R_n^{\mathrm{c}}}{R_n^{\mathrm{s}}}\le\frac{g(X^n,Y^n)}{f(S^k,D)}\Bigg), \tag{2.217}$$

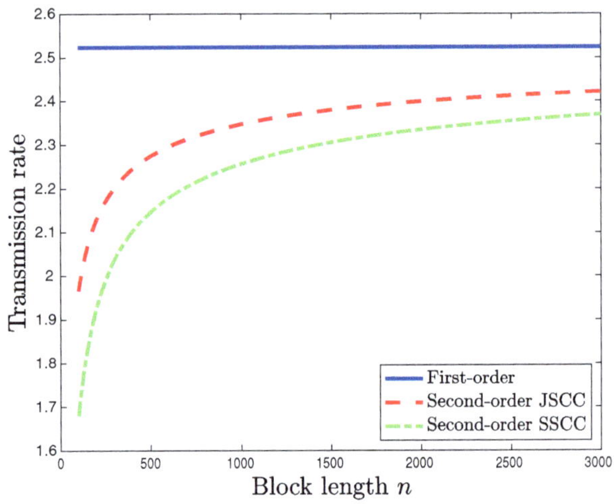

Fig. 2.9 Plot of first- and second-order asymptotics of JSCC for transmitting a GMS $P_S = \mathcal{N}(\cdot; 0, 1)$ over an AWGN channel with $\sigma^2 = 1$, $P = 3$, and $\varepsilon = 0.01$

it follows that

$$\mathbb{1}\left(R_n^{\mathrm{s}} < f(S^k, D) \text{ or } R_n^{\mathrm{c}} > g(X^n, Y^n)\right) \geq \mathbb{1}\left(\frac{R_n^{\mathrm{c}}}{R_n^{\mathrm{s}}} > \frac{g(X^n, Y^n)}{f(S^k, D)}\right), \tag{2.218}$$

and thus SSCC has large excess-distortion probability.

In the literature [32, Theorem 10], the following equivalent form of Theorem 2.5 usually appears. Given any $\varepsilon \in (0, 1)$, the optimal (n, k)-code for JSCC or SSCC with excess-distortion probability no greater than ε satisfies

$$nC(P_{Y|X}) - kR(P_S, D) = \sqrt{n}L_{\dagger}(P_S, P_{Y|X}, D, \varepsilon) + O(\log n). \tag{2.219}$$

where $\dagger \in \{\text{jscc}, \text{sscc}\}$.

2.5 Chapter Summary

This chapter presents second-order asymptotics for lossless source coding, lossy source coding, channel coding, and joint source-channel coding. Definitions, explanations, and proofs are available. Through the contents in this chapter, one could understand the notion, formulas, importance and history of second-order asymptotics. A take-home message is that second-order asymptotic analyses provide tight approximation to the performance of optimal codes for low-latency communication while classical first-order asymptotic Shannon theory are inaccurate.

2.6 Exercises

2.1 Fix the source distribution $P_X = [1/2, 1/4, 1/8, 1/8]$. Calculate the entropy $H(P_X)$, the source dispersion $V(P_X)$. Furthermore, plot $R^*(n, \varepsilon)$ up to second-order asymptotics for $n = 50$ to $n = 3000$ when ε is 0.01, 0.1, and 0.3.

2.2 Prove (2.57), which states that for any bounded distortion measure, the first-order asymptotics is the same under both the expected distortion and the excess-distortion probability criteria.

2.3 Fix any $p \in (0, 1)$. Consider a DMS with $P_X = [p, 1 - p]$. Calculate the distortion-tilted information density under the Hamming distortion measure and plot the second-order asymptotics with when $p = 0.2$, $D = 0.05$, and $\varepsilon = 0.01$ for $n = 50$ to $n = 3000$.

2.4 Set $P_X = [1/4, 1/4, 1/4, 1/4]$. Under the Hamming distortion measure, calculate the rate-distortion function and the distortion-dispersion function.

2.5 Prove the finite blocklength converse bound in Lemma 2.3 for lossy source coding using Claim 3 of Lemma 2.1.

2.6 Let $p = 0.8$ and let q be the solution of $H_b(p) + (1 - p) \log 2 = H_b(q)$. Let $\mathcal{X} = [6]$ and $\mathcal{Y} = [3]$. Consider the noisy channel $P_{Y|X}$ such that

$$P_{Y|X} = \begin{bmatrix} p & \frac{1-p}{2} & \frac{1-p}{2} \\ \frac{1-p}{2} & p & \frac{1-p}{2} \\ \frac{1-p}{2} & \frac{1-p}{2} & p \\ q & 1-q & 0 \\ 0 & q & 1-q \\ 1-q & 0 & q \end{bmatrix}. \tag{2.220}$$

Calculate the dispersion function of the above channel.

2.7 Plot second-order asymptotics for binary-input AWGN channels in Fig. 2.5 and compare the results with AWGN channels in Theorem 2.4 when $P = 3$ and $\varepsilon = 0.1$.

2.8 Fix any $(p, \tau) \in (0, 1)^2$. Calculate the second-order terms for transmitting a binary memoryless source with $P_X(1) = p$ over a BEC with parameter τ. Consider Hamming distortion measure and set D. Plot the second-order asymptotics of JSCC and SSCC when $p = 0.2$, $D = 0.05$, $\tau = 0.1$, and $\varepsilon = 0.01$.

References

1. C.E. Shannon, A mathematical theory of communication. Bell Syst. Tech. J. **27**(1), 379–423 (1948)

2. A.A. Yushkevich, On limit theorems connected with the concept of entropy of Markov chains. Uspekhi Matematicheskikh Nauk **8**(5), 177–180 (1953)
3. V. Strassen, Asymptotische Abschätzungen in Shannon's Informationstheorie, in *Transactions of the Third Prague Conference on Information Theory etc*, Czechoslovak Academy of Sciences (1962), pp. 689–723
4. M. Hayashi, Second-order asymptotics in fixed-length source coding and intrinsic randomness. IEEE Trans. Inf. Theory **54**(10), 4619–4637 (2008)
5. I. Kontoyiannis, S. Verdú, Optimal lossless data compression: Non-asymptotics and asymptotics. IEEE Trans. Inf. Theory **60**(2), 777–795 (2014)
6. S. Chen, M. Effros, V. Kostina, Lossless source coding in the point-to-point, multiple access, and random access scenarios. IEEE Trans. Inf. Theory **66**(11), 6688–6722 (2020)
7. L. Zhou, M. Motani, Finite blocklength lossy source coding for discrete memoryless sources. Found. Trend. Commun. Inf. Theory **20**(3), 157–389 (2023)
8. T.S. Han, *Information-Spectrum Methods in Information Theory* (Springer, Berlin, Heidelberg, 2003)
9. C.E. Shannon, Coding theorems for a discrete source with a fidelity criterion. IRE Nat. Conv. Rec **4**, 142–163 (1959)
10. A. Ingber, Y. Kochman, The dispersion of lossy source coding, in *IEEE 2011 Data Compression Conference* (2011), pp. 53–62
11. V. Kostina, S. Verdú, Fixed-length lossy compression in the finite blocklength regime. IEEE Trans. Inf. Theory **58**(6), 3309–3338 (2012)
12. M.T.C.A. J. Thomas, A. Thomas, *Elements of Information Theory* (Wiley-Interscience, Hoboken, 2006)
13. Z. Zhang, E.-H. Yang, V.K. Wei, The redundancy of source coding with a fidelity criterion-part one: Known statistics. IEEE Trans. Inf. Theory **43**(1), 71–91 (1997)
14. E.H. Yang, Z. Zhang, On the redundancy of lossy source coding with abstract alphabets. IEEE Trans. Inf. Theory **45**(4), 1092–1110 (1999)
15. I. Kontoyiannis, J. Zhang, Arbitrary source models and Bayesian codebooks in rate-distortion theory. IEEE Trans. Inf. Theory **48**(8), 2276–2290 (2002)
16. I. Csiszźr, On an extremum problem of information theory. Studia Sci. Math. Hungarica **9**(1), 57–72 (1974)
17. I. Csiszár, The method of types [information theory]. IEEE Trans. Inf. Theory **44**(6), 2505–2523 (1998)
18. S. Ihara, M. Kubo, Error exponent of coding for memoryless Gaussian sources with a fidelity criterion. IEICE Trans. Fund. Electron. Commun. Comput. Sci. **83-A**(10), 1891–1897 (2000)
19. C.A. Rogers, Covering a sphere with spheres. Mathematika **10**(02), 157–164 (1963)
20. V.Y.F. Tan, Asymptotic estimates in information theory with non-vanishing error probabilities. Found. Trend. Commun. Inf. Theory **11**(1–2), 1–184 (2014)
21. J.-L. Verger-Gaugry, Covering a ball with smaller equal balls in $\mathbb{R}^n$. Discrete Comput. Geometry **33**(1), 143–155 (2005)
22. L. Zhou, V.Y.F. Tan, M. Motani, Second-order and moderate deviation asymptotics for successive refinement. IEEE Trans. Inf. Theory **63**(5), 2896–2921 (2017)
23. V.Y.F. Tan, Moderate-deviations of lossy source coding for discrete and Gaussian sources, in *IEEE International Symposium on Information Theory Proceedings* (2012), pp. 920–924
24. A. Dembo, O. Zeitouni, *Large Deviations Techniques and Applications*, vol. 38 (Springer Science & Business Media, Berlin, 2009)
25. M. Hayashi, Information spectrum approach to second-order coding rate in channel coding. IEEE Trans. Inf. Theory **55**(11), 4947–4966 (2009)
26. Y. Polyanskiy, H.V. Poor, S. Verdú, Channel coding rate in the finite blocklength regime. IEEE Trans. Inf. Theory **56**(5), 2307–2359 (2010)
27. M. Tomamichel, V.Y.F. Tan, A tight upper bound for the third-order asymptotics for most discrete memoryless channels. IEEE Trans. Inf. Theory **59**(11), 7041–7051 (2013)
28. V.Y.F. Tan, M. Tomamichel, The third-order term in the normal approximation for the AWGN channel. IEEE Trans. Inf. Theory **61**(5), 2430–2438 (2015)

29. R.C. Yavas, V. Kostina, M. Effros, Third-order analysis of channel coding in the small-to-moderate deviations regime. IEEE Trans. Inf. Theory **70**(9), 6139–6170 (2024)
30. J. Scarlett, V.Y.F. Tan, G. Durisi, The dispersion of nearest-neighbor decoding for additive non-Gaussian channels. IEEE Trans. Inf. Theory **63**(1), 81–92 (2017)
31. D. Wang, A. Ingber, Y. Kochman, The dispersion of joint source-channel coding, in *Proceedings of the 49th Annual Allerton Conference on Communication, Control, and Computing*, Monticello, IL (2011), pp. 180–187
32. V. Kostina, S. Verdú, Lossy joint source-channel coding in the finite blocklength regime. IEEE Trans. Inf. Theory **59**(5), 2545–2575 (2013)

Chapter 3
Additive Arbitrary Noise Channel

This chapter studies the problem of transmitting messages through an additive arbitrary noise (AAN) channel. Although the studies in Chap. 2 built a solid foundation for low-latency communication [1–10], they share the assumption that additive noise is Gaussian. In most practical communication scenarios, it is challenging to obtain the exact distribution of channel noise. Therefore, it is essential to design code that has guaranteed performance without knowing the precise noise distribution.

To address the above problem, for the point-to-point (P2P) channel, Lapidoth [11] proposed a coding scheme using Gaussian codebooks and *nearest neighbor decoding* to combat AAN and showed that the capacity of the additive white Gaussian noise (AWGN) channel can be achieved for any AAN channel, where the noise power is the same as the AWGN channel. The problem of Lapidoth is named *mismatched channel coding* because the above coding scheme, which is optimal for AWGN channels, is not necessarily optimal for other AAN channels. Inspired by the asymptotic result of Lapidoth [11], Scarlett, Tan, and Durisi [12] derived the second-order asymptotics and generalized their results to the mismatched interference channel. In particular, the authors of [12] showed that the spherical codebook has strictly better low-latency performance than the i.i.d. Gaussian codebook.

This chapter presents the mismatched code of Lapidoth [11] and its low-latency performance in [12]. To better model practical scenarios, the assumption of unity noise power in [12] is slightly generalized to any positive real number.

L. Zhou, L. Bai, *Low-latency Communication*, Textbooks in Telecommunication Engineering, https://doi.org/10.1007/978-3-032-06627-5_3

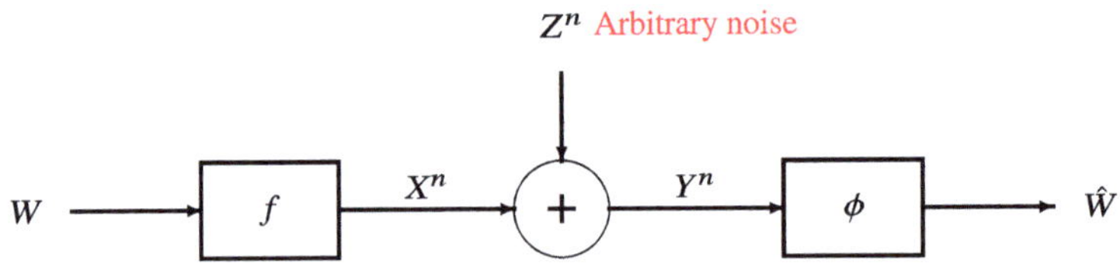

Fig. 3.1 System model for the additive arbitrary noise channel

3.1 Problem Formulation

3.1.1 System Model

Fix two integers $(n, M) \in \mathbb{N}^2$ and three alphabets $(\mathcal{X}, \mathcal{Y}, \mathcal{Z})$. As shown in Fig. 3.1, this chapter studies mismatched channel coding for data transmission over an AAN channel. Given the message $W \in [M]$, the transmitter generates an channel input $X^n \in \mathcal{X}^n$, which is corrupted by the additive noise $Z^n \in \mathcal{Z}^n$ and yields noisy channel output $Y^n \in \mathcal{Y}^n$ such that

$$Y^n = X^n + Z^n. \tag{3.1}$$

Using the decoder ϕ, the receiver estimates the transmitted message as $\hat{W}$.

The noise sequence Z^n is generated i.i.d. from a noise distribution satisfying the following moment constraints:

$$\sigma^2 := \mathbb{E}[Z^2],\ \zeta := \mathbb{E}[Z^4],\ \mathbb{E}[Z^6] < \infty. \tag{3.2}$$

As shown in the code design, only the second moment—the power of the noise—is required. The fourth moment ζ is used to characterize the low-latency performance, while the finiteness of the sixth moment guarantees the theoretical derivations. The above moment constraint is satisfied by common noise distributions, such as discrete, uniform, Gaussian, and Laplace distributions.

3.1.2 Coding Scheme

Fix any positive real number $P \in \mathbb{R}_+$. The mismatched code is defined as follows.

Definition 3.1 An (n, M, P)-code consists of:

- a random codebook C with M independent codewords $\{X^n(1), \ldots, X^n(M)\}$ subject to a power constraint P,

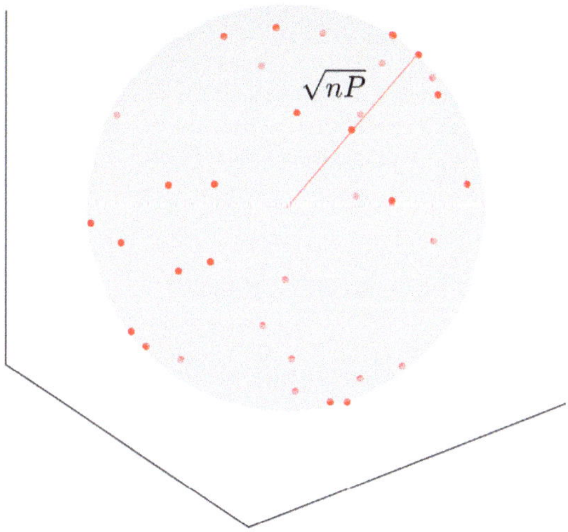

Fig. 3.2 Illustration of a spherical codebook when $n = 3$

- an encoder $f : [M] \to \mathcal{X}^n$ such that for each $w \in [M]$,

$$f(w) := X^n(w), \tag{3.3}$$

- a nearest neighbor (NN) decoder that applies the minimum Euclidean distance decoding rule, i.e.,

$$\phi(Y^n) := \hat{W} = \arg\min_{\bar{w}\in[M]} \|Y^n - X^n(\bar{w})\|^2. \tag{3.4}$$

The random codebook is either a spherical codebook or an i.i.d. Gaussian codebook. Fix any $m \in [M]$. In a spherical codebook, each codeword $X^n(m)$ is generated from the uniform distribution over the sphere with radius $\sqrt{nP}$, i.e.,

$$P_{X^n}^{\mathrm{sp}}(x^n(m)) := \frac{\delta(\|x^n(m)\|^2 - nP)}{S_n(\sqrt{nP})}, \tag{3.5}$$

where $\delta(\cdot)$ is the Dirac delta function and $S_n(r) = 2\pi^{n/2}\, r^{n-1}/\Gamma(n/2)$ is the surface area of a radius-r sphere where $\Gamma(\cdot)$ denotes the Gamma function. For i.i.d. spherical codebooks, the power constraint P is the maximal power constraint. Fig. 3.2 illustrates codewords in a spherical codebook with $n = 3$.

Analogously, in an i.i.d. Gaussian codebook, each codeword $X^n(m)$ is generated i.i.d. from the Gaussian distribution with mean zero and variance P, i.e.,

$$P_{X^n}^{\mathrm{iid}}(x^n(m)) := \prod_{i=1}^{n} \frac{1}{\sqrt{2\pi P}} \exp\left(\frac{x_i^2}{2P}\right). \tag{3.6}$$

For i.i.d. Gaussian codebooks, the power constraint P is the average power constraint.

In the rest of this section, an (n, M, P)-sp-code denotes an (n, M, P)-code using the spherical codebook, while an (n, M, P)-iid-code denotes an (n, M, P)-code using the i.i.d. Gaussian codebook.

3.1.3 Performance Metric

The reliability of the mismatched code is evaluated via the ensemble average error probability

$$\mathrm{P}_{\mathrm{e}}^{n} := \Pr\{\hat{W} \neq W\} \tag{3.7}$$

and the efficiency is evaluated via the transmission rate

$$R_n := \frac{\log M}{n}. \tag{3.8}$$

Note that the ensemble average error probability $\mathrm{P}_{\mathrm{e}}^{n}$ is averaged over the distributions of the random codebook $\mathcal{C}$, the messages $W \in [M]$ and the additive noise Z^n.

Analogously to the matched case in Chap. 2, the low-latency theoretical benchmark for the mismatched code is the maximal transmission rate subject to a constraint on the error probability. Specifically, for any $\varepsilon \in (0, 1)$ and $\dagger \in \{\mathrm{sp}, \mathrm{iid}\}$, the theoretical benchmark is

$$R_{\dagger}^{*}(n, \varepsilon) := \sup\big\{R_n : \exists \text{ an } (n, M, P) - \dagger\text{-code s.t. } \mathrm{P}_{\mathrm{e},n} \leq \varepsilon\big\}. \tag{3.9}$$

Although (3.9) shares the same form as the theoretical benchmark for the matched case in Chap. 2, a striking difference is that for the mismatched code, the converse is ensemble converse, which is used to show that the theoretical analysis of the given coding scheme is tight [13]. In contrast, for the matched case in Chap. 2, the converse is to show that the transmission rate is maximal among all possible coding schemes. Thus, for mismatched communication, the notion of converse is significantly different from the matched case.

It follows from [11, 12] that

$$R_{\dagger}^{*}(n, \varepsilon) = a + \frac{b_{\dagger}}{\sqrt{n}} + O\left(\frac{\log n}{n}\right). \tag{3.10}$$

In particular, Lapidoth showed that regardless of whether one uses the spherical or i.i.d. Gaussian codebook, the first-order rate equals to

$$a = C(P,\sigma^2) := \frac{1}{2}\log\left(1 + \frac{P}{\sigma^2}\right), \tag{3.11}$$

which is the capacity of an AWGN channel with the signal to noise ratio $\frac{P}{\sigma^2}$. The exact second-order term $b_\dagger$ and the order of the remainder term $O\left(\frac{\log n}{n}\right)$ were obtained by Scarlett, Tan and Durisi [12]. This chapter presents and proves the exact characterization of a and $b_\dagger$.

3.2 Second-Order Asymptotics and Discussions

Recall that $P \in \mathbb{R}_+$ is the power constraint, σ^2 is the second moment of the noise distribution and ζ is the fourth moment of the noise distribution. Define the following two dispersion functions:

$$V_{\mathrm{sp}}(P,\sigma^2,\zeta) := \frac{P^2(\zeta - \sigma^4) + 4\sigma^6 P}{4\sigma^4(P+\sigma^2)^2}, \tag{3.12}$$

$$V_{\mathrm{iid}}(P,\sigma^2,\zeta) := \frac{P^2(\zeta + \sigma^4) + 4\sigma^6 P}{4\sigma^4(P+\sigma^2)^2}. \tag{3.13}$$

Note that both dispersion functions increase in the fourth moment ζ.

Theorem 3.1 *For any* $\varepsilon \in (0,1)$ *and* $\dagger \in \{\mathrm{sp}, \mathrm{iid}\}$, *the maximal transmission rate* $R_\dagger^*(n,\varepsilon)$ *satisfies*

$$R_\dagger^*(n,\varepsilon) = C(P,\sigma^2) - \frac{\sqrt{V_\dagger(P,\sigma^2,\zeta)}Q^{-1}(\varepsilon)}{\sqrt{n}} + O\left(\frac{\log n}{n}\right). \tag{3.14}$$

Theorem 3.1 shows that the second-order rate of the mismatched code depends on the second- and fourth-moments of the noise distribution. When specialized to AWGN, it follows that $\zeta = 3\sigma^2$ and the maximal transmission rate using the spherical codebook $R_{\mathrm{sp}}^*(n,\varepsilon)$ equals the result in Theorem 2.4. Thus, the essence of mismatched coding is to apply the optimal code for Gaussian noise to data transmission over an AAN channel.

Furthermore, Theorem 3.1 shows that the low-latency performance is closely related to the fourth moment ζ of the noise distribution. The second moment σ^2 is used to design the codebook and characterizes the first-order rate. The fourth moment ζ determines the second-order term $b_\dagger$, which characterizes how fast the finite blocklength transmission rate approaches the asymptotic first-order rate $C(P,\sigma^2)$. For noise distributions with the same second moment σ^2, the larger the

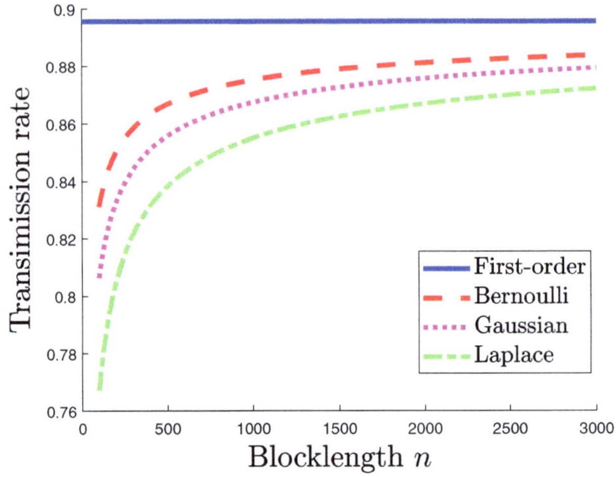

Fig. 3.3 Plot of first- and second-order rates of mismatched channel coding for different noise distributions with $P = 3$, $\sigma^2 = 0.6$ and $\varepsilon = 0.1$

fourth moment, the larger the dispersion function and thus the larger the transmission rate when $\varepsilon \in (0, 0.5)$. Figure 3.3 plots the first-order and second-order rates of the mismatched code for various noise distributions, including Gaussian, Laplace, and Bernoulli distributions. With the same second moment σ^2, the fourth moments of Gaussian, Bernoulli, and Laplace distributions are respectively $(3\sigma^4, \sigma^2, 6\sigma^4)$. Thus, the low-latency performance is best for Bernoulli noise and worst for Laplace noise.

Unlike Theorem 2.4 where the low-latency bound is optimal for a particular channel (i.e., AWGN), Theorem 3.1 is ensemble tight for a fixed coding scheme under AAN channels. Specifically, the converse proof of Theorem 3.1 provides an upper bound for the transmission rate of the mismatched code to ensure that the ensemble average error probability is bounded by ε. Thus, Theorem 3.1 does not show that the mismatched coding scheme has a universal optimal performance. Instead, when the noise distribution is known, it is possible to construct a better coding scheme with larger transmission rates. A suitable interpretation of Theorem 3.1 is that the mismatched code provides a robust way to combat additive arbitrary noise.

Comparing the dispersion term of the second-order rates using the i.i.d. Gaussian codebook (3.13) and the spherical codebook (3.12), it follows that

$$V_{\mathrm{sp}}(P, \sigma^2, \zeta) = V_{\mathrm{iid}}(P, \sigma^2, \zeta) - \frac{P^2\sigma^4}{2\sigma^4(P+\sigma^2)^2} \tag{3.15}$$

$$< V_{\mathrm{iid}}(P, \sigma^2, \zeta). \tag{3.16}$$

Thus, for any $\epsilon \in (0, 0.5)$, the spherical codebook has better low-latency performance than the i.i.d. Gaussian codebook by having a larger second-order

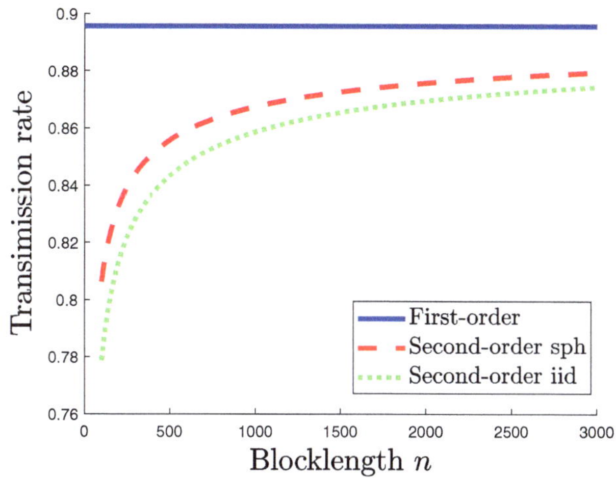

Fig. 3.4 Plot of the second-order asymptotics for (n, M, P)-sp-code and (n, M, P)-i.i.d. -code with power $P = 3$ and $\varepsilon = 0.1$ in a channel with additive noise $\sigma^2 = 1$ and $\zeta = 3$

transmission rate. In Fig. 3.4 plots the second-order rate for mismatched codes with both i.i.d. Gaussian codebooks and spherical codebooks when $P = 3$, $\varepsilon = 0.1$, $\sigma^2 = 1$ and $\zeta = 3$.

3.3 Proofs of Second-Order Asymptotics

This section proves Theorem 3.1 for both spherical codebooks and i.i.d. Gaussian codebooks.

3.3.1 Preliminaries

Given any $(x^n, y^n, z^n) \in (\mathbb{R}_+^n)^3$, define the *mismatched* information density as

$$\tilde{\imath}(x^n; y^n) := nC(P, \sigma^2) + \frac{\|y^n\|^2}{2(P+\sigma^2)} - \frac{\|y^n - x^n\|^2}{2\sigma^2}. \tag{3.17}$$

Furthermore, given a random vector $\bar{X}^n$ that is generated from the uniform distribution over the sphere with radius $\sqrt{nP}$ in (3.5), define the following probability

$$g(t, y^n) := \Pr\{\tilde{\imath}(\bar{X}^n; y^n) \geq t\}. \tag{3.18}$$

The following lemma generalizes the random coding union bound for the matched case [4, Theorem 16] to the mismatched setting.

Lemma 3.1 *There exists an (n, M, P)-sp-code such that the ensemble average error probability satisfies*

$$\mathrm{P}_{\mathrm{e}}^n \leq \mathbb{E}[\min\{1, M \Pr\{\tilde{\imath}(\bar{X}^n; Y^n) \geq \tilde{\imath}(X^n; Y^n)|X^n, Y^n\}\}],$$

where $(X^n, \bar{X}^n, Y^n) \sim P_{X^n} P_{X^n} P_{Y^n|X^n}$, P_{X^n} is the channel input distribution in (3.5) *and $P_{Y^n|X^n}$ is induced by the additive noise channel in* (3.1).

Proof The essence to prove Lemma 3.1 is to use the equivalence between NN decoding and maximum information density decoding. It follows from the NN decoder in (3.4) that given channel output Y^n, the message estimate is given by

$$\phi(Y^n) = \arg\min_{\bar{w}\in[M]} \|Y^n - X^n(\bar{w})\|^2 \tag{3.19}$$

$$= \arg\max_{\bar{w}\in[M]} nC(P, \sigma^2) + \frac{\|Y^n\|^2}{2(P+\sigma^2)} - \frac{\|Y^n - X^n(\bar{w})\|^2}{2\sigma^2} \tag{3.20}$$

$$= \arg\max_{\bar{w}\in[M]} \tilde{\imath}(X^n(\bar{w}); Y^n). \tag{3.21}$$

The result in (3.21) implies that minimizing the Euclidean distance between $X^n(\bar{w})$ and Y^n is equivalent to maximizing the mismatched information density $\tilde{\imath}(X^n(\bar{w}); Y^n)$ over all possible messages $\bar{w} \in [M]$. This way, the ensemble average error probability satisfies

$$\mathrm{P}_{\mathrm{e}}^n = \Pr\{\hat{W} \neq W\} \tag{3.22}$$

$$= \Pr\left\{\exists\, \bar{w} \in [M] :\ \bar{w} \neq W,\ \|Y^n - X^n(\bar{w})\|^2 \leq \|Y^n - X^n(W)\|^2\right\} \tag{3.23}$$

$$= \Pr\left\{\bigcup_{\bar{w}\neq W} \|Y^n - X^n(\bar{w})\|^2 \leq \|Y^n - X^n(w)\|^2\right\} \tag{3.24}$$

$$= \Pr\left\{\bigcup_{\bar{w}\neq W} \tilde{\imath}(X^n(\bar{w}); Y^n) \geq \tilde{\imath}(X^n(W); Y^n)\right\} \tag{3.25}$$

$$= \mathbb{E}\left[\Pr\left\{\bigcup_{\bar{w}\neq W} \tilde{\imath}(X^n(\bar{w}); Y^n) \geq \tilde{\imath}(X^n(W); Y^n)\middle|X^n(W), Y^n\right\}\right] \tag{3.26}$$

$$= \mathbb{E}\left[\min\left\{1, \Pr\left\{\bigcup_{\bar{w}\neq W} \tilde{\imath}(X^n(\bar{w}); Y^n) \geq \tilde{\imath}(X^n(W); Y^n)\middle|X^n(W), Y^n\right\}\right\}\right] \tag{3.27}$$

$$\leq \mathbb{E}\Big[\min\big\{1, (M-1)\Pr\{\tilde{\imath}(\bar{X}^n; Y^n) \geq \tilde{\imath}(X^n; Y^n)|X^n, Y^n\}\big\}\Big], \tag{3.28}$$

where (3.25) follows from the equivalence between NN decoding and maximal mutual information density decoding in (3.21), (3.26) follows from the law of total expectation, (3.27) follows since any probability is no greater than one, (3.28) follows from the union bound and the fact that for any $(w, \bar{w}) \in [M]^2$ with $w \neq w$, the joint distribution of $(X^n(w), X^n(\bar{w}), Y^n)$ is the same and given by $P_{X^n} P_{X^n} P_{Y^n|X^n}$. □

Recall the definition of the probability function $g(\cdot)$ in (3.18). The following lemma is similar to [14, Eq. (58)] and provides an upper bound for this probability term. A key distinction between Lemma 3.2 and [14] is the absence of the $1/\sqrt{n}$ term. This term is crucial for deriving the third-order asymptotics, but the following lemma suffices to obtain the second-order asymptotic result.

Lemma 3.2 *There exists a positive constant $G \in \mathbb{R}_+$ such that*

$$g(t, y^n) \leq G\exp(-t). \tag{3.29}$$

Proof Fix any $i \in [n]$. Let $\tilde{X}_i \sim \mathcal{N}(0, P)$, $\tilde{X}^n = (\tilde{X}_1, \ldots, \tilde{X}_n)$, $\tilde{Y}_i \sim \mathcal{N}(0, 1 + P)$, and $\tilde{Y}^n = (\tilde{Y}_1, \ldots, \tilde{Y}_n)$. Let $P_{\tilde{X}^n}$ denote the distribution of $\tilde{X}^n$ and let $\tilde{Y}^n = \tilde{X}^n + \tilde{Z}^n$, where $\tilde{Z}^n$ is generated i.i.d. from $\mathcal{N}(0, 1)$. Let $P_{\tilde{X}^n\tilde{Y}^n}$ be the joint distribution of $(\tilde{X}^n, \tilde{Y}^n)$ and let any other (conditional) distributions concerning $(\tilde{X}^n, \tilde{Y}^n)$ be induced by this joint distribution. In particular, $P_{\tilde{Y}^n|\tilde{X}^n}$ denotes the channel law for the Gaussian P2P channel. It follows from (3.17) that for any $(x^n, y^n) \in (\mathbb{R}^n)^2$,

$$\tilde{\imath}(x^n; y^n) = \log \frac{P_{\tilde{Y}^n|\tilde{X}^n}(y^n|x^n)}{P_{\tilde{Y}^n}(y^n)}. \tag{3.30}$$

Let $Y_{\mathrm{G}}^n = X^n + \tilde{Z}^n$ be the channel output of a Gaussian P2P channel when the input is spherical codeword X^n that is generated by P_{X^n} in (3.5). Let the joint distribution of (X^n, Y_{G}^n) be $P_{X^nY_{\mathrm{G}}}$ and let other (conditional) distributions concerning (X^n, Y_{G}^n) be induced by this joint distribution. For any $(x^n, y^n) \in (\mathbb{R}^n)^2$, define the following information density:

$$\imath_{\mathrm{G}}(x^n; y^n) := \log \frac{P_{Y_{\mathrm{G}}^n|X^n}(y^n|x^n)}{P_{Y_{\mathrm{G}}^n}(y^n)}. \tag{3.31}$$

It follows from the definition in (3.18) that g can be upper bounded as follows:

$$g(t, y^n) = \Pr_{P_{X^n}}\left\{\log \frac{P_{\tilde{Y}^n|\tilde{X}^n}(y^n|\bar{X}^n)}{P_{\tilde{Y}^n}(y^n)} \geq t\right\} \tag{3.32}$$

$$= \Pr_{P_{X^n}}\left\{\log \frac{P_{\tilde{Y}^n|\tilde{X}^n}(y^n|\bar{X}^n)}{P_{Y_{\mathrm{G}}^n}(y^n)} \geq t - \log \frac{P_{Y_{\mathrm{G}}^n}(y^n)}{P_{\tilde{Y}^n}(y^n)}\right\} \tag{3.33}$$

$$\leq \Pr_{P_{X^n}}\left\{\log\frac{P_{\tilde{Y}^n|\tilde{X}^n}(y^n|\bar{X}^n)}{P_{Y_{\mathrm{G}}^n}(y^n)} \geq t - \log G\right\} \tag{3.34}$$

$$= \Pr_{P_{X^n}}\left\{\log\frac{P_{Y_{\mathrm{G}}^n|X^n}(y^n|\bar{X}^n)}{P_{Y_{\mathrm{G}}^n}(y^n)} \geq t - \log G\right\} \tag{3.35}$$

$$= \Pr_{P_{X^n}}\{\imath_{\mathrm{G}}(\bar{X}^n; y^n) \geq t - \log G\} \tag{3.36}$$

$$= \int_{-\infty}^{\infty} P_{X^n}(\bar{x}^n)\mathbb{1}\{\imath_{\mathrm{G}}(\bar{x}^n; y^n) \geq t - \log G\}\mathrm{d}\bar{x}^n, \tag{3.37}$$

where (3.34) follows from [15, Lemma 1], which implies that $\frac{P_{Y_{\mathrm{G}}^n}(y^n)}{P_{\tilde{Y}^n}(y^n)} \leq G$, (3.35) follows since both $P_{Y_{\mathrm{G}}^n|X^n}$ and $P_{\tilde{Y}^n|\tilde{X}^n}$ denote the distribution of the Gaussian noise $\tilde{Z}^n$, and (3.36) follows from the definition in (3.31).

Bounding (3.37) follows by performing a change-of-measure. Recall that $\bar{X}^n$ follows the spherical distribution in (3.5) and Y_{G}^n is the Gaussian P2P channel with spherically distributed input. It follows that for any $(\bar{x}^n, y^n) \in (\mathbb{R}^n)^2$,

$$P_{X^n}(\bar{x}^n) = P_{X^n|Y_{\mathrm{G}}^n}(\bar{x}^n|y^n)\frac{P_{X^n}(\bar{x}^n)}{P_{X^n|Y_{\mathrm{G}}^n}(\bar{x}^n|y^n)} \tag{3.38}$$

$$= P_{X^n|Y_{\mathrm{G}}^n}(\bar{x}^n|y^n)\exp(-\imath_{\mathrm{G}}(\bar{x}^n; y^n)), \tag{3.39}$$

where (3.39) follows from the definition of information density in (3.31). Given any $\eta \in \mathbb{R}_+$, combining (3.39), (3.37) can be bounded as

$$g(t; y^n) = \int_{-\infty}^{\infty} P_{X^n|Y_{\mathrm{G}}^n}(\bar{x}^n|y^n)\exp(-\imath_{\mathrm{G}}(\bar{x}^n; y^n)) \times \mathbb{1}\{\imath_{\mathrm{G}}(\bar{x}^n; y^n) \geq t - \log G\}\mathrm{d}\bar{x}^n \tag{3.40}$$

$$= \mathbb{E}_{P_{X^n|Y_{\mathrm{G}}^n}}[\exp(-\imath_{\mathrm{G}}(\bar{x}^n; y^n))\mathbb{1}\{\imath_{\mathrm{G}}(\bar{x}^n; y^n) \geq t - \log G\}] \tag{3.41}$$

$$\leq \sum_{l=0}^{\infty}\exp(-(t - \log G + l\eta)) \times \Pr\left\{t - \log G + l\eta \leq \imath_{\mathrm{G}}(\bar{x}^n; y^n) \leq t - \log G + (l+1)\eta\right\} \tag{3.42}$$

$$\leq \sum_{l=0}^{\infty}\exp(-(t - \log G + l\eta)) \tag{3.43}$$

$$= \frac{\exp(-t + \log G)}{1 - \exp(-\eta)} \tag{3.44}$$

$$\leq G \exp(-t), \tag{3.45}$$

where (3.44) follows from the summation of a geometric sequence. □

Define two constants

$$c_1 := \zeta - \sigma^4 + 4\sigma^6 P, \tag{3.46}$$

$$c_2 := \zeta - \sigma^4. \tag{3.47}$$

Given any $\eta \in (0, 2P)$, define the following typical sets of powers:

$$\mathcal{P}_Y := \left\{ r \in \mathbb{R} : |r - (P + \sigma^2)| \leq \sqrt{c_1 \frac{\log n}{n}} \right\}, \tag{3.48}$$

$$\mathcal{P}_Z := \left\{ s \in \mathbb{R} : |s - \sigma^2| \leq \sqrt{c_2 \frac{\log n}{n}} \right\}, \tag{3.49}$$

$$\mathcal{Q} := \left\{ (r, s) \in \mathbb{R}^2 : r + P - s > \eta \right\}, \tag{3.50}$$

$$\mathcal{T} := (\mathcal{P}_Y \times \mathcal{P}_Z) \cap \mathcal{Q}. \tag{3.51}$$

Finally, the following lemma bounds the probability of having atypical output and noise powers.

Lemma 3.3 *Given either a spherical codeword X^n generated from* (3.5) *or an i.i.d. Gaussian codeword X^n generated from* (3.6) *and given Z^n generated i.i.d. from a distribution satisfying* (3.2)*, when $Y^n = X^n + Z^n$, it follows that*

$$\Pr\left\{ \left(\frac{\|Y^n\|^2}{n}, \frac{\|Z^n\|^2}{n} \right) \notin \mathcal{T} \right\} = O\left(\frac{1}{\sqrt{n}} \right). \tag{3.52}$$

3.3.2 *Achievability*

3.3.2.1 Spherical Codebooks

It follows from Lemma 3.1 that the ensemble average probability is upper bounded as

$$\mathrm{P}_{\mathrm{e}}^n \leq \mathbb{E}\Big[\min \left\{ 1, M \Pr\{\tilde{\imath}(\bar{X}^n; Y^n) \geq \tilde{\imath}(X^n; Y^n) | X^n, Y^n\} \right\} \Big] \tag{3.53}$$

$$= \mathbb{E}\Big[\min \left\{ 1, M g(\tilde{\imath}(X^n; Y^n), Y^n) \right\} \Big] \tag{3.54}$$

$$\leq \mathbb{E}\Big[\min\big\{1, MG\exp(-\tilde{\imath}(X^n; Y^n))\big\}\Big] \tag{3.55}$$

$$\leq \Pr\left\{MG\exp(-\tilde{\imath}(X^n; Y^n)) \geq \frac{1}{\sqrt{n}}\right\} + \frac{1}{\sqrt{n}}, \tag{3.56}$$

where (3.54) follows from the definition of the probability function $g(\cdot)$ in (3.18), (3.55) follows from the result in Lemma 3.2 that upper bounds the $g(\cdot)$ probability, and (3.56) follows because for any random variable J and integer $n \in \mathbb{N}$,

$$\mathbb{E}[\min\{1, J\}] = \mathbb{E}[1 \times \mathbb{1}(J > 1)] + \mathbb{E}[J \times \mathbb{1}(J \leq 1)] \tag{3.57}$$

$$\leq \Pr\{J > 1\} + \mathbb{E}\left[1 \times \mathbb{1}\Big(\frac{1}{\sqrt{n}} \leq J \leq 1\Big)\right] + \frac{1}{\sqrt{n}}\mathbb{E}\left[\mathbb{1}\Big(J \leq \frac{1}{\sqrt{n}}\Big)\right] \tag{3.58}$$

$$= \Pr\{J > 1\} + \Pr\left\{1 \geq J \geq \frac{1}{\sqrt{n}}\right\} + \frac{1}{\sqrt{n}}\Pr\left\{J < \frac{1}{\sqrt{n}}\right\} \tag{3.59}$$

$$= \Pr\left\{J \geq \frac{1}{\sqrt{n}}\right\} + \frac{1}{\sqrt{n}}\Pr\left\{J < \frac{1}{\sqrt{n}}\right\} \tag{3.60}$$

$$\leq \Pr\left\{J \geq \frac{1}{\sqrt{n}}\right\} + \frac{1}{\sqrt{n}}. \tag{3.61}$$

Thus, it follows from (3.56) and the definition of the mismatched information density (3.17) that

$$\mathrm{P}_{\mathrm{e}}^n \leq \Pr\left\{\log MG + \frac{1}{2}\log n \geq nC(P, \sigma^2) + \frac{\|Y^n\|^2}{2(P+\sigma^2)} - \frac{\|Y^n - X^n\|^2}{2\sigma^2}\right\} + \frac{1}{\sqrt{n}} \tag{3.62}$$

$$= \Pr\left\{\frac{\|Z^n\|^2}{2\sigma^2} - \frac{\|X^n + Z^n\|^2}{2(P+\sigma^2)} \geq nC(P, \sigma^2) - \log MG - \frac{1}{2}\log n\right\} + \frac{1}{\sqrt{n}} \tag{3.63}$$

$$= \Pr\Big\{(P+\sigma^2)\|Z^n\|^2 - \sigma^2\|X^n + Z^n\|^2 \geq 2\sigma^2(P+\sigma^2)(nC(P, \sigma^2) - \log MG - \frac{1}{2}\log n)\Big\} + \frac{1}{\sqrt{n}} \tag{3.64}$$

$$= \Pr\Big\{P\|Z^n\|^2 - nP\sigma^2 - 2\sigma^2\langle X^n, Z^n\rangle \geq 2\sigma^2(P+\sigma^2)(nC(P, \sigma^2) - \log MG - \frac{1}{2}\log n)\Big\} + \frac{1}{\sqrt{n}}, \tag{3.65}$$

where (3.63) follows since $Y^n = X^n + Z^n$ due to the channel law in (3.1) and (3.65) follows since $\|X^n\|^2 = nP$ when the spherical codebook is used (cf. (3.5)).

The rest of the proof bounds the probability term in (3.65) using the function version Berry-Esseen theorem (cf. Theorem 1.8). Let $\tilde{X}^n = (\tilde{X}_1, \ldots, \tilde{X}_n)$ be a sequence of n i.i.d. Gaussian random variables generated from the Gaussian distribution with mean zero and variance one. It follows from [16, Eq. (57)] that the spherical codeword X^n generated from (3.5) satisfies

$$X^n = \sqrt{nP}\frac{\tilde{X}^n}{\|\tilde{X}^n\|^2}. \tag{3.66}$$

Furthermore, for each $i \in [n]$ define the following zero-mean random variables

$$A_{1,i} := Z_i^2 - \sigma^2, \tag{3.67}$$

$$A_{2,i} := \sigma^2\sqrt{P}\tilde{X}_i Z_i, \tag{3.68}$$

$$A_{3,i} := \tilde{X}_i^2 - 1, \tag{3.69}$$

and let $\mathbf{A}_i := (A_{1,i}, A_{2,i}, A_{3,i})$ be the random vector. Note that $(\mathbf{A}_1, \ldots, \mathbf{A}_n)$ is a sequence of i.i.d. random vectors with zero mean and finite third absolute moments.

Given any three random variables $(a_1, a_2, a_3) \in \mathbb{R}^3$, define the function

$$f(a_1, a_2, a_3) := Pa_1 - \frac{2a_2}{\sqrt{1 + a_3}}. \tag{3.70}$$

It follows that

$$nf\left(\frac{1}{n}\sum_{i=1}^{n}\mathbf{A}_i\right) = P\|Z^n\|^2 - nP\sigma^2 - 2\sigma^2\langle X^n, Z^n\rangle, \tag{3.71}$$

and $f(\mathbf{0}_3) = 0$. For any $i \in [n]$, the covariance matrix of the random vector $\mathbf{A}_i$ satisfies

$$\mathsf{Cov}(\mathbf{A}_i) = \begin{bmatrix} \mathbb{E}[A_{1,i}^2] & \mathsf{Cov}(A_{1,i}, A_{2,i}) & \mathsf{Cov}(A_{1,i}, A_{3,i}) \\ \mathsf{Cov}(A_{2,i}, A_{1,i}) & \mathbb{E}[A_{2,i}^2] & \mathsf{Cov}(A_{2,i}, A_{3,i}) \\ \mathsf{Cov}(A_{3,i}, A_{1,i}) & \mathsf{Cov}(A_{3,i}, A_{2,i}) & \mathbb{E}[A_{3,i}^2] \end{bmatrix} \tag{3.72}$$

$$= \begin{bmatrix} \zeta - \sigma^4 & 0 & 0 \\ 0 & \sigma^6 P & 0 \\ 0 & 0 & 2 \end{bmatrix}. \tag{3.73}$$

Furthermore, the Jacobian matrix $\mathbf{J} := [J_1 \quad J_2 \quad J_3]$ of the three-parameter function $f(\cdot)$ around $\mathbb{E}[\mathbf{A}_i] = \mathbf{0}_3$ satisfies

$$J_1 := \frac{\partial f(\mathbf{a})}{\partial a_1}\bigg|_{\mathbf{a}=\mathbf{0}_3} = P, \tag{3.74}$$

$$J_2 := \frac{\partial f(\mathbf{a})}{\partial a_2}\bigg|_{\mathbf{a}=\mathbf{0}_3} = -2, \tag{3.75}$$

$$J_3 := \frac{\partial f(\mathbf{a})}{\partial a_3}\bigg|_{\mathbf{a}=\mathbf{0}_3} = 0. \tag{3.76}$$

Thus,

$$\mathbf{J}\,\mathsf{Cov}(\mathbf{A}_{1,1})\mathbf{J}^{\mathrm{T}} = P^2(\zeta - \sigma^4) + 4\sigma^6 P. \tag{3.77}$$

Recall the definition of the dispersion function $V_{\mathrm{sp}}(\cdot)$ in (3.12). Define the following constant:

$$\delta_n := \frac{2\sigma^2(P+\sigma^2)\big(nC(P,\sigma^2) - \log MG - \frac{1}{2}\log n\big)}{\sqrt{nP^2(\zeta - \sigma^4) + 4\sigma^6 P}} \tag{3.78}$$

$$= \frac{nC(P,\sigma^2) - \log MG - \frac{1}{2}\log n}{\sqrt{nV_{\mathrm{sp}}(P,\sigma^2,\zeta)}}. \tag{3.79}$$

It follows from the function version Berry-Esseen theorem (cf. Theorem 1.8) and the results in (3.65) and (3.71) that

$$\mathrm{P}_{\mathrm{e}}^n \le \Pr\left\{\sqrt{n}f\left(\frac{1}{n}\sum_{t\in[n]}\mathbf{X}_t\right) \ge \sqrt{P^2(\zeta-\sigma^4)+4\sigma^6 P}\,\delta_n\right\} + \frac{1}{\sqrt{n}} \tag{3.80}$$

$$\le \mathrm{Q}(\delta_n) + O\left(\frac{1}{\sqrt{n}}\right). \tag{3.81}$$

To make the error probability no greater than ε, the maximal value of M satisfies

$$\log M = nC(P,\sigma^2) - \sqrt{nV_{\mathrm{sp}}(P,\sigma^2,\zeta)}\mathrm{Q}^{-1}\left(\varepsilon + O\left(\frac{1}{\sqrt{n}}\right)\right) \tag{3.82}$$

$$= nC(P,\sigma^2) - \sqrt{nV_{\mathrm{sp}}(P,\sigma^2,\zeta)}\mathrm{Q}^{-1}(\varepsilon) + O(\log n), \tag{3.83}$$

where (3.83) follows from Taylor expansion of $\mathrm{Q}^{-1}(\cdot)$ around ε, analogously to (2.30). The achievability proof for spherical codebooks is now completed.

3.3.2.2 i.i.d. Gaussian Codebooks

The achievability proof of Theorem 3.1 with i.i.d. Gaussian codebooks is highly similar to the proof with spherical codebooks. This section emphasizes the differences.

Following (3.53) to (3.64), the error probability for i.i.d. Gaussian codebooks satisfies

$$\mathrm{P}_{\mathrm{e}}^n \leq \Pr\Big\{(P+\sigma^2)\|Z^n\|^2 - \sigma^2\|X^n+Z^n\|^2 \\ \geq 2\sigma^2(P+\sigma^2)\Big(nC(P,\sigma^2) - \log MG - \frac{1}{2}\log n\Big)\Big\} + \frac{1}{\sqrt{n}} \tag{3.84}$$

$$\leq \Pr\Big\{P\|Z^n\|^2 - \sigma^2\|X^n\|^2 - 2\sigma^2\langle X^n, Z^n\rangle \\ \geq 2\sigma^2(P+\sigma^2)\Big(nC(P,\sigma^2) - \log M - \frac{1}{2}\log n\Big)\Big\} + \frac{1}{\sqrt{n}}. \tag{3.85}$$

For each $i \in [n]$, define the random variable

$$\bar{A}_i := PZ_i^2 - \sigma^2 X_i^2 - 2\sigma^2 X_i Z_i, \tag{3.86}$$

where $X_i \sim \mathcal{N}(0, P)$ is the i-th element of an i.i.d. Gaussian codeword and Z_i is the i-th element of the noise sequence Z^n that is generated from a distribution P_Z satisfying the moment constraints in (3.2). Note that X_i and Z_i are independent. The expectation and variance of $\bar{A}_i$ are

$$\mathbb{E}[\bar{A}_i] = P\sigma^2 - \sigma^2 P - 0 = 0, \tag{3.87}$$

$$\begin{aligned}\mathsf{Var}[\bar{A}_i] &= \mathbb{E}[\bar{A}_i^2] && (3.88)\\ &= \mathbb{E}\big[P^2 Z_i^4 - 2P\sigma^2 X_i^2 Z_i^2 - 4P\sigma^2 X_i Z_i^3 + \sigma^4 X_i^4 \\ &\quad + 4\sigma^4 X_i^3 Z_i + 4\sigma^4 X_i^2 Z_i^2\big] && (3.89)\\ &= P^2\zeta - 2P^2\sigma^4 - 0 + 3P^2\sigma^4 + 0 + 4P\sigma^6 && (3.90)\\ &= P^2(\zeta + \sigma^4) + 4P\sigma^6, && (3.91)\end{aligned}$$

where (3.90) follows since $\mathbb{E}[X_i^3] = 0$ and $\mathbb{E}[X_i^4] = 3P^2$.

Thus, it follows from the result in (3.85) that

$$\mathrm{P}_{\mathrm{e}}^{n} \le \Pr\left\{ \sum_{i\in[n]} \tilde{A}_i \ge 2\sigma^2(P+\sigma^2)\Big(nC(P,\sigma^2) - \log M - \frac{1}{2}\log n\Big)\right\} + \frac{1}{\sqrt{n}} \tag{3.92}$$

$$\le \Pr\left\{\frac{1}{\sqrt{n}}\sum_{i=1}^{n}\frac{\bar{A}_i}{\sqrt{\mathsf{Var}[\bar{A}_i]}} \ge \frac{nC(P,\sigma^2) - \log M + \frac{1}{2}\log n}{\sqrt{nV_{\mathrm{iid}}(P,\sigma^2,\zeta)}}\right\} + \frac{1}{\sqrt{n}} \tag{3.93}$$

$$\le Q\left(\frac{nC(P,\sigma^2) - \log M - \frac{1}{2}\log n}{\sqrt{nV_{\mathrm{iid}}(P,\sigma^2,\zeta)}}\right) + O\Big(\frac{1}{\sqrt{n}}\Big), \tag{3.94}$$

where (3.93) follows from the definition of $V_{\mathrm{iid}}(P,\sigma^2,\zeta)$ in (3.13) and the result in (3.91) that

$$\frac{\mathsf{Var}[\bar{A}_i]}{V_{\mathrm{iid}}(P,\sigma^2,\zeta)} = \frac{(P^2(\zeta+\sigma^4)+4P\sigma^6)4\sigma^4(P+\sigma^2)}{P^2(\zeta+\sigma^4)+4\sigma^6 P} \tag{3.95}$$

$$= 4\sigma^4(P+\sigma^2)^2, \tag{3.96}$$

and (3.94) follows from the Berry-Esseen theorem

To make the error probability no greater than ε, the maximal value of M satisfies that

$$\log M = nC(P,\sigma^2) - \sqrt{nV_{\mathrm{iid}}(P,\sigma^2,\zeta)}\mathrm{Q}^{-1}\left(\varepsilon + O\Big(\frac{1}{\sqrt{n}}\Big)\right) \tag{3.97}$$

$$= nC(P,\sigma^2) - \sqrt{nV_{\mathrm{iid}}(P,\sigma^2,\zeta)}\mathrm{Q}^{-1}(\varepsilon) + O(\log n), \tag{3.98}$$

where (3.98) follows from Taylor expansion of $\mathrm{Q}^{-1}(\cdot)$ around ε. The achievability proof for i.i.d. Gaussian codebooks is now completed.

3.3.3 *Converse*

3.3.3.1 Spherical Codebooks

This section provides the ensemble converse proof for Theorem 3.1. In particular, lower bounds on the ensemble error probability are obtained, and the corresponding requirement on the transmission rate is obtained to ensure that the ensemble error probability is bounded by $\varepsilon \in (0, 1)$.

It follows from the definition of an (n, M, P)-sp-code in Definition 3.1 that the ensemble error probability satisfies

$$\mathrm{P}_\mathrm{e}^n = \Pr\left\{\exists\, \bar{W} \neq W :\ \|Y^n - X^n(\bar{W})\|^2 \leq \|Y^n - X^n(W)\|^2\right\} \tag{3.99}$$

$$= 1 - \Pr\left\{\forall\, \bar{W} \neq W :\ \|Y^n - X^n(\bar{W})\|^2 \geq \|Y^n - X^n(W)\|^2\right\} \tag{3.100}$$

$$= 1 - \mathbb{E}\Big[\Pr\left\{\forall\, \bar{W} \neq W :\ \|Y^n - X^n(\bar{W})\|^2 \geq \|Y^n - X^n(W)\|^2 | X^n(W), Y^n\right\}\Big] \tag{3.101}$$

$$= 1 - \mathbb{E}\Big[\Big(1 - \Pr\left\{\|Y^n - \bar{X}^n\|^2 \leq \|Y^n - X^n\|^2 | X^n, Y^n\right\}\Big)^{M-1}\Big], \tag{3.102}$$

where (3.99) follows since NN decoding is used, and (3.102) follows since each codeword is generated independently from the same distribution. Furthermore, $(X^n, \bar{X}^n, Y^n) \sim P_{X^n} P_{X^n} P_{Y^n|X^n}$, P_{X^n} is the channel input distribution in (3.5) and $P_{Y^n|X^n}$ is induced by the additive noise channel in (3.1).

For subsequent analyses, let $\hat{P}_Y := \frac{\|Y^n\|^2}{n}$ denote the power of the channel output and let $\hat{P}_Z := \frac{\|Z^n\|^2}{n}$ denote the power of the additive noise. This way, the probability term in (3.102) satisfies

$$\Pr\left\{\|Y^n - \bar{X}^n\|^2 \leq \|Y^n - X^n\|^2\right\} = \Pr\{\|Y^n - \bar{X}^n\|^2 \leq \|Z^n\|^2\} \tag{3.103}$$

$$= \Pr\left\{\|Y^n\|^2 + \|\bar{X}^n\|^2 - 2\langle \bar{X}^n, Y^n\rangle \leq \|Z^n\|^2\right\} \tag{3.104}$$

$$= \Pr\left\{2\langle \bar{X}^n, Y^n\rangle \geq n\hat{P}_Y + nP - n\hat{P}_Z\right\} \tag{3.105}$$

$$= \Pr\left\{X_1 \geq \frac{n\hat{P}_Y + nP - n\hat{P}_Z}{2\sqrt{n\hat{P}_Y}}\right\}, \tag{3.106}$$

where (3.103) follows since $Y^n = X^n + Z^n$ due to the channel model in (3.1), (3.105) follows since each spherical codeword has the power of P and thus $\|X^n\|^2 = nP$, and (3.106) follows from spherical symmetry where $\bar{X}_1$ is the first element of X^n. Specifically, since the random vector $\bar{X}^n$ is generated from the uniform distribution over the sphere (cf. (3.5)), the distribution of the inner product $\langle \bar{X}^n, Y^n\rangle$ depends on Y^n only from its norm $\hat{P}_Y$. Thus, one can take Y^n as $(\sqrt{n\hat{P}_Y}, 0, \ldots, 0)$ to ease the analysis.

Given any $(\hat{P}_Y, \hat{P}_Z) \in \mathbb{R}_+^2$, define the probability function

$$h(\hat{P}_Y, \hat{P}_Z) := \Pr\left\{\bar{X}_1 \geq \frac{n\hat{P}_Y + nP - n\hat{P}_Z}{2\sqrt{n\hat{P}_Y}}\right\}. \tag{3.107}$$

The error probability in (3.102) is equivalently given by

$$\mathrm{P}_{\mathrm{e}}^n = \mathbb{E}\big[1 - (1 - h(\hat{P}_Y, \hat{P}_Z))^{M-1}\big]. \tag{3.108}$$

Recall the definition of typical power set $\mathcal{T}$ in (3.51). Fix any $\gamma \in (0, 1)$ to be specified. When the blocklength n is sufficiently large, the error probability in (3.108) can be lower bounded as follows:

$$\mathrm{P}_{\mathrm{e}}^n \geq \mathbb{E}\left[1 - (1 - h(\hat{P}_Y, \hat{P}_Z))^{M-1}\mathbb{1}\left(h(\hat{P}_Y, \hat{P}_Z) \geq \frac{n}{M-1} \text{ and } (\hat{P}_Y, \hat{P}_Z) \in \mathcal{T}\right)\right] \tag{3.109}$$

$$\geq \Pr\left\{h(\hat{P}_Y, \hat{P}_Z) \geq \frac{n}{M-1} \text{ and } (\hat{P}_Y, \hat{P}_Z) \in \mathcal{T}\right\}(1 - \exp(-n\gamma)) \tag{3.110}$$

$$\geq \Pr\left\{h(\hat{P}_Y, \hat{P}_Z) \geq \frac{n}{M-1} \text{ and } (\hat{P}_Y, \hat{P}_Z) \in \mathcal{T}\right\} - \exp(-n\gamma), \tag{3.111}$$

where (3.110) follows since i) the following equality concerning the limit holds:

$$\lim_{a\to\infty}\left(1 - \frac{1}{a}\right)^a = \exp(-1), \tag{3.112}$$

ii) the codebook size M is exponentially in n and thus $\frac{n}{M-1} \to 0$ and iii) when $h(\hat{P}_Y, \hat{P}_Z) \geq \frac{n}{M-1}$, it follows that for any $\gamma \in (0, 1)$,

$$(1 - h(\hat{P}_Y, \hat{P}_Z))^{M-1} \leq \left(\left(1 - \frac{n}{M-1}\right)^{\frac{M-1}{n}}\right)^n \tag{3.113}$$

$$\to \exp(-n) \tag{3.114}$$

$$\leq \exp(-n\gamma). \tag{3.115}$$

The essence of the remaining proof bounds the probability $h(\hat{P}_Y, \hat{P}_Z)$ in (3.107). It follows from [17, Eq. (4)] that the pdf of $\bar{X}_1$ is given by

$$f_{\bar{X}_1}(x) = \frac{1}{\sqrt{\pi n P}}\frac{\Gamma(\frac{n}{2})}{\Gamma(\frac{n-1}{2})}\left(1 - \frac{x^2}{nP}\right)^{\frac{n-3}{2}}\mathbb{1}(x^2 \leq nP). \tag{3.116}$$

Note that $f_{\bar{X}_1}(x)$ is non-increasing in the absolute value $|x|$ of its parameter x. Fix any positive real number $\delta \in \mathbb{R}_+$. It follows from (3.107) that

$$h(\hat{P}_Y, \hat{P}_Z) \geq \Pr\left\{\frac{n\hat{P}_Y + nP - n\hat{P}_Z}{2\sqrt{n\hat{P}_Y}} \leq \bar{X}_1 \leq \frac{n\hat{P}_Y + nP - n\hat{P}_Z}{2\sqrt{n\hat{P}_Y}}\left(1 + \frac{\delta}{n}\right)\right\} \tag{3.117}$$

$$\geq \frac{\delta(\hat{P}_Y + P - \hat{P}_Z)}{2\sqrt{n\hat{P}_Y}} \frac{1}{\sqrt{\pi n P}} \frac{\Gamma(\frac{n}{2})}{\Gamma(\frac{n-1}{2})}$$

$$\times \left(1 - \frac{1}{nP}\left(\frac{n\hat{P}_Y + nP - n\hat{P}_Z}{2\sqrt{n\hat{P}_Y}}\left(1+\frac{\delta}{n}\right)\right)^2\right)^{\frac{n-3}{2}}. \tag{3.118}$$

Fix any $(\hat{P}_Y, \hat{P}_Z) \in \mathcal{T}$. It follows from the definition of $\mathcal{T}$ in (3.51) that i) $\hat{P}_Y$ and $\hat{P}_Z$ are both bounded, ii) $\hat{P}_Y + P - \hat{P}_Z$ is positive and iii) for any $\eta \in (0, 2P)$,

$$\frac{\eta}{2\sqrt{n(P+\sigma^2 - \sqrt{c_1 \log n/n})}} \leq \frac{\hat{P}_Y + P - \hat{P}_Z}{2\sqrt{n\hat{P}_Y}} \leq \frac{2P}{2\sqrt{n(P+\sigma^2 - \sqrt{c_1 \log n/n})}}. \tag{3.119}$$

Let

$$q_1^n := \frac{\eta\delta}{2\sqrt{n(P+\sigma^2 - \sqrt{c_1 \log n/n})}} \frac{1}{\sqrt{\pi n P}} \frac{\Gamma(\frac{n}{2})}{\Gamma(\frac{n-1}{2})}, \tag{3.120}$$

$$K_1 := \frac{3P^2}{P+\sigma^2}. \tag{3.121}$$

Since $\frac{\Gamma(\frac{n}{2})}{\Gamma(\frac{n-1}{2})} = O(\sqrt{n})$, it follows that $q_1^n = O(\sqrt{n})$. Furthermore, since $\eta < 2P$ and $\hat{P}_Y$ is asymptotically bounded by $P + \sigma^2$, for n sufficiently large, it follows that

$$\left(\frac{n\hat{P}_Y + nP - n\hat{P}_Z}{2\sqrt{n\hat{P}_Y}}\left(1+\frac{\delta}{n}\right)\right)^2 \leq \frac{n(\hat{P}_Y + P - \hat{P}_Z)^2}{4\hat{P}_Y}\left(1+\frac{\delta}{n}\right)^2 \tag{3.122}$$

$$\leq \frac{n(\hat{P}_Y + P - \hat{P}_Z)^2}{4\hat{P}_Y} + K_1. \tag{3.123}$$

Combining (3.118), (3.119) and (3.123), it follows that when the blocklength n is sufficiently large,

$$h(\hat{P}_Y, \hat{P}_Z) \geq q_1^n \left(1 - \frac{(\hat{P}_Y + P - \hat{P}_Z)^2}{4P\hat{P}_Y} - \frac{K_1}{nP}\right)^{\frac{n-3}{2}} \tag{3.124}$$

$$\geq q_1^n \left(\left(1 - \frac{(\hat{P}_Y + P - \hat{P}_Z)^2}{4P\hat{P}_Y}\right)\left(1 - \frac{K_2}{nP}\right)\right)^{\frac{n-3}{2}} \tag{3.125}$$

$$
\geq q_2^n \left(1 - \frac{(\hat{P}_Y + P - \hat{P}_Z)^2}{4P\hat{P}_Y}\right)^{\frac{n}{2}} \tag{3.126}
$$

$$
= q_2^n \exp\left(\frac{n}{2}\log\left(1 - \frac{(\hat{P}_Y + P - \hat{P}_Z)^2}{4P\hat{P}_Y}\right)\right), \tag{3.127}
$$

where in (3.125), K_2 is a constant such that $\frac{K_1}{nP} \leq \frac{K_2}{nP}\left(1 - \frac{(\hat{P}_Y+P-\hat{P}_Z)^2}{4P\hat{P}_Y}\right)$, and in (3.126), $q_2^n = q_1^n\left(1 - \frac{K_2}{nP}\right)^{\frac{n-3}{2}}\left(1 - \frac{(\hat{P}_Y+P-\hat{P}_Z)^2}{4P\hat{P}_Y}\right)^{-\frac{3}{2}}$.

Using (3.112), it follows that

$$
\lim_{n\to\infty}\left(1 - \frac{K_2}{nP}\right)^{\frac{n}{2}} = \exp\left(-\frac{K_2}{2P}\right). \tag{3.128}
$$

Thus, $q_2^n = O(\sqrt{n})$. Furthermore, the Taylor's expansion of $-\frac{1}{2}\log\left(1 - \frac{(\hat{P}_Y+P-\hat{P}_Z)^2}{4P\hat{P}_Y}\right)$ around $(\hat{P}_Y, \hat{P}_Z) = (P+\sigma^2, \sigma^2)$ satisfies

$$
\begin{aligned}
&-\frac{1}{2}\log\left(1 - \frac{(\hat{P}_Y + P - \hat{P}_Z)^2}{4P\hat{P}_Y}\right) \\
&= \frac{1}{2}\log\left(1 + \frac{P}{\sigma^2}\right) + \frac{1}{2(\sigma^2+P)}(\hat{P}_Y - (P+\sigma^2)) - \frac{1}{2\sigma^2}(\hat{P}_Z - \sigma^2) \\
&\quad + O(|\hat{P}_Y - (P+\sigma^2)|^2 + |\hat{P}_Z - \sigma^2|^2) \qquad (3.129) \\
&= C(P,\sigma^2) + \frac{\hat{P}_Y}{2(P+\sigma^2)} - \frac{\hat{P}_Z}{2\sigma^2} + O\left(\frac{\log n}{n}\right). \qquad (3.130)
\end{aligned}
$$

Thus,

$$
h(\hat{P}_Y, \hat{P}_Z) \geq q_2^n \exp\left(-n\left(C(P,\sigma^2) + \frac{\hat{P}_Y}{2(P+\sigma^2)} - \frac{\hat{P}_Z}{2\sigma^2} + O\left(\frac{\log n}{n}\right)\right)\right) \tag{3.131}
$$

$$
= \exp\left(-n\left(C(P,\sigma^2) + \frac{\hat{P}_Y}{2(P+\sigma^2)} - \frac{\hat{P}_Z}{2\sigma^2}\right) + O(\log n)\right). \tag{3.132}
$$

It follows from (3.111) and (3.132) with $\gamma = \frac{\log n}{n}$ that the ensemble error probability can be lower bounded as

$$\begin{aligned}
&\mathrm{P}_{\mathrm{e}}^{n} + \frac{1}{n} \\
&\geq \Pr\left\{h(\hat{P}_Y, \hat{P}_Z) \geq \frac{n}{M-1} \text{ and } (\hat{P}_Y, \hat{P}_Z) \in \mathcal{T}\right\} && (3.133)\\
&\geq \Pr\left\{\exp\left(-n\left(C(P,\sigma^2) + \frac{\hat{P}_Y}{2(P+\sigma^2)} - \frac{\hat{P}_Z}{2\sigma^2}\right) + O(\log n)\right) \geq \frac{n}{M-1}\right. \\
&\qquad\qquad \left.\text{and } (\hat{P}_Y, \hat{P}_Z) \in \mathcal{T}\right\} && (3.134)\\
&\geq \Pr\left\{-n\left(C(P,\sigma^2) + \frac{\hat{P}_Y}{2(P+\sigma^2)} - \frac{\hat{P}_Z}{2\sigma^2}\right) \geq -\log(M-1) + O(\log n)\right\} \\
&\quad - \Pr\{(\hat{P}_Y, \hat{P}_Z) \notin \mathcal{T}\} && (3.135)\\
&\geq \Pr\left\{\frac{\|Z^n\|^2}{2\sigma^2} - \frac{\|Y^n\|^2}{2(P+\sigma^2)} \geq nC(P,\sigma^2) - \log M + O(\log n)\right\} + O\left(\frac{1}{\sqrt{n}}\right) && (3.136)\\
&= \Pr\left\{P\|Z^n\|^2 - nP\sigma^2 - 2\sigma^2\langle X^n, Z^n\rangle\right. \\
&\qquad \left.\geq 2\sigma^2(P+\sigma^2)(nC(P,\sigma^2) - \log M + O(\log n))\right\} + O\left(\frac{1}{\sqrt{n}}\right), && (3.137)
\end{aligned}$$

where (3.135) follows by taking logarithm at both sides and using the inequality that $\Pr\{\mathcal{A}\cap\mathcal{B}\} \geq \Pr\{\mathcal{A}\} - \Pr\{\mathcal{B}^{\mathrm{c}}\}$ for any two events $(\mathcal{A}, \mathcal{B})$, (3.136) follows from the definitions that $\hat{P}_Y = \frac{\|Y^n\|^2}{n}$ and $\hat{P}_Z = \frac{\|Z^n\|^2}{n}$ and the fact that $\log(M-1) \geq \log M - \log 2$ for any $M \geq 2$, and (3.137) follows since $Y^n = X^n + Z^n$ due to the channel model in (3.1).

Note that (3.137) is exactly the same as (3.65) in the achievability part. Thus, using the function version Berry-Esseen theorem, it follows that

$$\mathrm{P}_{\mathrm{e}}^{n} \geq Q\left(\frac{nC(P,\sigma^2) - \log M + O(\log n)}{\sqrt{nV_{\mathrm{sp}}(P,\sigma^2,\zeta)}}\right) + O\left(\frac{1}{\sqrt{n}}\right). \tag{3.138}$$

For any real number $r > 0$, if M is chosen such that

$$\log M = nC(P,\sigma^2) - \sqrt{nV_{\mathrm{sp}}(P,\sigma^2,\zeta)}\,Q^{-1}\left(\varepsilon + r + \frac{1}{\sqrt{n}}\right) + O(\log n), \tag{3.139}$$

it follows that the error probability satisfies

$$\mathrm{P}_{\mathrm{e}}^n \geq \varepsilon + r > \varepsilon. \tag{3.140}$$

Therefore, the maximal value of M that ensures $\mathrm{P}_{\mathrm{e}}^n \leq \varepsilon$ satisfies

$$\log M = nC(P,\sigma^2) - \sqrt{nV_{\mathrm{sp}}(P,\sigma^2,\zeta)}\,Q^{-1}\Big(\varepsilon + O\Big(\frac{1}{\sqrt{n}}\Big)\Big) + O(\log n) \tag{3.141}$$

$$= nC(P,\sigma^2) - \sqrt{nV_{\mathrm{sp}}(P,\sigma^2,\zeta)}\,Q^{-1}\Big(\varepsilon\Big) + O(\log n). \tag{3.142}$$

The ensemble converse proof for spherical codebooks is now completed.

3.3.3.2 i.i.d. Gaussian Codebooks

Following the same logic as the converse proof of Theorem 3.1 for spherical codebooks in (3.102), the error probability for i.i.d. Gaussian codebooks satisfies

$$\mathrm{P}_{\mathrm{e}}^n = \mathbb{E}\Big[1 - \big(1 - \Pr\{\|Y^n - \bar{X}^n\|^2 \leq \|Y^n - X^n\|^2\}\big)^{M-1}\Big]. \tag{3.143}$$

Recall that $\hat{P}_Y = \frac{\|Y^n\|^2}{n}$ and $\hat{P}_Z = \frac{\|Z^n\|^2}{n}$ denote the powers of the channel output and the additive noise, respectively. Note that $Y^n = X^n + Z^n$ due to the channel law in (8.1). It follows that

$$\Pr\{\|Y^n - \bar{X}^n\|^2 \leq \|Y^n - X^n\|^2\} = \Pr\{\|Y^n - \bar{X}^n\|^2 \leq \|Z^n\|^2\} \tag{3.144}$$

$$= \Pr\{n\hat{P}_Y + \|\bar{X}^n\|^2 - 2\langle \bar{X}^n, Y^n\rangle \leq n\hat{P}_Z\}. \tag{3.145}$$

Since $\bar{X}^n$ is generated i.i.d. from the Gaussian distribution with mean zero and variance P, it follows from spherical symmetry that the distribution of $\langle \bar{X}^n, Y^n\rangle \sim \mathcal{N}(\cdot; 0, n\hat{P}_Y)$ depends on Y^n only through its power $\hat{P}_Y$. Thus, the probability term in (3.145) satisfies

$$\Pr\{\|Y^n - \bar{X}^n\|^2 \leq \|Y^n - X^n\|^2\} = \Pr\bigg\{\sum_{i=1}^n \Big(\bar{X}_i^2 - \sqrt{\hat{P}_Y}\Big)^2 \leq n\hat{P}_Z\bigg\} \tag{3.146}$$

$$= \Pr\Bigg\{-\frac{1}{n}\sum_{i=1}^n \frac{\Big(\bar{X}_i^2 - \sqrt{\hat{P}_Y}\Big)^2}{P} \geq -\frac{\hat{P}_Z}{P}\Bigg\} \tag{3.147}$$

$$=: h_{\mathrm{iid}}(\hat{P}_Y, \hat{P}_Z), \tag{3.148}$$

where (3.146) follows by choosing $Y^n = (\sqrt{\hat{P}_Y}, \ldots, \sqrt{\hat{P}_Y})$.

Recall the strong large deviations theorem in Theorem 1.4 (cf. [18, Theorem 3.7.4]). Fix any $s \in (0, 0.5)$. Let $P' := \frac{P}{1-2s}$ and $\mu := -\frac{2s\sqrt{\hat{P}_Y}}{1-2s}$. The moment generating function of $\frac{(\bar{X}_i - \sqrt{\hat{P}_Y})^2}{P}$ with parameter s satisfies

$$M(s) := \mathbb{E}\left[\exp\left(s\frac{(\bar{X}_i - \sqrt{\hat{P}_Y})^2}{P}\right)\right] \tag{3.149}$$

$$= \int \exp\left(\frac{s}{P}\left(\bar{x}_i^2 - 2\sqrt{\hat{P}_Y}\bar{x}_i + \hat{P}_Y\right)\right)\frac{1}{\sqrt{2\pi P}}\exp\left(-\frac{\bar{x}_i^2}{2P}\right)d\bar{x}_i \tag{3.150}$$

$$= \int \frac{1}{\sqrt{2\pi P}}\exp\left(-\frac{1}{2P}\left((1-2s)\bar{x}_i^2 + 4s\sqrt{\hat{P}_Y}\bar{x}_i - 2s\hat{P}_Y\right)\right)d\bar{x}_i \tag{3.151}$$

$$= \frac{1}{\sqrt{1-2s}}\int \frac{1}{\sqrt{2\pi P'}}\exp\left(-\frac{1}{2P'}\left(\bar{x}_i^2 + \frac{4s\sqrt{\hat{P}_Y}}{1-2s}\bar{x}_i - \frac{2s\hat{P}_Y}{1-2s}\right)\right)d\bar{x}_i \tag{3.152}$$

$$= \frac{1}{\sqrt{1-2s}}\int \frac{1}{\sqrt{2\pi P'}}\exp\left(-\frac{(\bar{x}_i-\mu)^2}{2P'}\right)\exp\left(\frac{\mu^2 + \frac{2s\hat{P}_Y}{1-2s}}{2P'}\right)d\bar{x}_i \tag{3.153}$$

$$= \frac{1}{\sqrt{1-2s}}\exp\left(\frac{\mu^2 + \frac{2s\hat{P}_Y}{1-2s}}{2P'}\right) \tag{3.154}$$

$$= \frac{1}{\sqrt{1-2s}}\exp\left(\frac{s\hat{P}_Y}{P(1-2s)}\right), \tag{3.155}$$

where (3.150) follows since $X_i \sim \mathcal{N}(0, P)$, (3.151) follows by combing the exponential terms, (3.152) follows from the definition of P', (3.153) follows from the definition of μ, (3.154) follows since the integral of any Gaussian distribution over the whole range equals one, and (3.155) follows from the definitions of P' and μ.

Thus, it follows from (3.155) and the strong large deviations in Theorem 1.4 that

$$h_{\mathrm{iid}}(\hat{P}_Y, \hat{P}_Z) = q_n \exp(-n\varphi(\hat{P}_Y, \hat{P}_Z)) \tag{3.156}$$

for some q_n satisfying $\log q(n) = O(\log n)$ and

$$\varphi(\hat{P}_Y, \hat{P}_Z) := \sup_{s\geq 0}\left\{-s\frac{\hat{P}_Z}{P} - \log M(-s)\right\} \tag{3.157}$$

$$= \sup_{s\geq 0}\left\{\frac{\hat{P}_Y s}{P(1+2s)}+\frac{1}{2}\log(1+2s)-s\frac{\hat{P}_Z}{P}\right\}. \tag{3.158}$$

It follows from algebra that the supremum in (3.158) is achieved by

$$s^* := \frac{P-2\hat{P}_Z+\sqrt{P^2+4\hat{P}_Y\hat{P}_Z}}{4\hat{P}_Z}. \tag{3.159}$$

Hence,

$$\begin{aligned}\varphi(\hat{P}_Y,\hat{P}_Z) &= \frac{1}{2}\log\left(1+\frac{P-2\hat{P}_Z+\sqrt{P^2+4\hat{P}_Y\hat{P}_Z}}{2\hat{P}_Z}\right)\\ &\quad+\frac{\hat{P}_Y\big(P-2\hat{P}_Z+\sqrt{P^2+4\hat{P}_Y\hat{P}_Z}\big)}{4P\hat{P}_Z\left(1+\frac{P-2\hat{P}_Z+\sqrt{P^2+4\hat{P}_Y\hat{P}_Z}}{4\hat{P}_Z}\right)}\\ &\quad-\frac{P-2\hat{P}_Z+\sqrt{P^2+4\hat{P}_Y\hat{P}_Z}}{4P}. \end{aligned}\tag{3.160}$$

It follows from Taylor's expansion of $\varphi(\hat{P}_Y,\hat{P}_Z)$ around $(\hat{P}_Y,\hat{P}_Z)=(P+\sigma^2,\sigma^2)$ that for any $(\hat{P}_Y,\hat{P}_Z)\in\mathcal{T}$,

$$\varphi(\hat{P}_Y,\hat{P}_Z)=C(P,\sigma^2)+\frac{\hat{P}_Y}{2(P+\sigma^2)}-\frac{\hat{P}_Z}{2\sigma^2}+O\left(\frac{\log n}{n}\right). \tag{3.161}$$

Combining (3.156) and (3.161) leads to a lower bound on $h_{\mathrm{iid}}(\hat{P}_Y,\hat{P}_Z)$, which is the same as the lower bound for $h(\hat{P}_Y,\hat{P}_Z)$ in (3.132). The rest of the proof is exactly the same as the spherical codebook as from (3.133) and thus omitted.

3.4 Chapter Summary

This chapter presents a mismatched code for channel coding, which can be used to transmit messages reliably over a point-to-point channel with additive arbitrary noise. The mismatched code uses either spherical codebooks or i.i.d. Gaussian codebooks with nearest neighbor decoding, which is optimal for Gaussian noise and depends on the noise distribution only through the second moment. Under mild moment conditions on the distribution of the additive noise, the low-latency performance of the mismatched code is obtained and shown to be ensemble tight. The condition is satisfied by common noise distribution including any discrete distribution, the Gaussian, Laplace, and exponential distributions. A key take-away message is that mismatched code is robust with performance guarantee, although it is not necessarily optimal when the channel noise distributions are perfectly known. Furthermore, although both spherical and i.i.d. Gaussian codebooks achieve the same asymptotic performance, the low-latency performance of the spherical codebook is better.

3.5 Exercises

3.1 What is mismatched channel coding? Why is it important?

3.2 What is ensemble converse for mismatched channel coding? How is it different from the usual converse argument for matched channel coding?

3.3 Calculate the value $\varphi(\hat{P}_Y, \hat{P}_Z)$ in (3.158).

3.4 Prove the result in Lemma 3.3 concerning the probability of having a typical powers for the channel noise and output.

3.5 Calculate the values of $R^*_{\text{iid}}(n, \varepsilon)$ and $R^*_{\text{sp}}(n, \varepsilon)$ when $n = 100$, $P = 1, \sigma^2 = 1$, $\zeta = 3$ and $\varepsilon = 0.1$.

3.6 Set $\sigma^2 = 0.3$. Calculate the fourth moment of Bernoulli, Gaussian, and Laplace noise distributions and compare the corresponding second-order rates.

References

1. V. Strassen, Asymptotische Abschätzungen in Shannon's Informationstheorie, in *Transactions of the Third Prague Conference on Information Theory etc*, Czechoslovak Academy of Sciences, Prague (1962), pp. 689–723
2. M. Hayashi, Second-order asymptotics in fixed-length source coding and intrinsic randomness. IEEE Trans. Inf. Theory **54**(10), 4619–4637 (2008)
3. M. Hayashi, Information spectrum approach to second-order coding rate in channel coding. IEEE Trans. Inf. Theory **55**(11), 4947–4966 (2009)
4. Y. Polyanskiy, H.V. Poor, S. Verdú, Channel coding rate in the finite blocklength regime. IEEE Trans. Inf. Theory **56**(5), 2307–2359 (2010)
5. A. Ingber, Y. Kochman, The dispersion of lossy source coding, in *2011 Data Compression Conference* (2011), pp. 53–62
6. V. Kostina, S. Verdú, Fixed-length lossy compression in the finite blocklength regime. IEEE Trans. Inf. Theory **58**(6), 3309–3338 (2012)
7. D. Wang, A. Ingber, Y. Kochman, The dispersion of joint source-channel coding, in *Proceedings of the 2011 49th Annual Allerton Conference on Communication, Control, and Computing*, Monticello, IL (2011), pp. 180–187
8. V. Kostina, S. Verdú, Lossy Joint Source-Channel Coding in the Finite Blocklength Regime. IEEE Trans. Inf. Theory **59**(5), 2545–2575 (2013)
9. V.Y.F. Tan, Asymptotic estimates in information theory with non-vanishing error probabilities. Found. and Trend. Commun. Inf. Theory **11**(1–2), 1–184 (2014)

10. L. Zhou, M. Motani, Finite blocklength lossy source coding for discrete memoryless sources. Found. Trend. Commun. Inf. Theory **20**(3), 157–389 (2023)
11. A. Lapidoth, Nearest neighbor decoding for additive non-Gaussian noise channels. IEEE Trans. Inf. Theory **42**(5), 1520–1529 (1996)
12. J. Scarlett, V.Y.F. Tan, G. Durisi, The dispersion of nearest-neighbor decoding for additive non-Gaussian channels. IEEE Trans. Inf. Theory **63**(1), 81–92 (2017)
13. R. Gallager, The random coding bound is tight for the average code (corresp.). IEEE Trans. Inf. Theory **19**(2), 244–246 (1973)
14. V.Y.F. Tan, M. Tomamichel, The third-order term in the normal approximation for the AWGN channel. IEEE Trans. Inf. Theory **61**(5), 2430–2438 (2015)
15. R.C. Yavas, V. Kostina, M. Effros, Gaussian multiple and random access channels: Finite-blocklength analysis. IEEE Trans. Inf. Theory **67**(11), 6983–7009 (2021)
16. E. MolavianJazi, J.N. Laneman, A second-order achievable rate region for Gaussian multi-access channels via a central limit theorem for functions. IEEE Trans. Inf. Theory **61**(12), 6719–6733 (2015)
17. A.J. Stam, Limit theorems for uniform distributions on spheres in high-dimensional Euclidean spaces. J. Appl. Probab. **19**(1), 221–228 (1982)
18. R.R. Bahadur, R.R. Rao, On deviations of the sample mean. Ann. Math. Statist. **31**(4), 1015–1027 (1960)

Chapter 4
Transmission of Arbitrary Source

This chapter studies the problem of transmitting an arbitrary memoryless source (AMS) over an additive arbitrary noise (AAN) channel. Traditionally, in a joint source-channel coding (JSCC) problem, one designs and optimizes coding schemes with the knowledge of the source distribution and the channel transition law. This way, the coding scheme is optimal for a particular source distribution and channel. However, in practical communication systems, obtaining the exact knowledge of the source distribution or the channel law is challenging. Thus, it is difficult and imperative to design code with performance guarantee, ignoring the exact source distribution and channel law.

Towards this goal, Lapidoth [1] pioneered the study of mismatched communication by using a fixed coding scheme to transmit a message over an AAN channel and proving the first-order asymptotic result. Such a code is named mismatched since the code is not optimal for the particular channel law (noise distribution). Corresponding results were also obtained by Lapidoth for lossy source coding [2, Theorem 3], where a fixed code is used to compress an AMS under quadratic distortion. Low-latency analysis of mismatched channel coding and mismatched lossy source coding was done by Scarlett, Tan, and Durisi [3] and by Zhou, Tan, and Motani [4], respectively. In particular, mismatched channel coding was presented in Chap. 3 and mismatched lossy source coding was summarized in [5, Chapter 6].

Subsequently, for JSCC, a mismatched coding scheme was proposed by Zhou, Tan, and Motani [6], who combined mismatched codes for channel coding and lossy source coding via the idea of *unequal error protection* [7, 8]. The coding scheme was inspired by the JSCC code for the matched case by Wang, Ingber, and Kochman [9]. The mismatched JSCC is proved with low-latency performance guarantee to transmit an AMS over an AAN channel. When specialized to a Gaussian memoryless source (GMS) and an additive white Gaussian noise (AWGN) channel, the results in [6] provide an alternative proof to Theorem 2.5 in Chap. 2, with structured code design. This chapter presents the mismatched JSCC code and its low-latency performance.

L. Zhou, L. Bai, *Low-latency Communication*, Textbooks in Telecommunication Engineering, https://doi.org/10.1007/978-3-032-06627-5_4

4.1 Problem Formulation

4.1.1 System Model

Fix two integers $(k, n) \in \mathbb{N}$ and five alphabets $(\mathcal{S}, \hat{\mathcal{S}}, \mathcal{X}, \mathcal{Y}, \mathcal{Z})$. Let $P_S \in \mathcal{P}(\mathcal{S})$ and $P_Z \in \mathcal{P}(\mathcal{Z})$ be two arbitrary distributions. Assume that the source sequence $S^k \in \mathcal{S}^k$ is generated i.i.d. from P_S. As shown in Fig. 4.1, the task of JSCC is to transmit a memoryless source over an additive noise channel reliably. Specifically, an encoder f converts the source sequence S^k into a channel input X^n, which is corrupted by additive noise Z^n and yields noisy output Y^n. At the receiver side, the decoder ϕ uses $Y^n = X^n + Z^n$ to generate an source estimate $\hat{S}^k$. The reliability is measured in the lossy sense by calculating the quadratic distortion $d(S^k, \hat{S}^k) = \|\hat{S}^k - S^k\|^2 = \sum_{i\in[k]}(\hat{S}_i - S_i)^2$.

Different from the matched case [9, 10] where the source distortion P_S and the noise distribution P_Z are known, this chapter studies the mismatched case where both (P_S, P_Z) are unknown. Instead, the statistics of source and channel noise distributions can be obtained via measurements. It is assumed that the second moment of both distributions is known and the sixth moment of both distributions is finite, i.e.,

$$\mathbb{E}[S^2] = \sigma_{\mathrm{s}}^2,\ \zeta_{\mathrm{s}} := \mathbb{E}[S^4],\ \mathbb{E}[S^6] < \infty, \tag{4.1}$$

$$\mathbb{E}[Z^2] = \sigma_{\mathrm{c}}^2,\ \zeta_{\mathrm{c}} := \mathbb{E}[Z^4],\ \mathbb{E}[Z^6] < \infty. \tag{4.2}$$

As shall be demonstrated, the second moments $(\sigma_{\mathrm{s}}^2, \sigma_{\mathrm{c}}^2)$ of both distributions are used for code design, the fourth moments $(\zeta_{\mathrm{s}}, \zeta_{\mathrm{c}})$ appear in low-latency performance, while the finiteness of sixth moment guarantees theoretical derivations using the function version of the Berry-Esseen theorem (cf. Theorem 1.8).

4.1.2 Coding Scheme

The mismatched JSCC code in [6] is summarized in Fig. 4.2. At a high level, the mismatched code consists of an encoder f that combines *source type partition* and *minimal Euclidean distance (MED) encoding*, and a decoder that uses *modified*

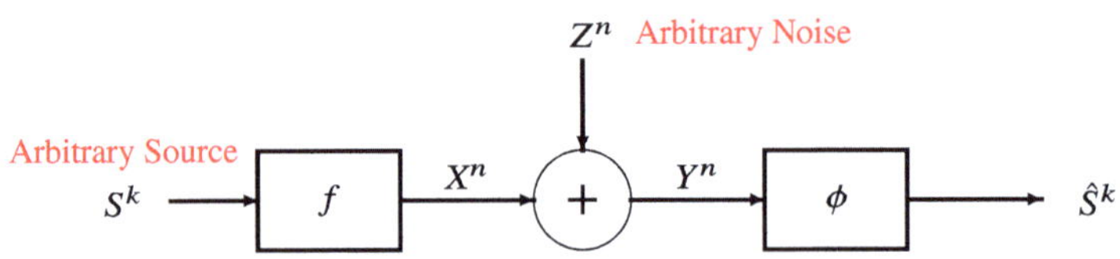

Fig. 4.1 System model for joint source-channel coding with additive arbitrary noise

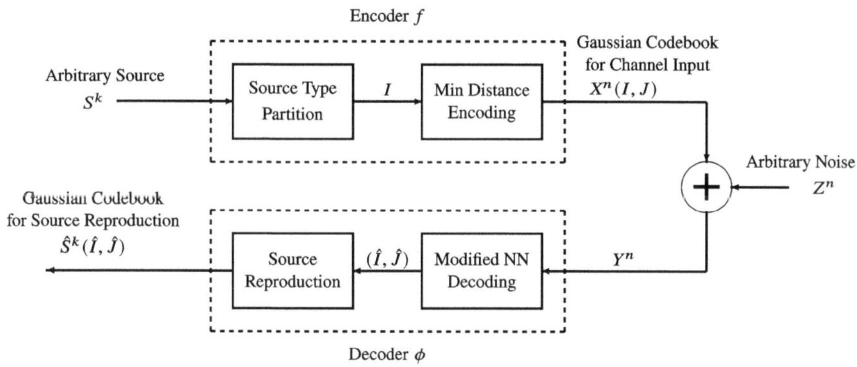

Fig. 4.2 Illustration of the mismatched JSCC code with an encoder f that uses modified minimum distance encoding and a decoder ϕ that uses modified nearest neighbor decoding

nearest neighbor (NN) decoding. Specifically, given a source sequence S^k, the encoder f first checks its type index I and subsequently maps it into a channel input $X^n(I, J)$, where J is the output of MED encoding. With additive noise corruption, noisy output Y^n is provided to the decoder, which uses a modified NN decoder to estimate (I, J) simultaneously and outputs the estimate of the source sequence as $\hat{S}(I, J)$. The key to the code is: i) how to partition the source sequence, ii) how to apply the modified NN decoding, and iii) how to design codebooks for channel input and source reproduction. These three key points are clarified in the remaining of this section.

4.1.2.1 Source Type Partition

The idea is to partition the source sequence into different types, depending on the power $\frac{\|S^k\|^2}{k} = \frac{1}{k}\sum_{i\in[k]} S_i^2$. This idea was used in [11] and in the proof of Theorem 2.2 for GMS.

Fix any $k \in \mathbb{N}$. Define the following constants:

$$\xi := \sqrt{\frac{\log k}{k}}, \tag{4.3}$$

$$N := \lceil 2k\xi \rceil, \tag{4.4}$$

$$\Upsilon(0) := (1 - \xi)\sigma_{\mathrm{S}}^2. \tag{4.5}$$

For each $i \in [N]$, define a real number and a power type:

$$\Upsilon(i) := \Big(1 - \xi + \frac{i}{k}\Big)\sigma_{\mathrm{S}}^2, \tag{4.6}$$

$$\mathcal{T}_i := \left\{s^k : k\Upsilon(i-1) \le \|s^k\|^2 < k\Upsilon(i)\right\}. \tag{4.7}$$

Finally, let

$$\mathcal{T} := \bigcup_{i\in[N]} \mathcal{T}_i. \tag{4.8}$$

Note that $\mathcal{T}_i$ collects source sequences s^k with powers in the range of $[\Upsilon(i-1), \Upsilon(i))$.

With the above definitions, given any sequence S^k, if $S^k \in \mathcal{T}_i$ for some $i \in [N]$, its power type is defined as i; otherwise, its power type is defined as 0.

4.1.2.2 Encoding and Decoding

For each $i \in [N]$, let $M_i \in \mathbb{N}$ be an integer and define the set

$$\mathcal{D} := \{(r, s) \in \mathbb{N}^2 : r \in [N],\ s \in [1 : M_r]\}. \tag{4.9}$$

The encoding and decoding operation is defined in the following mismatched code.

Definition 4.1 An (k, n)-code for mismatched JSCC consists of

1. a set of M_i source codewords $\{\hat{S}^k(i, j)\}_{j\in[M_i]}$ and a set of M_i channel codewords $\{X^n(i, j)\}_{j\in[M_i]}$ for each $i \in [N]$, which are known to the encoder and decoder;
2. an encoder f that combines source type partition and minimum Euclidean distance encoding. Specifically, if $S^k \notin \mathcal{T}$, the source type is 0 and the special codeword of $\mathbf{0}_n$ is transmitted. Otherwise, if $S^k \in \mathcal{T}_I$ and $\hat{S}^k(I, J)$ minimizes the Euclidean distance over all source codewords in the set $\{\hat{S}^k(I, \bar{j})\}_{\bar{j}\in[M_i]}$, i.e.,

$$J := \arg\min_{\bar{j}\in[M_i]} \|S^k - \hat{S}^k(I, \bar{j})\|^2; \tag{4.10}$$

3. a decoder ϕ that uses the modified nearest neighbor decoder rule. Specifically, the decoder declares generates the source estimate $\hat{S}^k(\hat{I}, \hat{J})$ if

$$(\hat{I}, \hat{J}) := \arg\min_{(\tilde{i},\tilde{j})\in\mathcal{D}} \|X^n(\tilde{i}, \tilde{j}) - Y^n\|^2 + 2\log M_{\tilde{i}}. \tag{4.11}$$

The above code applies the unequal error protection idea by assigning a different number of codewords M_i for different power types. In the special case of $N = 1$, the above code reduces to separate source-channel coding (SSCC) with M codewords in both source and channel codebooks.

4.1.2.3 Codebook Generation

The last thing to clarify is how the source and channel book is generated. Consistent with literature on mismatched communication [1–4, 6], two types of codebooks are considered: spherical codebooks and i.i.d. Gaussian codebooks. Fix $D \in (0, \sigma_{\mathrm{s}}^2)$ and $P \in \mathbb{R}_+$.

In a spherical codebook, each source codeword $\hat{S}^k$ (or channel codeword X^n) is generated independently and uniformly over a sphere with radius $\sqrt{k(\sigma_{\mathrm{s}}^2 - D)}$ (or $\sqrt{nP}$ where P is a positive number), i.e.,

$$\hat{S}^k \sim f_{\hat{S}^k}^{\mathrm{sp}}(\hat{s}^k) := \frac{\delta(\|\hat{s}^k\|^2 - k(\sigma_{\mathrm{s}}^2 - D))}{A_k(\sqrt{k(\sigma_{\mathrm{s}}^2 - D)})}, \tag{4.12}$$

$$X^n \sim f_{X^n}^{\mathrm{sp}}(x^n) := \frac{\delta(\|x^n\|^2 - nP)}{A_n(\sqrt{nP})}, \tag{4.13}$$

where $\delta(\cdot)$ is the Dirac delta function, $A_k(r) := k\pi^{k/2}r^{k-1}/\Gamma(\frac{k+2}{2})$ is the surface area of an k-dimensional sphere with radius r, and $\Gamma(\cdot)$ is the Gamma function.

In an i.i.d. Gaussian codebook, each source codeword $\hat{S}^k$ (or channel codeword X^n) is generated independently according to a product of univariate Gaussian distributions each with variance $\sigma_{\mathrm{s}}^2 - D$ (or P), i.e.,

$$\hat{S}^k \sim f_{\hat{S}^k}^{\mathrm{iid}}(\hat{s}^k) := \frac{\exp\left\{-\frac{\|\hat{s}^k\|^2}{2(\sigma_{\mathrm{s}}^2 - D)}\right\}}{(\sqrt{2\pi(\sigma_{\mathrm{s}}^2 - D)})^k}, \tag{4.14}$$

$$X^n \sim f_{X^n}^{\mathrm{iid}}(x^n) := \frac{\exp\left\{-\frac{\|x^n\|^2}{2P}\right\}}{(\sqrt{2\pi P})^n}. \tag{4.15}$$

4.1.3 Performance Metric

Fix any $D \in \mathbb{R}_+$ such that $D < \sigma_{\mathrm{s}}^2$. To evaluate the performance of an (n, k)-code, the following excess-distortion probability with respect to distortion level D, analogously to (2.54), is considered:

$$\mathrm{P_e}(k, n, D) := \Pr\{d(S^k, \phi(f(S^k))) > D\} \tag{4.16}$$

$$= \Pr\left\{S^k \notin \mathcal{T}\right\} + \sum_{i \in [N]} \Pr\left\{S^k \in \mathcal{T}_i,\ d(S^k, \phi(f(S^k))) > D\right\}, \tag{4.17}$$

where (4.17) follows from the constructed mismatched code in Definition 4.2.

Analogously to (2.209) for the matched case, given any $\varepsilon \in (0,1)$, the low-latency theoretical benchmark for mismatched JSCC is defined as follows:

$$R^*_{\mathrm{sp,sp}}(n, D, \varepsilon) := \sup\Bigg\{\frac{k}{n} : \exists \text{ an } (k,n)\text{-code for mismatched JSCC}$$
$$\text{with spherical source and channel codebooks s.t. } \mathrm{P_e}(k,n,D) \le \varepsilon\Bigg\}. \tag{4.18}$$

Note that the above definition differs from (2.209) for the matched case. Specifically, in the matched case, one needs to optimize code design to achieve best performance. In contrast, in (4.18), the coding scheme is fixed and the converse part is known as ensemble converse [12], which is used to show that the analysis of the given coding scheme is accurate. Similarly, one can define $R^*_{\dagger,\ddagger}(n, D, \varepsilon)$ for any $(\dagger, \ddagger) \in$ sp, iid when $\dagger$-type source codebook is used and $\ddagger$-type channel codebook is used. Analogously, when $N = 1$ and thus an SSCC code is used, the theoretical benchmark is defined as $R^{*,\mathrm{sscc}}_{\dagger,\ddagger}(n, D, \varepsilon)$.

4.2 Second-Order Asymptotics

4.2.1 Preliminaries

This section presents preliminaries for second-order asymptotics: the choice of M_i and preliminary definitions.

For $\dagger \in \{\mathrm{sp}, \mathrm{iid}\}$ and any source sequence $s^k \in \mathcal{S}^k$, by spherical symmetry, the non-excess-distortion probability $\Pr_{f^\dagger_{\hat{S}^k}}\{d(s^k, \hat{S}^k) \le D\}$ depends on s^k only through its norm $\|s^k\|$. Thus, for any s^k such that $\|s^k\|^2/k = p$, one can define

$$\Psi_\dagger(k, p) := \Pr_{f^\dagger_{\hat{S}^k}}\{d(s^k, \hat{S}^k) \le D\}. \tag{4.19}$$

For $\dagger \in \{\mathrm{sp}, \mathrm{iid}\}$, when a $\dagger$-type source codebook is used, for each $i \in [N]$, choose M_i such that

$$\log M_i = -\log \Psi_\dagger(k, \Upsilon(i)) + \log k. \tag{4.20}$$

Note that the above choice of M_i is robust since it only does not depend on source distribution.

Recall the definition of $(\sigma_{\mathrm{s}}^2, \sigma_{\mathrm{c}}^2, \zeta_{\mathrm{s}}, \zeta_{\mathrm{c}})$ in (4.2). Given any real number $P \in \mathbb{R}_+$, define the following functions:

$$V_{\mathrm{s}}(\sigma_{\mathrm{s}}^2, \zeta_{\mathrm{s}}) := \frac{\zeta_{\mathrm{s}} - \sigma^4}{4\sigma^4}, \tag{4.21}$$

$$V_{\mathrm{c}}^{\mathrm{sp}}(P, \zeta_{\mathrm{c}}) := \frac{P^2(\zeta_{\mathrm{c}} - \sigma_{\mathrm{c}}^4) + 4P\sigma_{\mathrm{c}}^6}{4\sigma_{\mathrm{c}}^4(P + \sigma_{\mathrm{c}}^2)^2}, \tag{4.22}$$

$$V_{\mathrm{c}}^{\mathrm{iid}}(P, \zeta_{\mathrm{c}}) := \frac{P^2(\zeta_{\mathrm{c}} + \sigma_{\mathrm{c}}^4) + 4P\sigma_{\mathrm{c}}^6}{4\sigma_{\mathrm{c}}^4(P + \sigma_{\mathrm{c}}^2)^2}. \tag{4.23}$$

Recall that for a GMS with mean zero and variance σ_{s}^2, the rate-distortion equals $R(\sigma_{\mathrm{s}}^2, D) = \frac{1}{2}\log\frac{\sigma_{\mathrm{s}}^2}{D}$ (cf. (2.67)), and for an AWGN channel with transmit power P and noise power of one, the capacity equals $C(P) = \frac{1}{2}\log(1 + P)$ (cf. (2.175)). Let

$$\rho(\sigma_{\mathrm{s}}^2, \sigma_{\mathrm{c}}^2, P, D) := \frac{C(P/\sigma_{\mathrm{c}}^2)}{R(\sigma_{\mathrm{s}}^2, D)}. \tag{4.24}$$

For any $\ddagger \in \{\mathrm{sp}, \mathrm{iid}\}$, define the following function

$$\begin{aligned} V_{\ddagger}(\sigma_{\mathrm{s}}^2, \zeta_{\mathrm{s}}, \sigma_{\mathrm{c}}^2, \zeta_{\mathrm{c}}, P, D) &:= \frac{\rho(\sigma_{\mathrm{s}}^2, \sigma_{\mathrm{c}}^2, P, D)V_{\mathrm{s}}(\sigma_{\mathrm{s}}^2, \zeta_{\mathrm{s}}) + V_{\mathrm{c}}^{\ddagger}(P, \zeta_{\mathrm{c}})}{(R(\sigma_{\mathrm{s}}^2, D))^2}, &(4.25)\\ &= \frac{C(P/\sigma_{\mathrm{c}}^2)V_{\mathrm{s}}(\sigma_{\mathrm{s}}^2, \zeta_{\mathrm{s}}) + R(\sigma_{\mathrm{s}}^2, D)V_{\mathrm{c}}^{\ddagger}(P, \zeta_{\mathrm{c}})}{(R(\sigma_{\mathrm{s}}^2, D))^3}. &(4.26) \end{aligned}$$

4.2.2 Results and Discussions

Theorem 4.1 *For any $\varepsilon \in [0, 1)$ and any $(\dagger, \ddagger) \in \{\mathrm{sp}, \mathrm{iid}\}^2$, the maximal transmission rate equals*

$$R_{\dagger,\ddagger}^*(n, D, \varepsilon) = \rho(\sigma_{\mathrm{s}}^2, \sigma_{\mathrm{c}}^2, P, D) - \sqrt{\frac{V_{\ddagger}(\sigma_{\mathrm{s}}^2, \zeta_{\mathrm{s}}, \sigma_{\mathrm{c}}^2, \zeta_{\mathrm{c}}, P, D)}{n}}\mathrm{Q}^{-1}(\varepsilon) + O\Big(\frac{\log n}{n}\Big). \tag{4.27}$$

Theorem 4.1 characterizes the low-latency performance of mismatched JSCC up to second-order asymptotics with the third-order term characterized. A key implication of Theorem 4.1 is that the coding scheme presented in Definition 4.2

is valid to transmit an arbitrary memoryless source over an additive arbitrary noise channel, since the low-latency performance guarantee of doing so is characterized.

If one cares about the first-order asymptotic rate in the Shannon theory sense, it follows that for any $(\dagger, \ddagger) \in \{\text{sp}, \text{iid}\}^2$,

$$\lim_{n\to\infty} R^*_{\dagger,\ddagger}(n, D, \varepsilon) = \rho(\sigma_\text{s}^2, \sigma_\text{c}^2, P, D). \tag{4.28}$$

Note that the left-hand side is a function of the excess-distortion probability and the types of source and channel codebook, while the right-hand side is simply a fixed number. The result in (4.28) implies that regardless of the source and channel codebooks, the asymptotic transmission rate remains the same for any tolerable excess-resolution probability. It is of even more interest to note that $\rho(\sigma_\text{s}^2, P, D) = \frac{C(P/\sigma_\text{c}^2)}{R(\sigma_\text{s}^2, D)}$ is the first-order asymptotic transmission rate to transmit a GMS over an AWGN channel. Thus, as long as the source power equals σ_s^2 and the additive memoryless noise power equals 1, the asymptotic performance to transmit an arbitrary memoryless over an additive arbitrary noise channel has the same asymptotic performance of transmitting a GMS over an AWGN channel, using the code in Definition 4.2. Such a result is very insightful and advocates the study of GMS and AWGN channels despite the exact distribution of source and channel noise is not necessarily Gaussian.

If one delves further and cares about low-latency performance, the second-order term in the transmission rate $R^*_{\dagger,\ddagger}(n, D, \varepsilon)$ scales in the order of $\frac{1}{\sqrt{n}}$ and equals

$$L^*_{\dagger,\ddagger}(n, D, \varepsilon) := \lim_{n\to\infty} \sqrt{n}\big(R^*_{\dagger,\ddagger}(n, D, \varepsilon) - \rho(\sigma_\text{s}^2, \sigma_\text{c}^2, P, D)\big) \tag{4.29}$$

$$= -\sqrt{V_\ddagger(\sigma_\text{s}^2, \zeta_\text{s}, \sigma_\text{c}^2, \zeta_\text{c}, P, D)}\mathrm{Q}^{-1}(\varepsilon). \tag{4.30}$$

In contrast to first-order asymptotics, the second-order term depends on the type of channel codebook. It follows from (4.22) and (4.23) that

$$V_\text{c}^\text{iid}(P, \zeta_\text{c}) = V_\text{c}^\text{sp}(P, \zeta_\text{c}) + \frac{P^2\sigma_\text{c}^4}{2\sigma_\text{c}^4(P+\sigma_\text{c}^2)^2}. \tag{4.31}$$

Thus, for any $\varepsilon \in (0, 0.5)$, $L^*_{\dagger,\text{iid}}(n, D, \varepsilon) < L^*_{\dagger,\text{sp}}(n, D, \varepsilon)$, which implies that i.i.d. Gaussian channel codebook leads to worse low-latency performance with small excess-distortion probability, although (4.28) implies that both channel codebooks achieve the same infinite-latency performance. This result disapproves the use of i.i.d. Gaussian codebooks for low-latency analysis of additive noise channels despite its simplicity. On the other hand, the type of source codebook does not affect the second-order term, implying that both source codebooks achieve the same low-latency performance. Thus, for simplicity, i.i.d. Gaussian codebook shall be used for data compression.

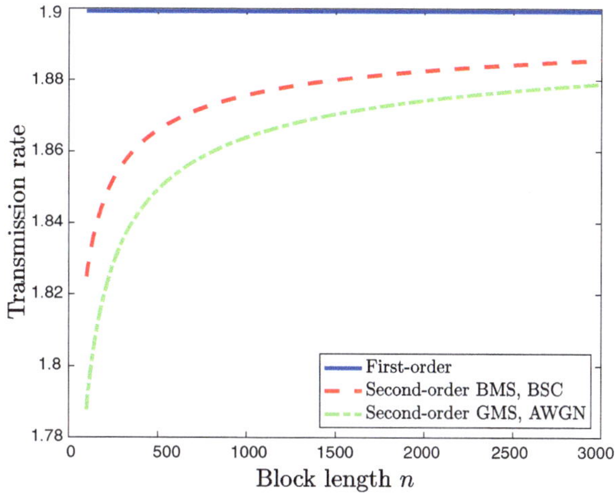

Fig. 4.3 Plot of first- and second-order asymptotics of mismatched JSCC and SSCC when $\varepsilon = 0.1$, $D = 0.05$, $P = 0.25$, $\sigma_s^2 = 0.5$ and $\sigma_c^2 = 0.3$. Two cases are considered: transmitting a BMS over a BSC and transmitting a GMS over an AWGN channel

Although the first-order asymptotic rate of the mismatched code is the same for different source and channel distributions satisfying (4.2), the second-order term depends on these distributions via the dispersion term $V_{\ddagger}(\sigma_s^2, \zeta_s, \sigma_c^2, \zeta_c, P, D)$. To illustrate, in Fig. 4.3, the first- and second-order asymptotics of the mismatched code for different source and channel distribution pairs. Two cases are considered: transmitting a binary memoryless source (BMS) over a BSC and transmitting a GMS over an AWGN channel. As observed, the second-order asymptotics in these two cases differ since the fourth moment of the source and channel noise distributions differ.

It is critical to note that Theorem 4.1 presents the analysis of a fixed code scheme, and the equality follows since an ensemble converse is derived, showing that the analysis of the fixed coding scheme is accurate. This is in stark contrast to the result in Theorem 2.5 for the matched case, where the equality in the coding theorem implies that optimal performance for particular source and channel noise distributions are obtained. Therefore, the result in Theorem 4.1 can be better understood as an achievability result when the source and channel noise distributions are unknown. When specialized to particular source and channel noise distribution, the achieved performance can be far from the optimal result for the matched case where the source and channel noise distributions are known. However, the advantage of the mismatched code is its robustness and simplicity. The robustness lies in it can be applied to any memoryless source and any additive noise with performance guarantee. The simplicity lies in the fact that it does not require exact knowledge of source and channel noise distributions.

When specialized to transmitting a GMS over an AWGN channel, the result in Theorem 4.1 is optimal (cf. Theorem 2.5), when spherical-type channel codebook is used. Thus, the essence of the mismatched code in Definition 4.2 is to apply optimal codes for GMS and AWGN channels to transmit an arbitrary memoryless source over an additive arbitrary noise channel. It might be possible other mismatched code can be constructed with performance guarantee under a particular distortion measure.

4.3 Loss of Separation

For the special case of $N = 1$, i.e., all the source sequences are mapped into the same type, the code in Definition 4.2 reduces to separate source-channel coding. For any $\ddagger \in \{\text{sp}, \text{iid}\}$, define

$$L_{\ddagger}^{*,\text{sscc}}(\varepsilon) := \frac{1}{R(\sigma_{\text{s}}^2, D)} \min_{\substack{(\varepsilon_1,\varepsilon_2)\in(0,1)^2:\\ \varepsilon_1+\varepsilon_2\le\varepsilon}} \Big(\sqrt{\rho(P, \sigma_{\text{s}}^2, D) V_{\text{s}}(\sigma_{\text{s}}^2, \zeta_{\text{s}})} \mathrm{Q}^{-1}(\varepsilon_1) + \sqrt{V_{\text{c}}^{\ddagger}(P, \zeta_{\text{c}})} \mathrm{Q}^{-1}(\varepsilon_2)\Big). \tag{4.32}$$

Similarly to Theorem 4.1, the following result holds for SSCC.

Theorem 4.2 *For any* $\varepsilon \in [0, 1)$ *and any* $(\dagger, \ddagger) \in \{\text{sp}, \text{iid}\}^2$*, the maximal transmission rate equals*

$$R_{\dagger,\ddagger}^{*,\text{sscc}}(n, D, \varepsilon) \ge \rho(\sigma_{\text{s}}^2, \sigma_{\text{c}}^2, P, D) + \frac{L_{\ddagger}^{*,\text{sscc}}(\varepsilon)}{\sqrt{n}} + O\Big(\frac{\log n}{n}\Big). \tag{4.33}$$

Theorem 4.2 implies that asymptotically as the blocklength tends to infinity, SSCC achieves the same first-order asymptotic rate $\rho(\sigma_{\text{s}}^2, \sigma_{\text{c}}^2, P, D)$ as JSCC. Furthermore, the second-order term of SSCC depends only on the type of the channel codebook. It follows from Theorems 4.1 and 4.2, in particular (4.30) and (4.32), that the second-order term of SSCC is generally smaller, unless the source dispersion $V_{\text{s}}(\sigma_{\text{s}}^2, \zeta_{\text{s}}) = 0$ or the channel dispersion $V_{\text{c}}^{\ddagger}(P, \zeta_{\text{c}}) = 0$. In Fig. 4.4, the first- and second-order asymptotics of transmitting a BMS over an AWGN channel using JSCC and SSCC mismatched codes are plotted. It is observed that SSCC is strictly inferior to JSCC in second-order asymptotics.

4.4 Proof of Second-Order Asymptotics

To prove Theorem 4.1, one needs to derive the results for all four possible combinations of source and channel codebooks, where each codebook can either be a spherical or an i.i.d. Gaussian codebook. This section first presents preliminary results, which is followed by achievability and converse proofs.

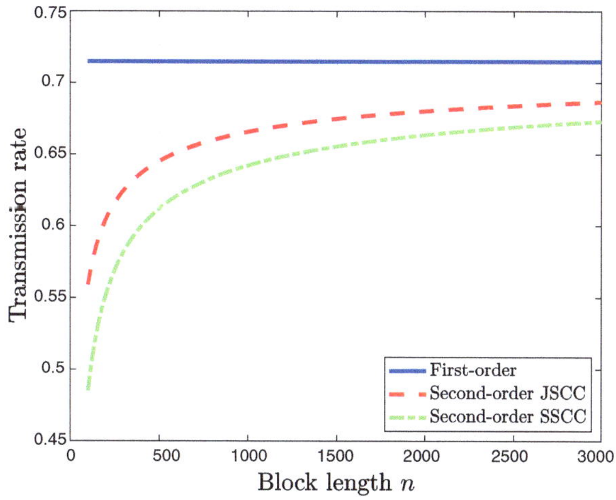

Fig. 4.4 Plot of first- and second-order asymptotics of mismatched JSCC and SSCC when $\varepsilon = 0.1$, $D = 0.05$, $P = 0.25$, $\sigma_s^2 = 0.3$ and $\sigma_c^2 = 0.1$. The case of transmitting a BMS over an AWGN channel is considered

4.4.1 *Preliminaries*

In the following, several key results are presented, including the decomposition of the excess-distortion probability, the analysis of the channel output, and theoretical benchmarks for mismatched compression. These results set the foundation for both achievability and converse analyses.

4.4.1.1 Bounds on Excess-Distortion Probability

Recall the mismatched JSCC code in Definition 4.2. Given any source sequence S^k, the encoder f declares an error if $S^k \notin \mathcal{T}$ and maps it into the codeword $X^n(I, J)$ if $S^k \in \mathcal{T}_I$ and $\hat{S}^k(I, J)$ minimizes the Euclidean distance with respect to S^k over all codewords in the sub-codebook $\{\hat{S}^k(I, \bar{j})\}_{\bar{j}\in[M_i]}$. Given the channel output Y^n, the channel decoder uses the modified nearest neighbor decoding (see (4.11)) to find $(\hat{I}, \hat{J})$ and declares $\hat{S}^k(\hat{I}, \hat{J})$ as the reproduced source sequence.

Given the above code, an excess-distortion event occurs if and only if one of the following events occurs:

1. $S^k \notin \mathcal{T}$;
2. $S^k \in \mathcal{T}_i$ for some $i \in [N]$ and one of the following events occurs:

   (a) The message pair (i, J) is transmitted correctly and the distortion is greater than D, i.e,

$$\mathcal{E}_1(i) := \{(\hat{I}, \hat{J}) = (i, J),\ d(S^k, \hat{S}^k(i, J)) > D\}, \tag{4.34}$$

(b) The message i is transmitted incorrectly and the distortion is greater than D, i.e.,

$$\mathcal{E}_2(i) := \{\hat{I} \neq i,\ d(S^k, \hat{S}^k(\hat{I}, \hat{J})) > D\}. \tag{4.35}$$

(c) The message i is transmitted correctly, the message J is transmitted incorrectly and the distortion is greater than D, i.e.,

$$\mathcal{E}_3(i) := \{\hat{I} = i, \hat{J} \neq J,\ d(S^k, \hat{S}^k(\hat{I}, \hat{J})) > D\}. \tag{4.36}$$

It follows from the definition of the excess-distortion probability in (4.17) and the definitions of error events above that

$$\mathrm{P_e}(k, n, D) = \Pr\left\{S^k \notin \mathcal{T}\right\} + \sum_{i\in[N]} \Pr\left\{\{S^k \in \mathcal{T}_i\} \cap \left(\mathcal{E}_1(i) \cup \mathcal{E}_2(i) \cup \mathcal{E}_3(i)\right)\right\}. \tag{4.37}$$

Thus, the excess-distortion probability can be upper bounded as

$$\begin{aligned}\mathrm{P_e}(k, n, D) \leq \Pr\left\{S^k \notin \mathcal{T}\right\} &+ \sum_{i\in[N]} \int_{s^k\in\mathcal{T}_i} \Pr\left\{d(s^k, \hat{S}^k(i, J)) > D\right\} f_{S^k}(s^k)\mathrm{d}s^k \\ &+ \sum_{i\in[N]} \int_{s^k\in\mathcal{T}_i} \Pr\{(\hat{I}, \hat{J}) \neq (i, J)\} f_{S^k}(s^k)\mathrm{d}s^k. \end{aligned} \tag{4.38}$$

where (4.38) follows from the the union bound and i) ignoring the requirement that $\{(\hat{I}, \hat{J}) = (i, J)\}$ in $\mathcal{E}_1(i)$, ii) ignoring the excess-distortion event $\{d(S^k, \hat{S}^K(\hat{I}, \hat{J})) > D\}$ in both $\mathcal{E}_2(i)$ and $\mathcal{E}_3(i)$, and iv) noting that $\{\hat{I} \neq i\} \cup \{\hat{I} = i,\ \hat{J} \neq J\}\} = \{(\hat{I}, \hat{J}) \neq (i, J)\}$. In (4.38), the first two probabilities are with respect to the joint distribution of the source sequence and source codebook, while the last probability is with respect to distributions of the channel codebook and the noise.

On the other hand, for each $i \in [N]$, the probability term in (4.37) is lower bounded as follows:

$$\begin{aligned}&\Pr\left\{\{S^k \in \mathcal{T}_i\} \cap \left(\mathcal{E}_1(i) \cup \mathcal{E}_2(i) \cup \mathcal{E}_3(i)\right)\right\} \\ &\geq \Pr\{\{S^k \in \mathcal{T}_i\} \cap \left(\mathcal{E}_2(i) \cup \mathcal{E}_3(i)\right)\} &(4.39)\\ &\geq \Pr\{\{S^k \in \mathcal{T}_i\} \cap \left(\{\hat{I} \neq i\} \cup \{\hat{I} = i,\ \hat{J} \neq J\}\right)\} \\ &\quad - \Pr\left\{\{S^k \in \mathcal{T}_i\} \bigcap \left(\{\{\hat{I} \neq i\} \cup \{\hat{I} = i,\ \hat{J} \neq J\}\} \cap \{d(S^k, \hat{S}^k(\hat{I}, \hat{J})) \leq D\}\right)\right\} &(4.40)\\ &\geq \Pr\{S^k \in \mathcal{T}_i,\ (\hat{I}, \hat{J}) \neq (i, J)\} - \Pr\{S^k \in \mathcal{T}_i,\ \hat{I} \neq i,\ d(S^k, \hat{S}^k(\hat{I}, \hat{J})) \leq D\} \\ &\quad - \Pr\{S^k \in \mathcal{T}_i,\ \hat{I} = i,\ \hat{J} \neq J,\ d(S^k, \hat{S}^k(\hat{I}, \hat{J})) \leq D\}. &(4.41)\end{aligned}$$

where (4.40) follows since $\Pr\{\mathcal{A} \cap (\mathcal{B} \cup \mathcal{C})\} \geq \Pr\{\mathcal{A} \cap \mathcal{B}\} - \Pr\{\mathcal{A} \cap \mathcal{B} \cap \mathcal{C}^C\}$ for any three sets $(\mathcal{A}, \mathcal{B}, \mathcal{C})$, and (4.41) follows by applying the union bound to the final term in (4.40) and using the fact that $\{\hat{I} \neq i\} \cup \{\hat{I} = i,\ \hat{J} \neq J\}\} = \{(\hat{I}, \hat{J}) \neq (i, J)\}$. In particular, in (4.40), $\mathcal{A} = \{\{S^k \in \mathcal{T}_i\}\}$, $\mathcal{B} = \big\{\{\hat{I} \neq i\} \cup \{\hat{I} = i,\ \hat{J} \neq J\}\big\}$ and $\mathcal{C} = \{d(S^k, \hat{S}^k(\hat{I}, \hat{J})) \leq D\}$.

Thus, it follows from (4.37) and (4.41) that the excess-distortion probability is lower bounded as follows:

$$
\begin{aligned}
\mathrm{P_e}(k, n, D) \geq \Pr\left\{S^k \notin \mathcal{T}\right\} &+ \sum_{i \in [N]} \Pr\{S^k \in \mathcal{T}_i,\ (\hat{I}, \hat{J}) \neq (i, J)\} \\
&- \Pr\{S^k \in \mathcal{T}_i,\ \hat{I} \neq i,\ d(S^k, \hat{S}^k(\hat{I}, \hat{J})) \leq D\} \\
&- \Pr\{S^k \in \mathcal{T}_i,\ \hat{I} = i,\ \hat{J} \neq J,\ d(S^k, \hat{S}^k(\hat{I}, \hat{J})) \leq D\}. \qquad (4.42)
\end{aligned}
$$

The upper and lower bounds on the excess-distortion probability in (4.38) and (4.42) are used in subsequent achievability and converse analyses, respectively.

4.4.1.2 Analysis of Error Probability of Modified NN Decoder

This section presents upper and lower bounds for the error probability of the channel decoding, given the source sequence and the source codebook. The results in this section hold regardless of the choices of source and channel codebooks.

Fix any $i \in [N]$. For simplicity, let $\hat{\mathbf{S}}_i := \{\hat{S}^k(i, \bar{j})\}_{\bar{j} \in [M_i]}$ denote the source codebook for power class $\mathcal{T}_i$, let $\hat{\mathbf{S}} := \cup_{i \in [N]} \hat{\mathbf{S}}_i$ denote the source codebook, and let $\mathbf{X} := \cup_{i \in [N]} \{X^n(i, \bar{j})\}_{\bar{j} \in [M_i]}$ denote the channel codebook. Furthermore, let $\hat{\mathbf{s}}_i$, $\hat{\mathbf{s}}$ and $\mathbf{x}$ be corresponding realizations.

Recall the definition of $\mathcal{D}$ in (4.9). For any $i \in [N]$, given a source sequence $S^k \in \mathcal{T}_i$ and a source codebook $\hat{\mathbf{S}}_i$, the output of the encoder f is (i, J), where

$$
J = \operatorname*{arg\,min}_{\bar{j} \in [M_i]} \|S^k - \hat{S}^k(i, \bar{j})\|^2, \qquad (4.43)
$$

and the output of the channel decoder is

$$
\begin{aligned}
(\hat{I}, \hat{J}) &= \operatorname*{arg\,min}_{(\tilde{i}, \tilde{j}) \in \mathcal{D}} \left(\|X^n(\tilde{i}, \tilde{j}) - Y^n\|^2 + 2 \log M_{\tilde{i}}\right) \qquad (4.44) \\
&= \operatorname*{arg\,min}_{(\tilde{i}, \tilde{j}) \in \mathcal{D}} \left(\|X^n(\tilde{i}, \tilde{j}) - (X^n(i, J) + Z^n)\|^2 + 2 \log M_{\tilde{i}}\right). \qquad (4.45)
\end{aligned}
$$

In (4.45), the output of the channel decoder $(\hat{I}, \hat{J})$ depends on the source sequence S^k and the source codebook $\mathbf{S}$ only through the type s of S^k and the sub-codebook

$\hat{\mathbf{S}}_i$, i.e., for any $s^k \in \mathcal{T}_i$ and any $\hat{\mathbf{s}}$,

$$\Pr\{(\hat{I}, \hat{J}) = (\hat{\imath}, \hat{\jmath})|S^k = s^k,\ \hat{\mathbf{S}} = \hat{\mathbf{s}}\} = \Pr\{(\hat{I}, \hat{J}) = (\hat{\imath}, \hat{\jmath})|S^k = s^k,\ \hat{\mathbf{S}}_i = \hat{\mathbf{s}}_i\}, \tag{4.46}$$

where the probability is calculated with respect to the distribution of the channel codebook $\mathbf{X}$.

Recall that $C(P) = \frac{1}{2}\log(1+P)$ (cf. (2.175)) and the distributions of $f^{\dagger}_{\tilde{X}^n}$ in (4.13) and (4.15). Given any (x^n, y^n), the mismatched information density (cf. (3.17)) is defined as

$$\imath(x^n; y^n) = nC\left(\frac{P}{\sigma_c^2}\right) + \frac{\|y^n\|^2}{2(P+\sigma_c^2)} - \frac{\|y^n - x^n\|^2}{2\sigma_c^2}. \tag{4.47}$$

For any $i \in [N]$ and any $\ddagger \in \{\text{sp}, \text{iid}\}$, define two quantities $\overline{h}_{\ddagger}(n, i)$ and $\underline{h}_{\ddagger}(n, i)$ as follows:

$$\overline{h}_{\ddagger}(n,i) := \min\left\{1, \mathbb{E}\left[\sum_{\tilde{i}\in[N]} M_{\tilde{i}} \Pr\left\{\imath(\bar{X}^n; Y^n) \le \imath(X^n; Y^n) - \log\frac{M_i}{M_{\tilde{i}}}\middle| X^n, Y^n\right\}\right]\right\}, \tag{4.48}$$

$$\underline{h}_{\ddagger}(n,i) := 1 - \left(1 - \mathbb{E}\left[\Pr\left\{\|\bar{X}^n - Y^n\|^2 \le \|X^n - Y^n\| \middle| X^n, Y^n\right\}\right]\right)^{M_i - 1}. \tag{4.49}$$

where the tuple $(\bar{X}^n, X^n, Y^n) \sim f^{\ddagger}_{X^n} f^{\ddagger}_{X^n} f_{Y^n|X^n}$.

For ease of notation, given $s^k \in \mathcal{T}_i$ and $\hat{\mathbf{s}}_i$, let

$$j(s^k, \hat{\mathbf{s}}_i) := \arg\min_{\bar{j}\in[M_i]} d(s^k, \hat{s}^k(i, \bar{j})). \tag{4.50}$$

The following lemma bounds the error probability of the channel decoder given a source sequence within a type class and a sub-codebook.

Lemma 4.1 *For any $i \in [N]$ and any $\ddagger \in \{\text{sp}, \text{iid}\}$, given any $s^k \in \mathcal{T}_i$ and any sub-codebook $\hat{\mathbf{s}}_i$, the error probability of the modified NN channel decoder satisfies*

$$\Pr\{(\hat{I}, \hat{J}) \neq (i, j(s^k, \hat{\mathbf{s}}_i))|S^k = s^k,\ \hat{\mathbf{S}}_i = \hat{\mathbf{s}}_i\} \le \overline{h}_{\ddagger}(n, i), \tag{4.51}$$

$$\Pr\{(\hat{I}, \hat{J}) \neq (i, j(s^k, \hat{\mathbf{s}}_i))|S^k = s^k,\ \hat{\mathbf{S}}_i = \hat{\mathbf{s}}_i\} \ge \underline{h}_{\ddagger}(n, i). \tag{4.52}$$

Lemma 4.1 holds regardless of the choice of the channel codebook. The upper bound in (4.51) is an extension of the random coding union (RCU) bound [13, Theorem 16] to the unequal message protection setting [7] while the lower bound in (4.52) is a proxy of the RCU bound in the other direction.

4.4.1.3 Results on Non-Excess-Distortion Probabilities

This section presents results concerning non-excess-distortion probabilities in [4]. Fix any $D \in (0, \sigma_{\mathrm{s}}^2)$. Let $r_1 := \sqrt{\sigma_{\mathrm{s}}^2 - D} - \sqrt{D}$ and $r_2 := \sqrt{\sigma_{\mathrm{s}}^2 - D} + \sqrt{D}$. For any $p \in \mathbb{R}_+$, define the following functions

$$R_{\mathrm{sp}}(p) := -\frac{1}{2}\log\left(1 - \frac{(p + \sigma_{\mathrm{s}}^2 - 2D)^2}{4p(\sigma_{\mathrm{s}}^2 - D)}\right), \tag{4.53}$$

$$s^*(p) := \max\left\{0, \frac{\sigma_{\mathrm{s}}^2 - 3D + \sqrt{(\sigma_{\mathrm{s}}^2 - D)^2 + 4pD}}{4D}\right\}, \tag{4.54}$$

$$R_{\mathrm{iid}}(p) := \frac{1}{2}\log(1 + 2s^*(p)) + \frac{s^*(p)p}{(1 + 2s^*(p))(\sigma_{\mathrm{s}}^2 - D)}d - \frac{s^*(p)D}{\sigma_{\mathrm{s}}^2 - D}. \tag{4.55}$$

Finally, given any integer $k \in \mathbb{N}$, let

$$\underline{g}(k, p) := \frac{\Gamma(\frac{k+2}{2})}{\sqrt{\pi}k\Gamma(\frac{k+1}{2})}\exp(-(k-1)R_{\mathrm{sp}}(p)), \tag{4.56}$$

$$\overline{g}(k, p) := \frac{\Gamma(\frac{k}{2})}{\Gamma(\frac{k-1}{2})\sqrt{\pi}}\exp(-(k-3)R_{\mathrm{sp}}(p)). \tag{4.57}$$

Recalling the definition of $\Psi_{\dagger}(\cdot)$ in (4.19). The following claims hold from [4].

Lemma 4.2

1. *Bounds on* $\Psi_{\mathrm{sp}}(k, p)$*:*
   (a) *If* $p \le r_1^2$ *or* $p \ge r_2^2$*, then* $\Psi_{\mathrm{sp}}(k, p) = 0$*;*
   (b) *If* $p \in (r_1^2, r_2^2)$*,* $\Psi_{\mathrm{sp}}(k, p) \ge \underline{g}(k, p)$*;*
   (c) *If* $p \in (r_1^2, r_2^2)$ *and* $p + \sigma_{\mathrm{s}}^2 - 2D \ge 0$*,* $\Psi_{\mathrm{sp}}(k, p) \le \overline{g}(k, p)$*.*
2. *Properties of* $R_{\mathrm{sp}}(p)$*:*
   (a) $R_{\mathrm{sp}}(p)$ *is increasing in* p *if* $p \ge |\sigma_{\mathrm{s}}^2 - 2D|$*;*
   (b) $R_{\mathrm{sp}}(\sigma_{\mathrm{s}}^2) = \frac{1}{2}\log\frac{\sigma_{\mathrm{s}}^2}{D}$*.*
3. *Bounds on* $\Psi_{\mathrm{iid}}(k, p)$*:*

$$\Psi_{\mathrm{iid}}(k, p) \sim \frac{\exp(-kR_{\mathrm{iid}}(p))}{s^*(p)\sqrt{\kappa(s^*(p), p)2\pi k}} \quad \text{as} \quad k \to \infty, \tag{4.58}$$

where

$$\kappa(s, p) := \frac{((\sigma_{\mathrm{s}}^2 - D)(1 + 2s) + 2p)^2}{(\sigma_{\mathrm{s}}^2 - D)(1 + 2s)^3}. \tag{4.59}$$

4. *Properties of* $R_{\mathrm{iid}}(p)$*:*
   (a) $R_{\mathrm{iid}}(p)$ *is increasing in* p *for all* $p \geq \max\{0, 2D - \sigma_{\mathrm{s}}^2\}$*;*
   (b) $R_{\mathrm{iid}}(\sigma_{\mathrm{s}}^2) = \frac{1}{2}\log\frac{\sigma_{\mathrm{s}}^2}{D}$.

It follows from the definitions in (4.53) and (4.55) that for any $p \in (r_1^2, r_2^2)$ such that $p \geq 2D - \sigma_{\mathrm{s}}^2$, $R_{\mathrm{sp}}(p)$ and $R_{\mathrm{iid}}(p)$ have the same Taylor expansions at $p = \sigma_{\mathrm{s}}^2$ as follows:

$$R_{\mathrm{sp}}(p) = R_{\mathrm{iid}}(p) = R(\sigma_{\mathrm{s}}^2, D) + \frac{p - \sigma_{\mathrm{s}}^2}{2\sigma_{\mathrm{s}}^2} + O\big((p - \sigma_{\mathrm{s}}^2)^2\big). \tag{4.60}$$

4.4.2 Achievability

Fix any $\varepsilon \in (0, 1)$ and $n \in \mathbb{N}$. For any any $\ddagger \in \{\mathrm{sp}, \mathrm{iid}\}$, let k be chosen such that

$$k = n\rho(\sigma_{\mathrm{s}}^2, \sigma_{\mathrm{c}}^2, P, D) - \sqrt{nV_{\ddagger}(\sigma_{\mathrm{s}}^2, \zeta_{\mathrm{s}}, \sigma_{\mathrm{c}}^2, \zeta_{\mathrm{c}}, P, D)}\mathrm{Q}^{-1}(\varepsilon) + O(\log n), \tag{4.61}$$

where $\rho(\sigma_{\mathrm{s}}^2, \sigma_{\mathrm{c}}^2, P, D)$ was defined in (4.24), $V_{\ddagger}(\sigma_{\mathrm{s}}^2, \zeta_{\mathrm{s}}, \sigma_{\mathrm{c}}^2, \zeta_{\mathrm{c}}, P, D)$ was defined in (4.26) and $O(\log n)$ is a multiple of $\log n$. Note that k scales linearly with n, i.e., $k = \Theta(n)$.

Using the definitions of ξ in (4.3) and N in (4.4), the number of power type classes is upper bounded as

$$N \leq 2k\xi + 1 = 2\sqrt{k\log k} + 1. \tag{4.62}$$

It follows from (4.7) that if $s^k \in \mathcal{T}_i$, when k is sufficiently large,

$$r_1^2 \leq \max\{|\sigma_{\mathrm{s}}^2 - 2D|, 2D - \sigma_{\mathrm{s}}^2\} \leq \frac{\|s^k\|^2}{k} \leq r_2^2. \tag{4.63}$$

Recall the distributions of $f_{\hat{S}^k}^{\dagger}$ in (4.12) and (4.14). For any $i \in [N]$ and $\dagger \in \{\mathrm{sp}, \mathrm{iid}\}$, let

$$f_{\hat{\mathbf{S}}_i}^{\dagger}(\hat{\mathbf{s}}_i) := \prod_{\bar{j}\in[M_i]} f_{\hat{S}^k}^{\dagger}(\hat{s}^k(i, \bar{j})), \tag{4.64}$$

$$f_{\hat{\mathbf{S}}}^{\dagger}(\hat{\mathbf{s}}) := \prod_{i\in[N]} f_{\hat{\mathbf{S}}_i}^{\dagger}(\hat{\mathbf{s}}_i) \tag{4.65}$$

denote the distributions of the source codebook for power class $\mathcal{T}_i$ and the source codebook. Recall the definition of $j(s^k, \hat{\mathbf{s}}_i)$ in (4.50). It follows from the result in (4.38) that for any $(\dagger, \ddagger) \in \{\mathrm{sp}, \mathrm{iid}\}^2$, the excess-distortion probability is upper bounded as

$$
\begin{aligned}
&\mathrm{P_e}(k, n, D) \\
&\le \Pr\left\{S^k \notin \mathcal{T}\right\} + \sum_{i \in [N]} \int_{s^k \in \mathcal{T}_i} (1 - \Pr_{f^{\dagger}_{\hat{S}^k}}\{d(s^k, \hat{S}^k) \le D\})^{M_i} f_{S^k}(s^k) \mathrm{d}s^k \\
&\quad + \sum_{i \in [N]} \int_{s^k \in \mathcal{T}_i} \left(\int_{\hat{\mathbf{s}}_i} \Pr\left\{(\hat{I}, \hat{J}) \neq (i, j(s^k, \hat{\mathbf{s}}_i))\right\} f_{\hat{\mathbf{S}}_i}(\hat{\mathbf{s}}_i) \mathrm{d}\hat{\mathbf{s}}_i \right) f_{S^k}(s^k) \mathrm{d}s^k \qquad (4.66) \\
&\le \Pr\left\{S^k \notin \mathcal{T}\right\} + \sum_{i \in [N]} \int_{s^k \in \mathcal{T}_i} (1 - \Pr_{f^{\dagger}_{\hat{S}^k}}\{d(s^k, \hat{S}^k) \le D\})^{M_i} f_{S^k}(s^k) \mathrm{d}s^k \\
&\quad + \sum_{i \in [N]} \int_{s^k \in \mathcal{T}_i} \overline{h}_{\ddagger}(n, i) f_{S^k}(s^k) \mathrm{d}s^k \qquad (4.67) \\
&\le \Pr\left\{S^k \notin \mathcal{T}\right\} + \sum_{i \in [N]} \int_{s^k \in \mathcal{T}_i} \exp(-M_i \Psi_{\dagger}(k, \|s^k\|^2/k)) f_{S^k}(s^k) \mathrm{d}s^k \\
&\quad + \sum_{i \in [N]} \int_{s^k \in \mathcal{T}_i} \overline{h}_{\ddagger}(n, i) f_{S^k}(s^k) \mathrm{d}s^k \qquad (4.68) \\
&\le \Pr\left\{S^k \notin \mathcal{T}\right\} + \exp(-k) + \sum_{i \in [N]} \Pr\{S^k \in \mathcal{T}_i\} \overline{h}_{\ddagger}(n, i), \qquad (4.69)
\end{aligned}
$$

where (4.66) follows since $d(s^k, \hat{S}^k(i, J)) > D$ implies that $d(s^k, \hat{S}^k(i, \bar{j})) > D$ for all $\bar{j} \in [M_i]$ and each source codeword $\hat{S}^k(i, \bar{j})$ is generated independently, (4.67) follows from (4.51), (4.68) follows from the inequality that $(1 - a)^M \le \exp(-Ma)$ for any $a \in [0, 1]$ and the definition of $\Psi_{\dagger}(\cdot)$ in (4.19), and (4.69) follows since i) $\Psi_{\dagger}(k, p)$ is decreasing in p for $p \ge \max\{\sigma_\mathrm{s}^2 - 2D, |\sigma_\mathrm{s}^2 - 2D|\}$ (which is implied by claims (ii)-(a) and (iv)-(a) in Lemma 4.2), (ii) $s^k \in \mathcal{T}_i$ implies that (4.63) holds and $\frac{\|s^k\|^2}{k} \le \Upsilon(i)$ (see (4.7)), and (iii) the choice of $\log M_i$ in (4.20).

The first term in (4.69) can be bounded by the Berry-Esseen Theorem. Let $V := \zeta_\mathrm{s} - \sigma_\mathrm{s}^4$ and $T := \mathbb{E}[|S^2 - \sigma_\mathrm{s}^2|^3]$. Note that $T < \infty$ since the sixth moment of the source distribution is finite (4.2). It follows that

$$
\Pr\left\{S^k \notin \mathcal{T}\right\} \le \Pr\left\{\frac{1}{k} \sum_{i \in [k]} (S_i^2 - \sigma_\mathrm{s}^2) > \xi \sigma_\mathrm{s}^2\right\} + \Pr\left\{\frac{1}{k} \sum_{i \in [k]} (S_i^2 - \sigma_\mathrm{s}^2) < -\xi \sigma_\mathrm{s}^2\right\} \qquad (4.70)
$$

$$\leq 2\mathrm{Q}\Big(\sqrt{\frac{\log k}{V}}\sigma_{\mathrm{s}}^2\Big) + \frac{2T}{\sqrt{k}V^{3/2}} \tag{4.71}$$

$$\leq 2\exp\Big(-\frac{\sigma_{\mathrm{s}}^4 \log k}{2V}\Big) + \frac{2T}{\sqrt{k}V^{3/2}}, \tag{4.72}$$

where (4.70) follows since $s^k \notin \mathcal{T}$ implies $\frac{\|s^k\|^2}{k} < (1-\xi)\sigma_{\mathrm{s}}^2$ or $\frac{\|s^k\|^2}{k} > (1+\xi)\sigma_{\mathrm{s}}^2$, (4.71) follows from the Berry-Esseen theorem and definition of ξ in (4.3), and (4.72) follows from $\mathrm{Q}(a) \leq \exp(-\frac{a^2}{2})$. The upper bound in (4.72) scales in the order of $O(\frac{1}{\sqrt{k}})$, which tends to 0 as $k \to \infty$.

The following lemma bounds the last term in (4.69).

Lemma 4.3 *For any* $(\dagger, \ddagger) \in \{\mathrm{sp}, \mathrm{iid}\}^2$,

$$\sum_{i\in[N]} \Pr\{S^k \in \mathcal{T}_i\}\overline{h}_{\ddagger}(n,i) \leq \varepsilon + O\Big(\frac{1}{\sqrt{n}}\Big). \tag{4.73}$$

The proof of Lemma 4.3 is deferred to the end of this subsection.

For any $(\dagger, \ddagger) \in \{\mathrm{sp}, \mathrm{iid}\}^2$, combining (4.69), (4.72) and (4.73) and using the definitions of $\rho(\sigma_{\mathrm{s}}^2, \sigma_{\mathrm{c}}^2, P, D)$ in (4.24) and $V_{\ddagger}(\sigma_{\mathrm{s}}^2, \zeta_{\mathrm{s}}, \sigma_{\mathrm{c}}^2, \zeta_{\mathrm{c}}, P, D)$ in (4.26), it follows from the choice of k in (4.61) that

$$\limsup_{n\to\infty} \mathrm{P_e}(k, n, D) \leq \varepsilon. \tag{4.74}$$

The achievability proof is now completed.

Proof of Lemma 4.3 Using the definitions of the mismatched information density in (4.47) and $\overline{h}_{\ddagger}(n,i)$ in (4.48), for any $\ddagger \in \{\mathrm{sp}, \mathrm{iid}\}$, it follows that

$$\overline{h}_{\ddagger}(n,i) - \frac{1}{\sqrt{n}} \leq \Pr\bigg\{\sum_{\tilde{i}\in[N]} M_{\tilde{i}} K_0 \exp\Big(-\imath(X^n;Y^n) + \log\frac{M_i}{M_{\tilde{i}}}\Big) \geq \frac{1}{\sqrt{n}}\bigg\} \tag{4.75}$$

$$= \Pr\bigg\{N K_0 M_i \exp(-\imath(X^n;Y^n)) \geq \frac{1}{\sqrt{n}}\bigg\} \tag{4.76}$$

$$= \Pr\bigg\{\frac{(P+\sigma_{\mathrm{c}}^2)\|Z^n\|^2 - \sigma_{\mathrm{c}}^2\|X^n+Z^n\|^2}{2(P+\sigma_{\mathrm{c}}^2)\sigma_{\mathrm{c}}^2} \geq nC(P/\sigma_{\mathrm{c}}^2) - \log M_i - \log(NK_0\sqrt{n})\bigg\}, \tag{4.77}$$

where (4.75) follows similarly to (2.197), which follows from the inequality that $\mathbb{E}[\min\{1, J\}] \leq \Pr\{J > \frac{1}{\sqrt{n}}\} + \frac{1}{\sqrt{n}}$ for any random variable J and K_0 is a finite constant defined in [14, Eq. (58)], and (4.77) follows from the definition of the mismatched information density $\imath(x^n; y^n)$ in (4.47).

Recall the definitions of $V_{\mathrm{s}}(\zeta_{\mathrm{s}}, \sigma_{\mathrm{s}}^2)$ in (4.21), $V_{\mathrm{c}}^{\mathrm{sp}}(\zeta_{\mathrm{c}}, P)$ in (4.22), $V_{\mathrm{c}}^{\mathrm{iid}}(\zeta_{\mathrm{c}}, P)$ in (4.23), $\rho(\sigma_{\mathrm{s}}^2, \sigma_{\mathrm{c}}^2, P, D)$ in (4.24) and the choice of k in (4.61). First consider the case where spherical codebooks are used for both source and channel codebooks, i.e., $\dagger = \mathrm{sp}$ and $\ddagger = \mathrm{sp}$. For simplicity, let

$$\begin{aligned} V_1 &:= 4\sigma_{\mathrm{s}}^4\sigma_{\mathrm{c}}^4(P+\sigma_{\mathrm{c}}^2)^2\big(\rho(\sigma_{\mathrm{s}}^2, \sigma_{\mathrm{c}}^2, P, D)V_{\mathrm{s}}(\zeta_{\mathrm{s}}, \sigma_{\mathrm{s}}^2) + V_{\mathrm{c}}^{\mathrm{sp}}(\zeta_{\mathrm{c}}, P)\big) \\ &= \rho(\sigma_{\mathrm{s}}^2, \sigma_{\mathrm{c}}^2, P, D)(P+\sigma_{\mathrm{c}}^2)^2\sigma_{\mathrm{c}}^4(\zeta_{\mathrm{s}} - \sigma_{\mathrm{s}}^4) + \sigma_{\mathrm{s}}^4(4P\sigma_{\mathrm{c}}^2 + P^2(\zeta_{\mathrm{c}} - \sigma_{\mathrm{c}}^4)). \end{aligned} \tag{4.78}$$

Using the choice of M_i in (4.20), the definition of $\overline{h}_{\ddagger}(n, i)$ in (4.48) and the result in (4.77), it follows that

$$\sum_{i\in[N]} \Pr\{S^k \in \mathcal{T}_i\}\overline{h}_{\mathrm{sp}}(n, i) - \frac{1}{\sqrt{n}}$$

$$\begin{aligned} &\le \sum_{i\in[N]} \mathbb{E}\Bigg[\Pr\Bigg\{\frac{(P+\sigma_{\mathrm{c}}^2)\|Z^n\|^2 - \sigma_{\mathrm{c}}^2\|X^n + Z^n\|^2}{2(P+\sigma_{\mathrm{c}}^2)\sigma_{\mathrm{c}}^2} \ge nC(P/\sigma_{\mathrm{c}}^2) \\ &\qquad + \log \Psi_{\mathrm{sp}}(k, \Upsilon(i)) - \log k - \log(NK_0\sqrt{n})\Bigg\}\Bigg| S^k \in \mathcal{T}_i\Bigg] \end{aligned} \tag{4.79}$$

$$\begin{aligned} &\le \sum_{i\in[N]} \mathbb{E}\Bigg[\Pr\Bigg\{\frac{(P+\sigma_{\mathrm{c}}^2)\|Z^n\|^2 - \sigma_{\mathrm{c}}^2\|X^n + Z^n\|^2}{2(P+\sigma_{\mathrm{c}}^2)\sigma_{\mathrm{c}}^2} \ge nC(P/\sigma_{\mathrm{c}}^2) \\ &\qquad + \log \underline{g}\left(k, \frac{\|S^k\|^2 + \sigma_{\mathrm{s}}^2}{k}\right) - \log k - \log(NK_0\sqrt{n})\Bigg\}\Bigg| S^k \in \mathcal{T}_i\Bigg] \end{aligned} \tag{4.80}$$

$$\begin{aligned} &\le \sum_{i\in[N]} \mathbb{E}\Bigg[\Pr\Bigg\{\frac{(P+\sigma_{\mathrm{c}}^2)\|Z^n\|^2 - \sigma_{\mathrm{c}}^2\|X^n + Z^n\|^2}{2(P+\sigma_{\mathrm{c}}^2)\sigma_{\mathrm{c}}^2} \ge nC(P/\sigma_{\mathrm{c}}^2) \\ &\qquad - (k-1)R_{\mathrm{sp}}\left(\frac{\|S^k\|^2 + \sigma_{\mathrm{s}}^2}{k}\right) + \log\frac{\Gamma(\frac{k+2}{2})}{\sqrt{\pi}\Gamma(\frac{k+1}{2})} \\ &\qquad - \log k - \log(NK_0\sqrt{n})\Bigg\}\Bigg| S^k \in \mathcal{T}_i\Bigg] \end{aligned} \tag{4.81}$$

$$= \sum_{i\in[N]} \int_{s^k\in\mathcal{T}_i} \Pr\left\{ \frac{(P+\sigma_{\mathrm{c}}^2)\|Z^n\|^2 - \sigma_{\mathrm{c}}^2\|X^n+Z^n\|^2}{2(P+\sigma_{\mathrm{c}}^2)\sigma_{\mathrm{c}}^2} \geq nC(P/\sigma_{\mathrm{c}}^2) - kR(\sigma_{\mathrm{s}}^2, D) \right.$$

$$\left. - \frac{\|s^k\|^2 - (k-1)\sigma_{\mathrm{s}}^2}{2\sigma_{\mathrm{s}}^2} + O(\log n) \right\} f_{S^k}(s^k)\mathrm{d}s^k \tag{4.82}$$

$$= \Pr\left\{ S^k \in \mathcal{T}, \frac{P\|Z^n\|^2 - n\sigma_{\mathrm{c}}^2 P - 2\sigma_{\mathrm{c}}^2\langle X^n, Z^n\rangle}{2(P+\sigma_{\mathrm{c}}^2)\sigma_{\mathrm{c}}^2} \geq nC(P/\sigma_{\mathrm{c}}^2) \right.$$

$$\left. - kR(\sigma_{\mathrm{s}}^2, D) - \frac{\|S^k\|^2 - k\sigma_{\mathrm{s}}^2}{2\sigma_{\mathrm{s}}^2} + O(\log n) \right\} \tag{4.83}$$

$$\leq \Pr\left\{ \frac{P\|Z^n\|^2 - n\sigma_{\mathrm{c}}^2 P - 2\sigma_{\mathrm{c}}^2\langle X^n, Z^n\rangle}{2(P+\sigma_{\mathrm{c}}^2)\sigma_{\mathrm{c}}^2} + \frac{\|S^k\|^2 - k\sigma_{\mathrm{s}}^2}{2\sigma_{\mathrm{s}}^2} \geq nC(P/\sigma_{\mathrm{c}}^2) \right.$$

$$\left. - kR(\sigma_{\mathrm{s}}^2, D) + O(\log n) \right\} \tag{4.84}$$

$$= \Pr\left\{ \sigma_{\mathrm{s}}^2\big(P\|Z^n\|^2 - n\sigma_{\mathrm{c}}^2 P - 2\sigma_{\mathrm{c}}^2\langle X^n, Z^n\rangle\big) + (P+\sigma_{\mathrm{c}}^2)\big(\|S^k\|^2 - k\sigma_{\mathrm{s}}^2\big) \right.$$

$$\left. \geq 2\sigma_{\mathrm{s}}^2(P+\sigma_{\mathrm{c}}^2)\sigma_{\mathrm{c}}^2\big(nC(P/\sigma_{\mathrm{c}}^2) - kR(\sigma_{\mathrm{s}}^2, D) + O(\log n)\big) \right\} \tag{4.85}$$

$$\leq \varepsilon + O\left(\frac{1}{\sqrt{n}}\right), \tag{4.86}$$

where (4.80) follows since $\frac{\|s^k\|^2+\sigma_{\mathrm{s}}^2}{k} \geq \Upsilon(i)$ when $s^k \in \mathcal{T}_i$, $\Psi_{\mathrm{sp}}(k,p) \geq \underline{g}(k,p)$ (see Claim (i) in Lemma 4.2) and $\underline{g}(k,p)$ is decreasing in $p \geq |\sigma_{\mathrm{s}}^2 - 2D|$, (4.81) follows from the definition of $\underline{g}(\cdot)$ in (4.56), (4.82) follows by i) using the Taylor expansion of $R_{\mathrm{sp}}(p)$ at $p = \sigma_{\mathrm{s}}^2$ (see (4.60)) and the bound on N in (4.62), and ii) noting that $\Gamma(\frac{k+2}{2})/(k\Gamma(\frac{k+1}{2})) = O(\frac{1}{\sqrt{k}})$, $k = \Theta(n)$, $\left|\frac{\|s^k\|^2}{k} - \sigma_{\mathrm{s}}^2\right|^2 \leq (\xi + \frac{1}{k})^2\sigma^4 = O(\frac{\log k}{k})$ since $s^k \in \Upsilon(i)$ (see (4.6)) for some $i \in [N]$ and $\xi = \sqrt{\frac{\log k}{k}}$ (see (4.3)), and (4.86) follows by applying the Berry-Esseen theorem for functions of independent random variables and using the choice of k in (4.61).

Next consider the case where the spherical source coding and i.i.d. Gaussian channel codebook are used. Compared with the case where $(\dagger, \ddagger) = (\mathrm{sp}, \mathrm{sp})$, the proof when $(\dagger, \ddagger) = (\mathrm{sp}, \mathrm{iid})$ is exactly the same until (4.82). It follows that

$$\begin{aligned}
&\sum_{i\in[N]} \Pr\{S^k \in \mathcal{T}_i\}\overline{h}_{\mathrm{iid}}(n,i)\\
&\le \Pr\left\{S^k \in \mathcal{T}_i,\ \frac{(P+\sigma_{\mathrm{c}}^2)\|Z^n\|^2-\sigma_{\mathrm{c}}^2\|X^n+Z^n\|^2}{2(P+\sigma_{\mathrm{c}}^2)\sigma_{\mathrm{c}}^2} \ge nC(P/\sigma_{\mathrm{c}}^2)-kR(\sigma_{\mathrm{s}}^2,D)\right.\\
&\qquad\left. -\frac{\|S^k\|^2-(k+1)\sigma_{\mathrm{s}}^2}{2\sigma_{\mathrm{s}}^2}+O(\log n)\right\} &(4.87)\\
&\le \Pr\left\{\frac{(P+\sigma_{\mathrm{c}}^2)\|Z^n\|^2-\sigma_{\mathrm{c}}^2\|X^n+Z^n\|^2}{2(P+\sigma_{\mathrm{c}}^2)\sigma_{\mathrm{c}}^2} \ge nC(P/\sigma_{\mathrm{c}}^2)-kR(\sigma_{\mathrm{s}}^2,D)\right.\\
&\qquad\left. -\frac{\|S^k\|^2-(k+1)\sigma_{\mathrm{s}}^2}{2\sigma_{\mathrm{s}}^2}+O(\log n)\right\} &(4.88)\\
&= \Pr\left\{\sum_{i\in[n]} \sigma_{\mathrm{s}}^2(PZ_i^2-\sigma_{\mathrm{c}}^2X_i^2-2\sigma_{\mathrm{c}}^2X_iZ_i)+(P+\sigma_{\mathrm{c}}^2)\sigma_{\mathrm{c}}^2(S_i^2-\sigma_{\mathrm{s}}^2)\right.\\
&\qquad\left.\ge 2\sigma_{\mathrm{s}}^2(P+\sigma_{\mathrm{c}}^2)\sigma_{\mathrm{c}}^2\big(nC(P/\sigma_{\mathrm{c}}^2)-kR(\sigma_{\mathrm{s}}^2,D)+O(\log n)\big)\right\} &(4.89)\\
&\le \varepsilon + O\left(\frac{1}{\sqrt{n}}\right), &(4.90)
\end{aligned}$$

where (4.89) follows since the channel input X^n is generated i.i.d. from $\mathcal{N}(0, P)$ when the channel codebook is an i.i.d. Gaussian codebook, and (4.90) follows similarly to the proof of (4.86) by applying the Berry-Esseen theorem for functions of independent random variables.

Finally, consider the case where the source codebook is an i.i.d. Gaussian codebook and the channel codebook is either a spherical or an i.i.d. Gaussian codebook. Using (4.20) and (4.51), similarly to arguments leading to (4.82), it follows that

$$\begin{aligned}
&\sum_{i\in[N]} \Pr\{S^k \in \mathcal{T}_i\}\overline{h}_{\ddagger}(n,i)\\
&\le \sum_{i\in[N]}\int_{s^k\in\mathcal{T}_i} \Pr\left\{\frac{(P+\sigma_{\mathrm{c}}^2)\|Z^n\|^2-\sigma_{\mathrm{c}}^2\|X^n+Z^n\|^2}{2(P+\sigma_{\mathrm{c}}^2)\sigma_{\mathrm{c}}^2} \ge nC(P/\sigma_{\mathrm{c}}^2)-kR(\sigma_{\mathrm{s}}^2,D)\right.\\
&\qquad\left. -\frac{\|s^k\|^2-(k+1)\sigma_{\mathrm{s}}^2}{2\sigma_{\mathrm{s}}^2}+O(\log n)\right\} f_{S^k}(s^k)\mathrm{d}s^k. &(4.91)
\end{aligned}$$

When the channel codebook is a spherical codebook, the rest of the proof is exactly the same as (4.83)–(4.86). When the channel codebook is an i.i.d. Gaussian codebook, the rest of the proof is exactly the same as (4.87)–(4.90). □

4.4.3 Ensemble Converse

In the ensemble converse proof, for any $\varepsilon \in (0,1)$ and $(\dagger, \ddagger) \in \{\mathrm{sp}, \mathrm{iid}\}^2$, assume for the sake of contradiction that there exists a sequence of (k,n)-codes such that $\mathrm{P_e}(k,n,D) \leq \varepsilon$ and

$$k = n\rho(\sigma_\mathrm{s}^2, \sigma_\mathrm{c}^2, P, D) - \sqrt{nV_\ddagger(\sigma_\mathrm{s}^2, \zeta_\mathrm{s}, \sigma_\mathrm{c}^2, \zeta_\mathrm{c}, P, D)}\mathrm{Q}^{-1}(\varepsilon+\tau) + O(\log n) \tag{4.92}$$

for some $\tau > 0$ and the $O(\log n)$ term is a multiple of $\log n$.

Recall the lower bound on the excess-distortion probability in (4.41) and the definition of $j(s^k, \hat{\mathbf{s}}_i)$ in (4.50). Fix any $i \in [N]$. For any $(\dagger, \ddagger) \in \{\mathrm{sp}, \mathrm{iid}\}^2$, each term in the first sum in (4.41) is lower bounded as follows:

$$\begin{aligned}
&\Pr\{S^k \in \mathcal{T}_i, (\hat{I}, \hat{J}) \neq (i, J)\} \\
&= \int_{s^k \in \mathcal{T}_i} \left(\int_{\hat{\mathbf{s}}_i} \Pr\left\{(\hat{I}, \hat{J}) \neq (i, j(s^k, \hat{\mathbf{s}}_i)) \middle| S^k = s^k,\ \hat{\mathbf{S}}_i = \hat{\mathbf{s}}_i\right\} f_{\hat{\mathbf{S}}_i}(\hat{\mathbf{s}}_i) \right) f_S^k(s^k) \mathrm{d}s^k && (4.93) \\
&\geq \int_{s^k \in \mathcal{T}_i} \underline{h}_\ddagger(n,i) f_S^k(s^k) \mathrm{d}s^k && (4.94) \\
&= \Pr\Bigg\{S^k \in \mathcal{T}_i, \frac{(P+\sigma_\mathrm{c}^2)\|Z^n\|^2 - \sigma_\mathrm{c}^2\|X^n+Z^n\|^2}{2(P+\sigma_\mathrm{c}^2)\sigma_\mathrm{c}^2} \\
&\geq nC(P/\sigma_\mathrm{c}^2) - \log M_i + O(\log n)\Bigg\}, && (4.95)
\end{aligned}$$

where (4.94) follows by using the result in (4.52), and (4.95) follows similarly to [3, Eq. (74)].

Using (4.92) and (4.95), similarly to steps proving Lemma 4.3, it follows that for any $(\dagger, \ddagger) \in \{\mathrm{sp}, \mathrm{iid}\}^2$,

$$\Pr\left\{S^k \notin \mathcal{T}\right\} + \sum_{i \in [N]} \Pr\{S^k \in \mathcal{T}_i, (\hat{I}, \hat{J}) \neq (i, J)\} \geq \varepsilon + \tau + O\left(\frac{1}{\sqrt{n}}\right). \tag{4.96}$$

The following lemma bounds show that the last two terms in (4.42), the lower bound of the excess-distortion probability, are negligible.

Lemma 4.4 *For any* $(\dagger,\ddagger)\in\{\mathrm{sp},\mathrm{iid}\}^2$ *and for all* $i\in[N]$,

$$\Pr\left\{d(S^k,\hat{S}^k(\hat{I},\hat{J}))\le D\middle|S^k\in\mathcal{T}_i,\ \hat{I}\neq i\right\}$$

$$\le\exp\left(-k\Big(\frac{1}{2}\log\frac{\sigma_{\mathrm{s}}^2}{D}+O\Big(\sqrt{\frac{\log k}{k}}\Big)\Big)\right), \tag{4.97}$$

$$\Pr\left\{d(S^k,\ddot{S}^k(\ddot{I},\ddot{J}))\le D|S^k\in\mathcal{T}_i,\ \hat{I}=i,\ \hat{J}\neq J\right\}$$

$$\le\exp\left(-k\Big(\frac{1}{2}\log\frac{\sigma_{\mathrm{s}}^2}{D}+O\Big(\sqrt{\frac{\log k}{k}}\Big)\Big)\right). \tag{4.98}$$

Combining the results in (4.42), (4.96) and Lemma 4.4, one concludes for any k satisfying (4.92), for any $(\dagger,\ddagger)\in\{\mathrm{sp},\mathrm{iid}\}^2$,

$$\liminf_{n\to\infty}\mathrm{P_e}(k,n,D)\ge\varepsilon+\tau, \tag{4.99}$$

This violates the condition that $\mathrm{P_e}(k,n,D)\le\varepsilon$. Since $\tau>0$ is arbitrary, the converse proof is completed.

Proof of Lemma 4.4 For simplicity, given $s^k\in\mathcal{T}_i$ and $\hat{\mathbf{s}}_i$, let

$$\Pr\{(\hat{I},\hat{J})=(\hat{i},\hat{j})|s^k,\hat{\mathbf{s}}_i\}:=\Pr\{(\hat{I},\hat{J})=(\hat{i},\hat{j})|S^k=s^k,\hat{\mathbf{S}}_i=\hat{\mathbf{s}}_i\}. \tag{4.100}$$

First consider (4.97). For any $\dagger\in\{\mathrm{sp},\mathrm{iid}\}$, it follows that

$$\begin{aligned}
&\Pr\{S^k\in\mathcal{T}_i,\ \hat{I}\neq i,\ d(S^k,\hat{S}^k(\hat{I},\hat{J}))\le D\}\\
&=\int_{s^k\in\mathcal{T}_i}\Bigg(\sum_{\hat{i}\in[N]}\sum_{\hat{j}\in[M_{\hat{i}}]}\int_{\hat{\mathbf{s}}}\Pr\{(\hat{I},\hat{J})=(\hat{i},\hat{j})|s^k,\hat{\mathbf{s}}\}\\
&\qquad\times\mathbb{1}\big(\hat{i}\neq i,\ d(s^k,\hat{S}^k(\hat{i},\hat{j}))\le D\big)f_{\hat{\mathbf{S}}}^{\dagger}(\hat{\mathbf{s}})\mathrm{d}\hat{\mathbf{s}}\Bigg)f_S^k(s^k)\mathrm{d}s^k
\end{aligned} \tag{4.101}$$

$$\begin{aligned}
&=\int_{s^k\in\mathcal{T}_i}\Bigg(\sum_{\hat{i}\in[N]\setminus\{i\}}\sum_{\hat{j}\in[M_{\hat{i}}]}\Bigg\{\int_{\hat{\mathbf{s}}_i}\Pr\{(\hat{I},\hat{J})=(\hat{i},\hat{j})|s^k,\hat{\mathbf{s}}_i\}\\
&\qquad\times\Big(\int_{\hat{\mathbf{s}}_{\hat{i}}}\mathbb{1}\big(d(s^k,\hat{S}^k(\hat{i},\hat{j}))\le D\big)f_{\hat{\mathbf{S}}_{\hat{i}}}^{\dagger}(\hat{\mathbf{s}}_{\hat{i}})\mathrm{d}\hat{\mathbf{s}}_{\hat{i}}\Big)f_{\hat{\mathbf{S}}_i}^{\dagger}(\hat{\mathbf{s}}_i)\mathrm{d}\hat{\mathbf{s}}_i\Bigg\}\Bigg)f_S^k(s^k)\mathrm{d}s^k
\end{aligned} \tag{4.102}$$

$$= \int_{s^k\in\mathcal{T}_i}\bigg(\int_{\hat{\mathbf{s}}_i}\sum_{\hat{i}\in[N]\backslash\{i\}}\sum_{\hat{j}\in[M_{\hat{i}}]}\Pr\{(\hat{I},\hat{J})=(\hat{i},\hat{j})|s^k,\hat{\mathbf{s}}_i\}$$

$$\times \Pr_{f^{\dagger}_{\hat{S}^k}}\{d(s^k,\hat{S}^k(\hat{i},\hat{j}))\le D\}f^{\dagger}_{\hat{\mathbf{S}}_i}(\hat{\mathbf{s}}_i)\mathrm{d}\hat{\mathbf{s}}_i\bigg)f^k_S(s^k)\mathrm{d}s^k \tag{4.103}$$

$$= \int_{s^k\in\mathcal{T}_i}\bigg(\int_{\hat{\mathbf{s}}_i}\sum_{\hat{i}\in[N]\backslash\{i\}}\sum_{\hat{j}\in[M_{\hat{i}}]}\Pr\{(\hat{I},\hat{J})=(\hat{i},\hat{j})|s^k,\hat{\mathbf{s}}_i\}f^{\dagger}_{\hat{\mathbf{S}}_i}(\hat{\mathbf{s}}_i)\mathrm{d}\hat{\mathbf{s}}_i\bigg)$$

$$\times \Psi_{\dagger}\Big(k,\frac{\|s^k\|^2}{k}\Big)f^k_S(s^k)\mathrm{d}s^k \tag{4.104}$$

$$\le \int_{s^k\in\mathcal{T}_i}\bigg(\int_{\hat{\mathbf{s}}_i}\sum_{\hat{i}\in[N]\backslash\{i\}}\sum_{\hat{j}\in[M_{\hat{i}}]}\Pr\{(\hat{I},\hat{J})=(\hat{i},\hat{j})|s^k,\hat{\mathbf{s}}_i\}f^{\dagger}_{\hat{\mathbf{S}}_i}(\hat{\mathbf{s}}_i)\mathrm{d}\hat{\mathbf{s}}_i\bigg)$$

$$\times \Psi_{\dagger}\big(k,\Upsilon(0)\big)f^k_S(s^k)\mathrm{d}s^k \tag{4.105}$$

$$\le \Pr\{S^k\in\mathcal{T}_i,\ \hat{I}\neq i\}\exp\bigg(-k\Big(\frac{1}{2}\log\frac{\sigma_{\mathrm{s}}^2}{D}+O\Big(\sqrt{\frac{\log k}{k}}\Big)\Big)\bigg), \tag{4.106}$$

where (4.102) follows since i) $(\hat{I},\hat{J})$ is independent of all sub-codebooks $\hat{\mathbf{s}}_{\bar{i}}$ for $\bar{i}\in[N]\backslash\{i\}$ (see Sect. 4.4.1.2), and ii) therefore the whole codebook $\hat{\mathbf{s}}$ can be decomposed into sub-codebooks and integrate over each codebook separately, (4.104) follows by using the definition of $\Psi_{\dagger}(\cdot)$ in (4.19) and noting that each source codeword is generated independently according to the same distribution (see (4.12) or (4.14)), (4.105) follows since i) $\Psi_{\dagger}\big(k,p\big)$ is decreasing in p for $p\ge\max\{|\sigma_{\mathrm{s}}^2-2D|,2D-\sigma_{\mathrm{s}}^2\}$ and ii) for all $s^k\in\mathcal{T}_i$, it follows that $\frac{\|s^k\|^2}{k}\ge\Upsilon(i-1)\ge\Upsilon(0)$, and (4.106) follows from the bounds on $\Psi_{\dagger}(\cdot)$ in Lemma 4.2 and the Taylor expansion of $R_{\dagger}(p)$ at σ_{s}^2 (see (4.60)) similarly to (4.82) and noting the definitions of $\Upsilon(0)$ in (4.5) and ξ in (4.3).

Now consider (4.98). Recall the definition of $j(s^k,\hat{\mathbf{s}}_i)$ in (4.50). The following lemma is critical.

Lemma 4.5 *For any $i\in[N]$, given any $s^k\in\mathcal{T}_i$ and any sub-codebook $\hat{\mathbf{s}}_i$, then for any $(\hat{i},\hat{j})$ such that $\hat{i}=i$ and $\hat{j}\neq j(s^k,\hat{\mathbf{s}}_i)$, it follows that* $\Pr\{(\hat{I},\hat{J})=(i,\hat{j})|s^k,\hat{\mathbf{s}}_i\}$ *depends only on i. For brevity, denote this quantity as $q_i :=$* $\Pr\{(\hat{I},\hat{J})=(i,\hat{j})|s^k,\hat{\mathbf{s}}_i\}$.

It follows from Lemma 4.5 that for each $i \in [N]$,

$$\begin{aligned}
&\Pr\{S^k \in \mathcal{T}_i,\ \hat{I}=i,\ \hat{J}\neq J)\}\\
&=\int_{s^k\in\mathcal{T}_i}\left(\int_{\hat{\mathbf{s}}_i}\sum_{\hat{j}\in[M_i]\setminus\{j(s^k,\hat{\mathbf{s}}_i)\}}\Pr\{(\hat{I},\hat{J})=(i,\hat{j})|s^k,\hat{\mathbf{s}}_i\}f_{\hat{\mathbf{S}}_i}(\hat{\mathbf{s}}_i)\mathrm{d}\hat{\mathbf{s}}_i\right)f_{S^k}(s^k)\mathrm{d}s^k \qquad (4.107)\\
&=\int_{s^k\in\mathcal{T}_i}\left(\int_{\hat{\mathbf{s}}_i}\sum_{\hat{j}\in[M_i]:\ \hat{j}\neq j(s^k,\hat{\mathbf{s}}_i)}q_i f_{\hat{\mathbf{S}}_i}(\hat{\mathbf{s}}_i)\mathrm{d}\hat{\mathbf{s}}_i\right)f_{S^k}(s^k)\mathrm{d}s^k \qquad (4.108)\\
&=\int_{s^k\in\mathcal{T}_i}(M_i-1)q_i f_{S^k}(s^k)\mathrm{d}s^k. \qquad (4.109)
\end{aligned}$$

Furthermore,

$$\begin{aligned}
&\Pr\{S^k\in\mathcal{T}_i,\hat{I}=i,\hat{J}\neq J,\ d(S^k,\hat{S}^k(\hat{I},\hat{J}))\leq D\}\\
&=\int_{s^k\in\mathcal{T}_i}\left(\int_{\hat{\mathbf{s}}_i}\sum_{\hat{j}\in[M_i]}\Pr\left\{(\hat{I},\hat{J})=(i,\hat{j})|s^k,\hat{\mathbf{s}}_i\right\}\mathbb{1}\right.\\
&\qquad\times\left(d(s^k,\hat{s}^k(i,\hat{j}))\leq D,\ \hat{j}\neq j(s^k,\hat{\mathbf{s}}_i)\right)\\
&\qquad\qquad\left.\times f_{\hat{\mathbf{S}}_i}^{\dagger}(\hat{\mathbf{s}}_i)\mathrm{d}\hat{\mathbf{s}}_i\right)f_S^k(s^k)\mathrm{d}s^k \qquad (4.110)\\
&=\int_{s^k\in\mathcal{T}_i}\left(\int_{\hat{\mathbf{s}}_i}\sum_{\hat{j}\in[M_i]\setminus\{j(s^k,\hat{\mathbf{s}}_i)\}}\Pr\left\{(\hat{I},\hat{J})=(i,\hat{j})|s^k,\hat{\mathbf{s}}_i\right\}\right.\\
&\qquad\left.\times\mathbb{1}(d(s^k,\hat{s}^k(i,\hat{j}))\leq D)f_{\hat{\mathbf{S}}_i}^{\dagger}(\hat{\mathbf{s}}_i)\mathrm{d}\hat{\mathbf{s}}_i\right)f_S^k(s^k)\mathrm{d}s^k \qquad (4.111)\\
&=\int_{s^k\in\mathcal{T}_i}\left(\int_{\hat{\mathbf{s}}_i}\sum_{\hat{j}\in[M_i]\setminus\{j(s^k,\hat{\mathbf{s}}_i)\}}q_i\mathbb{1}(d(s^k,\hat{s}^k(i,\hat{j}))\leq D)f_{\hat{\mathbf{S}}_i}^{\dagger}(\hat{\mathbf{s}}_i)\mathrm{d}\hat{\mathbf{s}}_i\right)f_S^k(s^k)\mathrm{d}s^k \qquad (4.112)\\
&\leq\int_{s^k\in\mathcal{T}_i}\left(\int_{\hat{\mathbf{s}}_i}\sum_{\hat{j}\in[M_i]}q_i\mathbb{1}(d(s^k,\hat{s}^k(i,\hat{j}))\leq D)\times f_{\hat{\mathbf{S}}_i}^{\dagger}(\hat{\mathbf{s}}_i)\mathrm{d}\hat{\mathbf{s}}_i\right)f_S^k(s^k)\mathrm{d}s^k \qquad (4.113)\\
&=\int_{s^k\in\mathcal{T}_i}M_iq_i\left(\Pr_{f_{\hat{S}^k}^{\dagger}}\{d(s^k,\hat{S}^k(i,\hat{j}))\leq D\}\right)f_S^k(s^k)\mathrm{d}s^k \qquad (4.114)\\
&\leq\int_{s^k\in\mathcal{T}_i}M_iq_i\Psi_{\dagger}(k,\Upsilon(0))f_S^k(s^k)\mathrm{d}s^k \qquad (4.115)
\end{aligned}$$

$$= \frac{M_i}{M_i - 1} \Psi_{\dagger}(k, \Upsilon(0)) \Pr\{S^k \in \mathcal{T}_i,\ \hat{I} = i,\ \hat{J} \neq J)\} \tag{4.116}$$

$$\leq \Pr\{S^k \in \mathcal{T}_i,\ \hat{I} = i,\ \hat{J} \neq J\} \exp\Big(- k\Big(\frac{1}{2}\log\frac{\sigma_s^2}{D} + O\Big(\sqrt{\frac{\log k}{k}}\Big)\Big)\Big), \tag{4.117}$$

where (4.112) follows from Lemma 4.5, (4.115) follows from similar steps leading to (4.105), (4.116) follows from (4.109), and (4.117) follows from i) the fact that $\frac{M_i}{M_i - 1} \to 1$ as $k \to \infty$ and ii) a Taylor expansion similar to (4.106).

The proof of Lemma 4.4 is now completed. □

4.5 Chapter Summary

This chapter presents a mismatched code that can be used to transmit an arbitrary memoryless source over an additive arbitrary noise channel, reliably in a lossy manner under quadratic loss. A mild condition is that the source distribution and the channel noise distribution have finite sixth moment, which is satisfied by most discrete distributions and continuous distributions. It is critical to note that the mismatched code in this chapter requires no knowledge of the exact source and channel noise distributions, in contrast to the usual information theoretical analyses. Such a code is thus robust and practical. The second-order asymptotics of this mismatched code is analyzed and shown to be ensemble optimal. A key take-away message is that mismatched code is robust with performance guarantee, although it is not necessarily optimal when the source and channel noise distributions are perfectly known.

4.6 Exercises

4.1 What is mismatched JSCC? What is the difference between the mismatched and the matched case?

4.2 Why is the coding scheme in Definition 4.2 an JSCC code? Can the code be degenerated to SSCC?

4.3 Let $(p, \xi) \in (0, 0.5)^2$. Consider a binary memoryless source with $P_S = [p, 1 - p]$ and a BSC with crossover probability ξ. Calculate the second-order asymptotic bound in Theorem 4.1 when either spherical codebook or the i.i.d. Gaussian channel codebook are used and plot the bound for $n = 50$ to $n = 3000$ when the power $P = 0.5$, the error probability $\varepsilon = 0.01$, the distortion level $D = 0.02$, and the parameters $p = 0.3$, $\xi = 0.2$.

4.4 Let $(p, P, \sigma) \in (0, 0.5) \times \mathbb{R}_+^2$. Consider a binary memoryless source with $P_S = [p, 1-p]$ and an AWGN channel with power P and noise power σ^2. Calculate the second-order asymptotic bound in Theorem 4.1 when spherical source codebook or i.i.d. Gaussian channel codebook are used and plot the bound for $n = 50$ to $n = 3000$ when the power $P = 3$, the noise power $\sigma_{\mathrm{c}}^2 = 1$, the error probability $\varepsilon = 0.1$, the distortion level $D = 0.05$, and the parameter $p = 0.2$.

4.5 Show that the mutual information density in (4.47) is the accumulated information density in (2.186) for an AWGN channel (cf. Definition 2.7) with $P_X = \mathcal{N}(\cdot; 0, P)$ and the additive noise $Z \sim \mathcal{N}(\cdot; 0, \sigma_{\mathrm{c}}^2)$.

4.6 Prove Lemma 4.1 regarding the performance of the modified NN decoder.

4.7 Justify the results in (4.86) and (4.90) concerning the upper bound on the excess-distortion probability, using Theorem 1.8, the function version Berry-Esseen theorem.

4.8 Justify the result in (4.96) concerning the lower bound on the excess-distortion probability, using Theorem 1.8, the function version Berry-Esseen theorem.

4.9 Proof the result in Lemma 4.5.

References

1. A. Lapidoth, Nearest neighbor decoding for additive non-Gaussian noise channels. IEEE Trans. Inf. Theory **42**(5), 1520–1529 (1996)
2. A. Lapidoth, On the role of mismatch in rate distortion theory. IEEE Trans. Inf. Theory **43**(1), 38–47 (1997)
3. J. Scarlett, V.Y.F. Tan, G. Durisi, The dispersion of nearest-neighbor decoding for additive non-Gaussian channels. IEEE Trans. Inf. Theory **63**(1), 81–92 (2017)
4. L. Zhou, V.Y.F. Tan, M. Motani, Refined asymptotics for rate-distortion using Gaussian codebooks for arbitrary sources. IEEE Trans. Inf. Theory **65**(5), 3145–3159 (2019)
5. L. Zhou, M. Motani, Finite blocklength lossy source coding for discrete memoryless sources. Found. Trend. Commun. Inf. Theory **20**(3), 157–389 (2023)
6. L. Zhou, V.Y.F. Tan, M. Motani, The dispersion of mismatched joint source-channel coding for arbitrary sources and additive channels. IEEE Trans. Inf. Theory **65**(4), 2234–2251 (2019)
7. Y.Y. Shkel, V.Y. F. Tan, S.C. Draper, Unequal message protection: Asymptotic and non-asymptotic tradeoffs. IEEE Trans. Inf. Theory **61**(10), 5396–5416 (2015)
8. I. Csiszár, On the error exponent of source-channel transmission with a distortion threshold. IEEE Trans. Inf. Theory **28**(11), 823–838 (1982)
9. D. Wang, A. Ingber, Y. Kochman, The dispersion of joint source-channel coding, in *Proceedings of the 2011 49th Annual Allerton Conference on Communication, Control, and Computing*, Monticello, IL (2011), pp. 180–187
10. V. Kostina, S. Verdú, Lossy Joint Source-Channel Coding in the Finite Blocklength Regime. IEEE Trans. Inf. Theory **59**(5), 2545–2575 (2013)
11. L. Zhou, V.Y.F. Tan, M. Motani, Second-order and moderate deviation asymptotics for successive refinement. IEEE Trans. Inf. Theory **63**(5), 2896–2921 (2017)
12. R. Gallager, The random coding bound is tight for the average code (corresp.). IEEE Trans. Inf. Theory **19**(2), 244–246 (1973)

13. Y. Polyanskiy, H.V. Poor, S. Verdú, Channel coding rate in the finite blocklength regime. IEEE Trans. Inf. Theory **56**(5), 2307–2359 (2010)
14. V.Y.F. Tan, T. Tomamichel, The third-order term in the normal approximation for the AWGN channel. IEEE Trans. Inf. Theory **61**(5), 2430–2438 (2015)

Chapter 5
Multiterminal Compression

This chapter studies the problem of multiterminal data compression, where multiple encoders or decoders exist. A typical example is the successive refinement (SR) problem [1] with two encoders and two decoders. This problem generalizes the lossy source coding problem in Chap. 2 by having one additional encoder that further compresses the source and having an additional decoder that generates a finer source estimate using outputs of both encoders. Multiterminal data compression is much more efficient than repeatedly applying the coding scheme for the point-to-point lossy source coding problem and finds applications in image compression, video compression, and clinical diagnosis.

The SR problem has been widely studied. For any discrete memoryless source, the rate-distortion region, which consists of rate pairs that ensure vanishing excess-distortion probability, was characterized by Rimoldi [1]. Koshelev [2] and Equitz-Cover [3] studied the sufficient and necessary condition for successive refinability, under which the sum rate of two encoders equals the rate-distortion function of point-to-point lossy source coding with the distortion level of the second decoder of the SR problem. The successive refinability implies that there is no loss in the compression rate when one serves two decoders with different distortion levels. Any Gaussian memoryless source (GMS) under the quadratic distortion measure is successively refinable [3]. For low-latency communication, the theoretical benchmark of the SR problem was characterized by Zhou, Tan, and Motani [4] and by No, Ingber, and Weissman [5] under two performance criteria for any DMS under bounded distortion measure and any GMS under quadratic distortion.

Although the above analyses of the SR problem are comprehensive, there is one common assumption: perfect knowledge of the source distribution. However, obtaining the exact source distribution before data compression is challenging. Thus, it is desired to propose a universal coding scheme that could be used to compress an arbitrary memoryless source (AMS). For the point-to-point setting, such coding scheme was proposed by Lapidoth [6], where optimal coding scheme for a GMS is used to compress AMS under quadratic distortion. Since the coding scheme

L. Zhou, L. Bai, *Low-latency Communication*, Textbooks in Telecommunication Engineering, https://doi.org/10.1007/978-3-032-06627-5_5

is not necessarily optimal for the compressed source, the coding scheme is named mismatched. The asymptotic performance of mismatched lossy source coding was characterized by Lapidoth [6, Theorem 3], and the low-latency performance was obtained by Zhou, Tan, and Motani [7]. For the SR problem, the corresponding result was characterized by Bai, Wu, and Zhou [8]. This chapter introduces the mismatched coding scheme and the low-latency theoretical benchmark for the SR problem in [8].

5.1 Problem Formulation

Fix integers $(n, M_1, M_2) \in \mathbb{N}^2$, three alphabets $(\mathcal{X}, \hat{\mathcal{X}}_1, \hat{\mathcal{X}}_2)$ and two positive real numbers $(D_1, D_2) \in \mathbb{R}_+^2$ such that $D_2 < D_1$. As shown in Fig. 5.1, the SR problem has two layers of encoders and decoders. For each $i \in [2]$, the encoder f_i has access to a source sequence $X^n \in \mathcal{X}^n$ and compresses it into a message $S_i \in [M_i]$. Two decoders are asymmetric and classified into weak and strong decoders. The weak decoder ϕ_1 aims to reproduce the source sequence X^n within distortion level D_1 using S_1 while the strong decoder ϕ_2 aims to reproduce X^n within a finer distortion level $D_2 < D_1$ with the additional access to S_2. The reproduced source sequences of decoders (ϕ_1, ϕ_2) are $\hat{X}_1^n$ and $\hat{X}_2^n$, respectively.

The source sequence X^n is generated i.i.d. from P_X defined on the alphabet $\mathcal{X}$ such that the second moment satisfies

$$\mathbb{E}[X^2] = \sigma^2. \tag{5.1}$$

The second moment can be measured via source symbol observations and is used to construct the mismatched code. Without loss of generality, assume that the desired distortion levels at the two decoders satisfy $\sigma^2 > D_1 > D_2$. This chapter generalizes the mismatched lossy source coding [6, Theorem 3] to the multiterminal SR setting [1]. Specifically, the mismatched SR code uses the minimal Euclidean distance encoding with spherical or i.i.d. Gaussian codebooks to compress an AMS X^n under quadratic distortion. Recall that for any two sequences $(x^n, y^n) \in (\mathbb{R}^n)^2$,

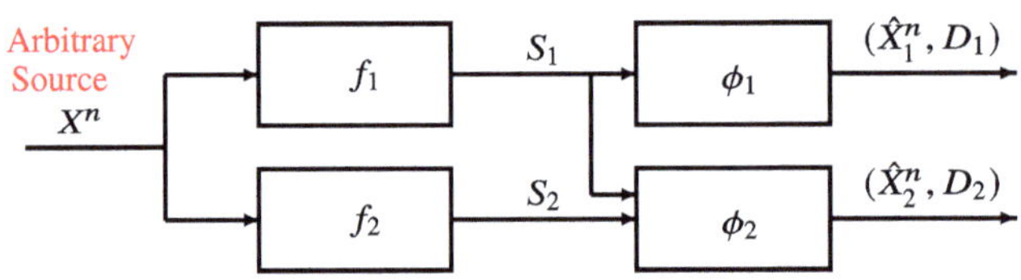

Fig. 5.1 System model for successive refinement of an arbitrary memoryless source

the quadratic distortion measure is defined as follows:

$$d(x^n, y^n) := \frac{1}{n}\|x^n - y^n\|^2 = \frac{1}{n}\sum_{i\in[n]}(x_i - y_i)^2. \tag{5.2}$$

5.1.1 Coding Scheme

Definition 5.1 An (n, M_1, M_2)-code for the mismatched SR problem consists of

- A set of M_1 codewords $\{Y^n(i)\}_{i\in[M_1]}$ known by encoders (f_1, f_2) and decoders (ϕ_1, ϕ_2)
- A set of M_2 codewords $\{Z^n(i, j)\}_{j\in[M_2]}$ known by the encoder f_2 and the decoder ϕ_2 for each $i \in [M_1]$,
- Two encoders f_1 and f_2 that use successive minimum Euclidean distance encoding to compress the source sequence X^n, i.e.,

$$f_1(X^n) := \arg\min_{i\in[M_1]} d(X^n, Y^n(i)), \tag{5.3}$$

$$f_2(X^n) := \arg\min_{j\in[M_2]} d(X^n, Z^n(f_1(X^n), j)), \tag{5.4}$$

- Two decoders ϕ_1 and ϕ_2 that operate as follows:

$$\phi_1(f_1(X^n)) = Y^n(f_1(X^n)), \tag{5.5}$$

$$\phi_2(f_1(X^n), f_2(X^n)) = Z^n(f_1(X^n), f_2(X^n)). \tag{5.6}$$

For simplicity, throughout the chapter, $\hat{X}_1^n$ and $Y^n(f_1(X^n))$ are interchangeably used to denote the reproduced source sequence of the first decoder ϕ_1, and similarly $\hat{X}_2^n$ and $Z^n(f_1(X^n), f_2(X^n))$ are used interchangeably.

To specify the codewords in Definition 5.1, define the following distributions that are functions of the two parameters: a vector $c^n = (c_1, \ldots, c_n) \in \mathbb{R}^n$ and a positive real number $P \in \mathbb{R}_+$.

- A uniform distribution over the surface of a sphere with center c^n and radius $\sqrt{nP}$ is defined as follows: for any $u^n = (u_1, \ldots, u_n) \in \mathbb{R}^n$,

$$f_{\mathrm{sp}}(u^n|c^n, P) = \frac{\mathbb{1}(\|u^n - c^n\|^2 - nP)}{S_n(\sqrt{nP})}, \tag{5.7}$$

where $\mathbb{1}(\cdot)$ is the indicator function, $S_n(r) = n\pi^{n/2}r^{n-1}/\Gamma(\frac{n+2}{2})$ is the surface area of an n-dimensional sphere with radius r, and $\Gamma(\cdot)$ is the Gamma function.

- A product Gaussian distribution is defined as follows: for any $u^n = (u_1, \ldots, u_n) \in \mathbb{R}^n$,

$$f_{\text{iid}}(u^n|c^n, P) = \prod_{i\in[n]} \frac{\exp\left(-\frac{(u_i-c_i)^2}{2P}\right)}{\sqrt{2\pi P}}. \tag{5.8}$$

Let $P_Y := \sigma^2 - \lambda D_1$ and $P_Z := \lambda D_1 - D_2$. Fix any positive real number $\lambda \in \mathbb{R}_+$. Both spherical and i.i.d. Gaussian codebooks are considered. Since both encoders can use either spherical or i.i.d. Gaussian codebooks, there are in total four different combinations of the codebooks. For ease of notation, throughout the paper, the type of the codebook used by encoder f_1 is denoted by $\dagger$ and the type of the codebook used by encoder f_2 is denoted by $\ddagger$. Thus, $(\dagger, \ddagger) \in \{\text{sp}, \text{iid}\}^2$, where "sp" is short for "spherical." For any $(\dagger, \ddagger) \in \{\text{sp}, \text{iid}\}^2$, encoder f_1 uses a random codebook with M_1 independent codewords $(Y^n(1), \ldots, Y^n(M_1))$, where each codeword $Y^n(i)$ is generated according to the distribution $f_\dagger(Y^n(i)|\mathbf{0}^n, P_Y)$; for each $i \in [M_1]$ and $Y^n(i)$, encoder f_2 uses a random codebook with M_2 independent codewords $(Z^n(i, 1), \ldots, Z^n(i, M_2))$, where each codeword $Z^n(i, j)$ is generated according to the distribution $f_\ddagger(Z^n(i, j)|Y^n(i), P_Z)$. For $n = 2$, the encoders' codebooks are illustrated in Fig. 5.2. In a spherical codebook, the codewords of f_1 distribute uniformly over the surface of the sphere with center of the origin and the radius of $\sqrt{nP_Y} = \sqrt{n(\sigma^2 - \lambda D_1)}$. When $f_1(X^n) = i$, given the codeword $Y^n(i)$, the codewords for f_2 distribute uniformly over the surface of the sphere with center $Y^n(i)$ and the radius $\sqrt{nP_Z} = \sqrt{n(\lambda D_1 - D_2)}$. In an i.i.d. Gaussian codebook, the codewords of f_1 are generated independently and each codeword is generated i.i.d. from the Gaussian distribution with mean 0 and variance $\sigma^2 - \lambda D_1$. When $f_1(X^n) = i$, given the codeword $Y^n(i)$, the codewords for encoder f_2 are generated similarly from a Gaussian product distribution.

5.1.2 Performance Criterion

To evaluate the performance of the mismatched coding scheme, there are two different performance criteria: the joint excess-distortion probability (JEP) that measures the probability that either decoder fails to decode X^n within the desired distortion level, and the separate excess-distortion probabilities (SEP) that measure the decoding failure of two decoders separately. The SEP are defined as follows:

$$\mathrm{P}_{\mathrm{e},n}^{\dagger}(D_1|M_1, M_2) := \Pr\{d(X^n, \hat{X}_1^n) > D_1\}, \tag{5.9}$$

$$\mathrm{P}_{\mathrm{e},n}^{\ddagger}(D_2|M_1, M_2) := \Pr\{d(X^n, \hat{X}_2^n) > D_2\}. \tag{5.10}$$

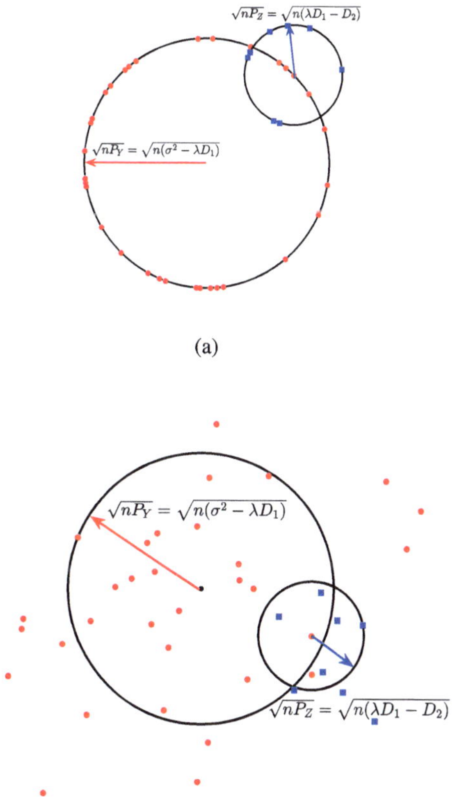

Fig. 5.2 Illustration of codewords when both encoders use the same kind of codebooks. The red dots denote the codewords of encoder f_1 and the blue squares denote the codewords of encoder f_2. (**a**) Both encoders use spherical codebooks. (**b**) Both encoders use i.i.d. Gaussian codebooks

The JEP is defined analogously:

$$\mathrm{P}_{\mathrm{e},n}^{\dagger,\ddagger}(D_1, D_2|M_1, M_2) := \Pr\left\{d(X^n, \hat{X}_1^n) > D_1 \text{ or } d(X^n, \hat{X}_2^n) > D_2\right\}. \tag{5.11}$$

Note that the above probability terms average not only over the distribution of the source sequence X^n but also over the random codebooks. These definitions are consistent with existing studies on mismatched communication [6, 7, 9, 10].

5.1.3 The Rate-Distortion Region

The rate-distortion region collects rate pairs of two encoders so that the mismatched code in Definition 5.1 ensures vanishing JEP and SEP. Fix any $(\dagger, \ddagger) \in \text{sp, iid}$.

Definition 5.2 A rate pair $(R_1, R_2) \in \mathbb{R}_+^2$ is said to be $(D_1, D_2|\dagger, \ddagger)$-achievable if there exists a sequence of (n, M_1, M_2)-codes using $(\dagger, \ddagger)$ codebooks such that

$$\limsup_{n\to\infty} \frac{1}{n} \log M_1 \le R_1, \tag{5.12}$$

$$\limsup_{n\to\infty} \frac{1}{n} \log M_1 M_2 \le R_1 + R_2, \tag{5.13}$$

and

$$\lim_{n\to\infty} \mathrm{P}_{\mathrm{e},n}^{\dagger,\ddagger}(D_1, D_2|M_1, M_2) = 0. \tag{5.14}$$

The convex closure of the set of all $(D_1, D_2|\dagger, \ddagger)$-achievable rate pairs is called the rate-distortion region and denoted as $\mathcal{R}_{\dagger,\ddagger}(D_1, D_2)$.

As a corollary of the results in Theorem 5.1 by letting $n \to \infty$, one has the following inner bound to $\mathcal{R}_{\dagger,\ddagger}(D_1, D_2)$:

$$\mathcal{R}_{\mathrm{inner}} := \left\{(R_1, R_2) : \ R_1 \ge \frac{1}{2} \log \frac{\sigma^2}{D_1}, \ R_1 + R_2 \ge \frac{1}{2} \log \frac{\sigma^2}{D_2}\right\}. \tag{5.15}$$

The same inner bound holds under SEP since

$$\mathrm{P}_{\mathrm{e},n}^{\dagger,\ddagger}(D_1, D_2|M_1, M_2) \ge \max_{i\in[2]} \mathrm{P}_{\mathrm{e},n}^{\dagger}(D_i|M_1, M_2). \tag{5.16}$$

Note that $\mathcal{R}_{\mathrm{inner}}$ is the rate-distortion region for a GMS under the quadratic distortion measure [3, 4]. One might wonder whether the ensemble converse for mismatched lossy source coding in [6, Theorem 3] implies the ensemble converse for mismatched successive refinement. Unfortunately, the answer doesn't directly follow since the codewords for the second encoder of mismatched successive refinement are not equivalent to the codewords for the encoder in mismatched rate-distortion, a figure illustration of which are available Figs. 5.3 and 5.4 where both encoders use either spherical or i.i.d. Gaussian codebooks. It is worthwhile future work to investigate whether the rate-distortion region $\mathcal{R}_{\mathrm{inner}}$ is ensemble tight.

5.1.4 *Definition of Theoretical Benchmark*

This section defines the second-order asymptotics. Different from the point-to-point (P2P) settings of previous chapters, for multiterminal problems, the theoretical benchmark is a multidimensional region instead of a value. As shown in the previous section, the rate-distortion region is the first-order term that collects all rate pairs with vanishing JEP and SEP. This corresponds to the rate-distortion function for the

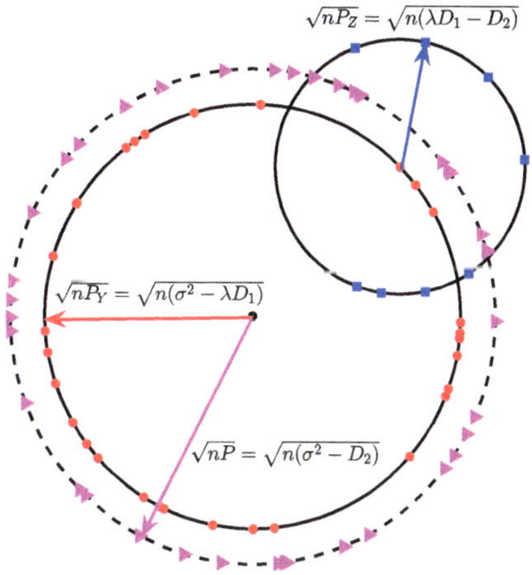

Fig. 5.3 Illustration of the spherical codebooks for $n = 2$. The magenta triangle denote the codeword of encoder f, the red dots denote the codewords of encoder f_1, and the blue squares denote the codewords of encoder f_2. In point-to-point source coding, the codewords are generated uniformly over the surface of the sphere denoted as dash circle with center of origin and the radius $\sqrt{nP} = \sqrt{n(\sigma^2 - D_2)}$. However, in successive refinement setting, the codewords of encoder f_1 are generated uniformly over the surface of the sphere with center of origin and the radius $\sqrt{nP_Y} = \sqrt{n(\sigma^2 - \lambda D_1)}$, and the codewords of encoder f_2 are generated uniformly over the surface of the sphere centered on the given codeword of encoder f_1 with radius $\sqrt{nP_Z} = \sqrt{n(\lambda D_1 - D_2)}$. Thus, the codewords of encoder f_2 are not uniformly distributed over the surface of the sphere, which makes it different from the spherical codebook used in the P2P mismatched lossy source coding problem

P2P lossy source coding in Chap. 2. To characterize the low-latency performance, the second-order asymptotics refers to bound the finite blocklength rate pairs subject to a constraint on JEP or SEP. In particular, the finite blocklength rate pairs approach the boundary rate pair in the rate-distortion region as the blocklength increases. The task is to characterize the speed of approach in the order of the blocklength and the corresponding coefficient. For most problems, the second-order scale is in the order of $\frac{1}{\sqrt{n}}$, and the task is to characterize the coefficient for the second-order term. This chapter defines and characterizes the second-order region for the mismatched SR problem.

Fix any boundary rate pair (R_1^*, R_2^*) of the rate-distortion region $\mathcal{R}_{\text{inner}}(D_1, D_2)$. The second-order asymptotics under the JEP criterion characterizes how a finite blocklength rate pair approaches $(R_1^*, R_2^*)$when a non-vanishing JEP is tolerated.

Definition 5.3 Given any $\varepsilon \in (0, 1)$, a pair $(L_1, L_2) \in \mathbb{R}_+^2$ is said to be second-order $(R_1^*, R_2^*, D_1, D_2, \varepsilon|\dagger, \ddagger)$-achievable under JEP if there exists a sequence of (n, M_1, M_2)-codes per Definition 5.1 such that

$$\limsup_{n\to\infty} \frac{1}{\sqrt{n}}(\log M_1 - nR_1^*) \le L_1, \tag{5.17}$$

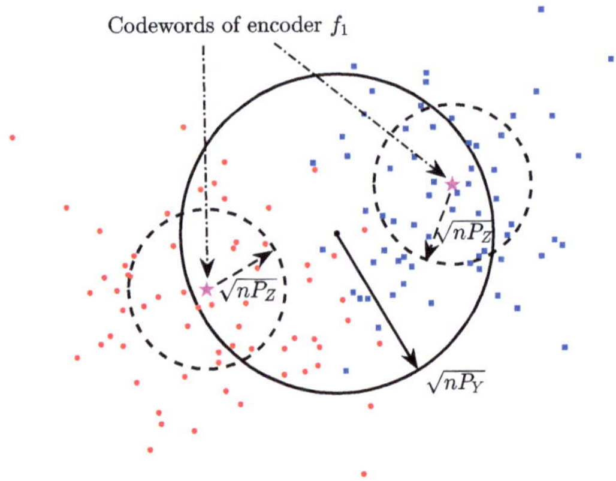

Fig. 5.4 Illustration of i.i.d. Gaussian codebooks for $n = 2$. The magenta stars denote the codewords of encoder f_1, and the red dots and blue squares denote the codewords of encoder f_2. Assume that $M_1 = 2$. The codewords $Y^n(1), Y^n(2)$ of encoder f_1 are generated i.i.d. from the Gaussian distribution with mean 0 and variance P_Y. Given each codeword $Y^n(i)$, the codewords $\{Z^n(i,j)\}_{j\in[M_2]}$ of encoder f_2 are generated i.i.d. from the Gaussian distribution with mean $Y^n(i)$ and variance P_Z. Although the marginal distribution of each codeword $Z^n(i,j)$ is Gaussian with mean 0 and variance $P_Y + P_Z = \sigma^2 - D_2$, the dependence of the codewords make it different from the i.i.d. Gaussian codebook in the P2P mismatched lossy source coding problem

$$\limsup_{n\to\infty} \frac{1}{\sqrt{n}}(\log M_1 M_2 - nR_1^* - nR_2^*) \le L_2, \tag{5.18}$$

and

$$\limsup_{n\to\infty} \mathrm{P}_{\mathrm{e},n}^{\dagger,\ddagger}(D_1, D_2|M_1, M_2) \le \varepsilon. \tag{5.19}$$

The closure of the set of all second-order $(R_1^*, R_2^*, D_1, D_2, \varepsilon|\dagger, \ddagger)$-achievable rate pairs is named the second-order rate region and denoted as $\mathcal{L}(R_1^*, R_2^*, D_1, D_2, \varepsilon|\dagger, \ddagger)$.

An equivalent form of (5.17) and (5.18) are as follows,

$$\log M_1 \le nR_1^* + \sqrt{n}L_1 + o(\sqrt{n}), \tag{5.20}$$

$$\log M_1 M_2 \le n(R_1^* + R_2^*) + \sqrt{n}L_2 + o(\sqrt{n}), \tag{5.21}$$

where (L_1, L_2) are second-order rate pair. This form parallels the P2P case in Chap. 2.

Under SEP, one only defines the second-order rate region for the corner point of rate-distortion region, which corresponds to the minimal rates of both encoders. This is because the second-order coding rates are not well defined for other points on the boundary of the rate-distortion region for at least one encoder.

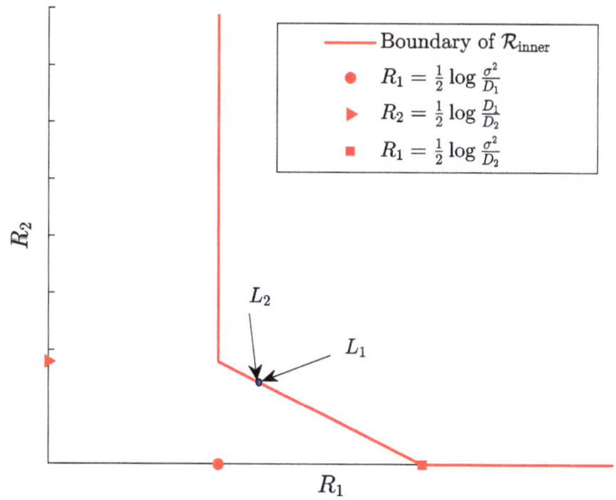

Fig. 5.5 Illustration of the first-order achievable rate region and the pair (L_1, L_2)

Definition 5.4 Given any $(\varepsilon_1, \varepsilon_2) \in (0, 1)^2$, a pair $(L_1, L_2) \in \mathbb{R}_+^2$ is said to be second-order $(D_1, D_2, \varepsilon_1, \varepsilon_2|\dagger, \ddagger)$-achievable under SEP if there exists a sequence of (n, M_1, M_2)-codes such that

$$\limsup_{n\to\infty} \frac{1}{\sqrt{n}}\left(\log M_1 - \frac{n}{2}\log\frac{\sigma^2}{D_1}\right) \leq L_1, \tag{5.22}$$

$$\limsup_{n\to\infty} \frac{1}{\sqrt{n}}\left(\log M_1 M_2 - \frac{n}{2}\log\frac{\sigma^2}{D_2}\right) \leq L_2, \tag{5.23}$$

and

$$\limsup_{n\to\infty} \mathrm{P}_{\mathrm{e},n}^{\dagger}(D_1|M_1, M_2) \leq \varepsilon_1, \tag{5.24}$$

$$\limsup_{n\to\infty} \mathrm{P}_{\mathrm{e},n}^{\ddagger}(D_2|M_1, M_2) \leq \varepsilon_2. \tag{5.25}$$

The closure of the set of all second-order $(D_1, D_2, \varepsilon_1, \varepsilon_2|\dagger, \ddagger)$-achievable rate pairs is named the second-order rate region and denoted as $\mathcal{L}_{\mathrm{SEP}}(D_1, D_2, \varepsilon_1, \varepsilon_2|\dagger, \ddagger)$.

The illustration of a pair (L_1, L_2) is provided in Fig. 5.5.

5.2 Second-Order Asymptotics

This section presents achievable second-order asymptotics under both JEP and SEP for all four combinations of spherical and i.i.d. Gaussian codebooks used by both encoders.

In addition to the second moment constraint in (5.1), the fourth moment of the source distribution P_X is used to characterize the performance and the sixth moment of the source distribution is required finite to guarantee the theoretical analysis, i.e.,

$$\zeta := \mathbb{E}[X^4],\ \mathbb{E}[X^6] < \infty. \tag{5.26}$$

The mismatched source dispersion for the P2P setting in [7] is also critical:

$$\mathrm{V}(\sigma^2, \zeta) := \frac{\zeta - \sigma^4}{4\sigma^4} = \frac{\mathrm{Var}[\mathrm{X}^2]}{4(\mathbb{E}[X^2]^2)}. \tag{5.27}$$

5.2.1 Results Under JEP

Theorem 5.1 *Given any $\varepsilon \in (0, 1)$, for any $\lambda \in (\frac{D_2}{D_1}, 1]$ and $(\dagger, \ddagger) \in \{\mathrm{sp}, \mathrm{iid}\}^2$, there exists a sequence of (n, M_1, M_2)-codes using Gaussian codebooks such that*

$$\log M_1 = \frac{n}{2}\log\frac{\sigma^2}{\lambda D_1} + \sqrt{n\mathrm{V}(\sigma^2, \zeta)}Q^{-1}(\varepsilon) + O(\log n), \tag{5.28}$$

$$\log M_1 M_2 = \frac{n}{2}\log\frac{\sigma^2}{D_2} + \sqrt{n\mathrm{V}(\sigma^2, \zeta)}Q^{-1}(\varepsilon) + O(\log n), \tag{5.29}$$

and

$$\lim_{n\to\infty} \mathrm{P}_{\mathrm{e},n}^{\dagger,\ddagger}(D_1, D_2|M_1, M_2) \le \varepsilon. \tag{5.30}$$

Theorem 5.1 generalizes the achievability results in [7, Theorem 1] for the P2P mismatched lossy source coding problem to the SR setting. For any $\lambda \in (\frac{D_2}{D_1}, 1]$, when the target rate pair is $(R_1^*, R_2^*) = (\frac{1}{2}\log\frac{\sigma^2}{\lambda D_1}, \frac{1}{2}\log\frac{\lambda D_1}{D_2})$, the pair $(L_1, L_2) = (\sqrt{\mathrm{V}(\sigma^2, \zeta)}Q^{-1}(\varepsilon), \sqrt{\mathrm{V}(\sigma^2, \zeta)}Q^{-1}(\varepsilon))$ is second-order achievable. This result generalizes [7, Theorem 1] by showing that all four combinations of spherical and i.i.d. Gaussian codebooks achieve the same second-order asymptotics.

At the first glance, Theorem 5.1 is not in the same form when presenting second-order asymptotics as [4, Theorem 20] for a GMS in the matched SR case. With proper choice of the parameter λ in Theorem 5.1, the following inner bound to the second-order coding region $\mathcal{L}(R_1^*, R_2^*, D_1, D_2, \varepsilon|\dagger, \ddagger)$ is obtained.

- Case (i): Let $\lambda \in (\frac{D_2}{D_1}, 1)$, $R_1^* = \frac{1}{2}\log\frac{\sigma^2}{\lambda D_1} > \frac{1}{2}\log\frac{\sigma^2}{D_1}$, $R_1^* + R_2^* = \frac{1}{2}\log\frac{\sigma^2}{D_2}$. It follows that

$$\mathcal{L}(R_1^*, R_2^*, D_1, D_2, \varepsilon|\dagger, \ddagger) \supseteq \left\{(L_1, L_2) : L_2 \ge \sqrt{\mathrm{V}(\sigma^2, \zeta)}Q^{-1}(\varepsilon)\right\}. \tag{5.31}$$

- Case (ii): Let $\lambda = 1$, $R_1^* = \frac{1}{2}\log\frac{\sigma^2}{D_1}$ and $R_1^* + R_2^* > \frac{1}{2}\log\frac{\sigma^2}{D_2}$. It follows that

$$\mathcal{L}(R_1^*, R_2^*, D_1, D_2, \varepsilon|\dagger, \ddagger) \supseteq \left\{(L_1, L_2) : L_1 \geq \sqrt{\mathrm{V}(\sigma^2, \zeta)}Q^{-1}(\varepsilon)\right\}. \tag{5.32}$$

- Case (iii): Let $\lambda = 1$, $R_1^* = \frac{1}{2}\log\frac{\sigma^2}{D_1}$ and $R_1^* + R_2^* = \frac{1}{2}\log\frac{\sigma^2}{D_2}$. It follows that

$$\mathcal{L}(R_1^*, R_2^*, D_1, D_2, \varepsilon|\dagger, \ddagger) \supseteq \left\{(L_1, L_2) : \min\{L_1, L_2\} \geq \sqrt{\mathrm{V}(\sigma^2, \zeta)}Q^{-1}(\varepsilon)\right\}. \tag{5.33}$$

Recall that the power of each codeword for encoder f_1 and encoder f_2 are $P_Y = \sigma^2 - \lambda D_1$ and $P_Z = \lambda D_1 - D_2$, respectively. If one uses a codebook with $\lambda = 1$ as in [8], only the results for cases (ii) and (iii) can be achieved. When $\lambda < 1$, the power of each codeword for encoder f_1 increases and the power of each codewords for encoder f_2 decreases. The proof of Theorem 1 shows that $R_1^* = \frac{1}{2}\log\frac{\sigma^2}{\lambda D_1} > \frac{1}{2}\log\frac{\sigma^2}{D_1}$ and $R_2^* = \frac{1}{2}\log\frac{\lambda D_1}{D_2} < \frac{1}{2}\log\frac{D_1}{D_2}$ is asymptotically achievable, as a consequence of the proposed refined second-order asymptotics. By introducing the parameter $\lambda \in (\frac{D_2}{D_1}, 1]$, all boundary points of the rate-distortion region can be covered, which allows one to derive achievable second-order asymptotics for all cases of interest, especially for case (i) when the rate of encoder f_1 is larger than $\frac{1}{2}\log\frac{\sigma^2}{D_1}$.

When specialized to a GMS, $\mathrm{V}(\sigma^2, \zeta) = \frac{1}{2}$. This implies that the proof of Theorem 5.1 is an alternative achievability proof for the matched SR problem [4, Theorem 20] with structured code design. Specifically, the achievability proof in [4, Theorem 20] used a covering lemma in [11] that bounds the minimal number of codewords required to cover a ball but did not specify how these codewords should be constructed. In contrast, the proof of Theorem 5.1 specifies that the same second-order asymptotic performance can be achieved with an arbitrary combination of spherical and i.i.d. Gaussian codebooks using successive minimum Euclidean distance encoding. For the source distribution $P_X = \mathcal{N}(0, 1)$, the distortion level $D_1 = 0.5$, $D_2 = 0.35$, the first and second order achievable regions are illustrated in Fig. 5.6. When the blocklength is large enough, the second-order region is close to the rate-distortion region.

Furthermore, as illustrated in Fig. 5.7, if the parameter λ is fixed to be 1, the rate-distortion region can't be achieved even in first-order sense. Such a result confirms the necessity of introducing the parameter λ as a power control factor.

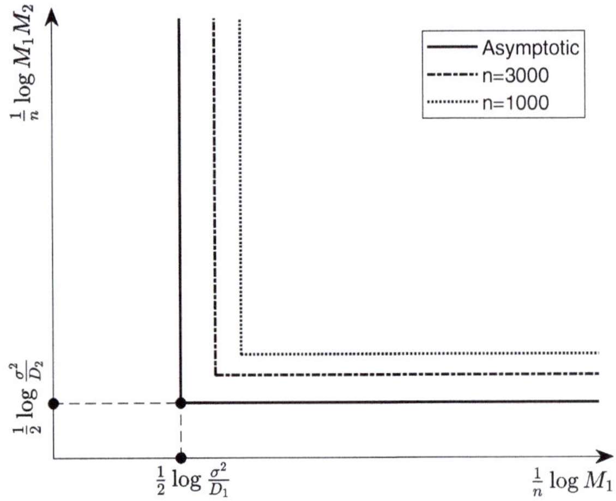

Fig. 5.6 Comparison of the rate-distortion regions and the achievable second-order region with different blocklength

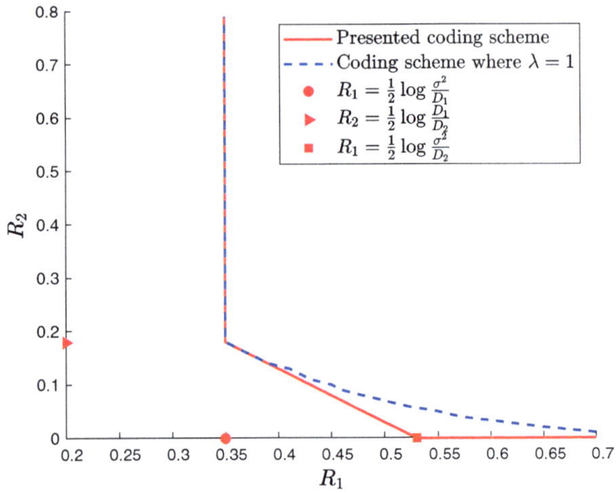

Fig. 5.7 Comparison of the achievable rate regions between the proposed coding scheme and the coding scheme with $\lambda = 1$

5.2.2 *Results Under SEP*

Theorem 5.2 *Given any $\varepsilon_1, \varepsilon_2 \in (0,1)^2$, for any $(\dagger, \ddagger) \in \{\mathrm{sp}, \mathrm{iid}\}^2$,*

$$\mathscr{L}_{\mathrm{SEP}}(D_1, D_2, \varepsilon_1, \varepsilon_2|\dagger, \ddagger) \supseteq \Big\{(L_1, L_2) : L_1 \geq \sqrt{\mathrm{V}(\sigma^2, \zeta)}\mathrm{Q}^{-1}\big(\min\{\varepsilon_1, \varepsilon_2\}\big), \\ L_2 \geq \sqrt{\mathrm{V}(\sigma^2, \zeta)}\mathrm{Q}^{-1}\big(\min\{\varepsilon_1, \varepsilon_2\}\big)\Big\}. \tag{5.34}$$

Under SEP, the minimum of the tolerable excess-distortion probabilities $(\varepsilon_1, \varepsilon_2)$ dominates the second-order rates of both encoders. This is because under the mismatched coding scheme, only a non-excess-distortion event of decoder ϕ_1 could guarantee a non-excess-distortion event of decoder ϕ_2 with proper rate choices due to the proposed layered coding scheme.

Theorem 5.2 generalizes the achievability part of the second-order asymptotic results in [5] for the matched SR problem to the mismatched scenario. When specialized to a GMS, Theorem 5.2 provides an alternative proof for [5, Theorem 7] with a structured codebook. It is important to note that the results in [5, Theorem 7] holds also when $\min\{\varepsilon_1, \varepsilon_2\}$ dominates the second-order coding rate since a similar layered coding scheme based on sphere covering was used.

The concept *successive refinability* [2, 3, 5] is defined as follows up to second-order. Simply speaking, if a source and distortion measure triplet are successively refinable, there is no loss of optimality of separating the compression process into two shares: one for a coarse reconstruction and the other for finer reconstruction. The sum rate of separating the compression process is the same as the optimal code of not separating the compression process, if the same distortion level is required.

Definition 5.5 Let $\varepsilon_1 = \varepsilon_2 = \varepsilon \in (0, 1)$. The source is strongly successively refinable at (D_1, D_2, ε) if there exists a $(D_1, D_2, \varepsilon_1, \varepsilon_2|\dagger, \ddagger)$-achievable pair (L_1, L_2) such that

$$\frac{1}{n}\log M_1 = R(P_X, D_1) + \sqrt{\frac{L_1}{n}} + o\Big(\frac{1}{\sqrt{n}}\Big), \tag{5.35}$$

$$\frac{1}{n}\log M_1 M_2 = R(P_X, D_2) + \sqrt{\frac{L_2}{n}} + o\Big(\frac{1}{\sqrt{n}}\Big), \tag{5.36}$$

where $R(P_X, D_i)$ for $i \in [2]$ is the rate-distortion function in (2.59). Furthermore, if the source is strongly successively refinable at (D_1, D_2, ε) for all $R^*(n, D_1, \varepsilon) < R^*(n, D_2, \varepsilon)$ (cf. (2.96)), then the source is strongly successively refinable.

Following the proof of Theorem 5.1 and letting $\lambda = 1$ and $\varepsilon = \min\{\varepsilon_1, \varepsilon_2\}$, there exists a sequence of (n, M_1, M_2) codes such that

$$\log M_1 = \frac{n}{2}\log\frac{\sigma^2}{D_1} + \sqrt{n\mathrm{V}(\sigma^2, \zeta)}\mathrm{Q}^{-1}(\varepsilon) + O(\log n), \tag{5.37}$$

$$\log M_1 M_2 = \frac{n}{2}\log\frac{\sigma^2}{D_2} + \sqrt{n\mathrm{V}(\sigma^2, \zeta)}\mathrm{Q}^{-1}(\varepsilon) + O(\log n), \tag{5.38}$$

and

$$\lim_{n\to\infty} \mathrm{P}_{\mathrm{e},n}^{\dagger}(D_1|M_1, M_2) \le \varepsilon, \tag{5.39}$$

$$\lim_{n\to\infty} \mathrm{P}_{\mathrm{e},n}^{\ddagger}(D_2|M_1, M_2) \le \varepsilon. \tag{5.40}$$

Recall that the rate-distortion function for a GMS equals to $\frac{1}{n}\log\frac{\sigma^2}{D_i}$ with $i \in [2]$ (cf. (2.90)). Theorem 5.2 implies that when $\varepsilon_1 = \varepsilon_2 = \varepsilon$ for any $\varepsilon \in (0, 1)$, any memoryless source satisfying (5.1) and (5.26) is strongly successive refinable under the quadratic distortion measure using the mismatched coding scheme. In other words, the sum rate in (5.38) equals the rate required to achieve (5.40) without the first layer of encoders and decoders, up to second-order asymptotics.

5.3 Proofs of Second-Order Asymptotics

To prove Theorem 5.1, one needs to derive the achievable second-order asymptotics for all four possible combinations of the codebooks.

5.3.1 Preliminaries

5.3.1.1 Definitions

Recall that $P_Y = \sigma^2 - \lambda D_1$ and $P_Z = \lambda D_1 - D_2$. For any $\varepsilon \in (0, 1)$, let

$$\mathrm{V} := \mathrm{Var}[X^2] = \zeta - \sigma^4, \tag{5.41}$$

$$a_n := \sqrt{\mathrm{V}\frac{\log n}{n}}, \tag{5.42}$$

$$b_n := \sqrt{\frac{\mathrm{V}}{n}}\mathrm{Q}^{-1}(\varepsilon). \tag{5.43}$$

In subsequent analysis, for simplicity, let the random variable W denote the normalized ℓ_2 norm of source sequence, i.e., $\frac{\|X^n\|^2}{n}$. Let f_W be the distribution of random variable W and w be particular realization of W. Similarly, define the random variable L as the quadratic distortion of the first layer, i.e., $L = d(X^n, \hat{X}_1^n)$. Furthermore, let f_L be denote the induced distribution of the random variable L and l be a particular realization of L.

For any (x^n, y^n) and $(\dagger, \ddagger) \in \{\mathrm{sp}, \mathrm{iid}\}^2$, define following non-excess-distortion probabilities

$$\Psi_{\dagger}(n, w, D) := \Pr_{f_{\dagger}(\cdot|\mathbf{0}^n, P_Y)}\{d(x^n, Y^n) \le D\}, \tag{5.44}$$

$$\Phi_{\ddagger}(n, x^n, y^n) := \Pr_{f_{\ddagger}(\cdot|y^n, P_Z)}\{d(x^n, Z^n) \le D_2\}, \tag{5.45}$$

where $f_{\cdot}(\cdot)$ is the distribution of codewords (cf. Definition 5.1). Define two constants

$$\beta_1 := \sqrt{P_Z} - \sqrt{D_2}, \tag{5.46}$$

$$\beta_2 := \sqrt{P_Z} + \sqrt{D_2}. \tag{5.47}$$

It follows that

$$\beta_1^2 \leq |P_Z - D_2|, \tag{5.48}$$

where (5.48) can be justified as follows. Note that $\beta_1^2 = \lambda D_1 - 2\sqrt{P_Z}\sqrt{D_2}$. If $\lambda D_1 \geq 2D_2$, then $\sqrt{P_Z} \geq \sqrt{D_2}$ and $|P_Z - D_2| = \lambda D_1 - 2D_2$, and thus

$$\beta_1^2 = \lambda D_1 - 2\sqrt{P_Z}\sqrt{D_2} \tag{5.49}$$

$$\leq \lambda D_1 - 2D_2 \tag{5.50}$$

$$= |P_Z - D_2|. \tag{5.51}$$

If $\lambda D_1 < 2D_2$, then $\sqrt{P_Z} < \sqrt{D_2}$ and $|P_Z - D_2| = 2D_2 - \lambda D_1$, and thus

$$\beta_1^2 - |P_Z - D_2| = \lambda D_1 - 2\sqrt{P_Z}\sqrt{D_2} - 2D_2 + \lambda D_1 \tag{5.52}$$

$$= 2(P_Z - \sqrt{P_Z}\sqrt{D_2}) \tag{5.53}$$

$$= 2\sqrt{P_Z}(\sqrt{P_Z} - \sqrt{D_2}) \tag{5.54}$$

$$< 0. \tag{5.55}$$

Note that (5.44) is valid since $\Pr_{f_{\dagger}^{*}(\cdot|\mathbf{0}^n, P_Y)}\{d(x^n, Y^n) \leq D\}$ depends on x^n only through its normalized ℓ_2 norm $w = \frac{\|x^n\|^2}{n}$.

Given any $(s, w, P, D) \in \mathbb{R}_+^4$, define the following functions

$$R_{\mathrm{iid}}(s, w, P, D) := \frac{1}{2}\log(1 + 2s) + \frac{sw}{(1 + 2s)P} - \frac{sD}{P}, \tag{5.56}$$

$$s^*(w, P, D) := \max\left\{0, \frac{P - 2D + \sqrt{P^2 + 4wD}}{4D}\right\}, \tag{5.57}$$

$$R_{\mathrm{iid}}(w, P, D) := R_{\mathrm{iid}}(s^*(w, P, D), w, P, D). \tag{5.58}$$

Note that $R_{\mathrm{iid}}(w, P_Y, D_1)$ is the exponential decay rate of the non-excess-distortion probability with respect to the distortion level D_1 for a source sequence x^n with power $w = \frac{1}{n}||x^n||^2$ when its reproduction sequence is generated from $f_{\mathrm{iid}}(Y^n|\mathbf{0}^n, P_Y)$ (cf. (5.8)), i.e.,

$$\lim_{n\to\infty} -\frac{1}{n}\log \Pr_{f_{\mathrm{iid}}(Y^n|\mathbf{0}^n, P_Y)}\{d(x^n, Y^n) \leq D_1\}$$

$$= R_{\mathrm{iid}}(w, P_Y, D_1). \tag{5.59}$$

Furthermore, $R_{\mathrm{iid}}(l, P_Z, D_2)$ is the the exponential decay rate of the non-excess-distortion probability with respect to the distortion level D_2 for any source sequence x^n whose quadratic distortion with respect to the output codeword $\hat{x}_1^n$ of encoder f_1 is $l := d(x^n, \hat{x}_1^n)$ when its reproduction sequence is generated from $f_{\mathrm{iid}}(Z^n|\hat{x}_1^n, P_Z)$, i.e.,

$$\begin{aligned}&\lim_{n\to\infty} -\frac{1}{n}\log \Pr_{f_{\mathrm{iid}}(Z^n|\hat{x}_1^n, P_Z)}\{d(x^n, Z^n) \le D_2\}\\&= R_{\mathrm{iid}}(l, P_Z, D_2).\end{aligned} \tag{5.60}$$

5.3.1.2 Lemmas

Similarly to [12, Theorem 37], $\Phi_{\mathrm{sp}}(n, x^n, y^n)$ depends on (x^n, y^n) only through their quadratic distortion $l := d(x^n, y^n)$ and the following lemma can be obtained.

Lemma 5.1 *Recall that* $|x|^+ = \max\{x, 0\}$. *For any* (x^n, y^n) *such that* $\sqrt{l} \in [|\beta_1|^+, \beta_2]$

$$\begin{aligned}\Phi_{\mathrm{sp}}(n, x^n, y^n) &\ge \frac{\Gamma(\frac{n+2}{2})}{\sqrt{\pi}n\Gamma(\frac{n+1}{2})}\left(1 - \frac{(l + P_Z - D_2)^2}{4lP_Z}\right)^{\frac{n-1}{2}}\\&=: \underline{h}(n, l),\end{aligned} \tag{5.61}$$

and otherwise $\Phi_{\mathrm{sp}}(n, x^n, y^n) = 0$.

The following function is also useful in the proof with i.i.d. Gaussian codebooks.

$$\kappa(s, w, P) := \frac{(P(1 + 2s) + 2w)^2}{P(1 + 2s)^3}. \tag{5.62}$$

Recall the definition of $R_{\mathrm{iid}}(w, P, D)$, $s^*(w, P, D)$ and $\kappa(s, w, P)$ in (5.57), (5.58) and (5.62), respectively. Similar to Zhou et al. [7, cf. (89)], using the strong large deviations theorem in [13, Theorem 3.7.4], noting that $\Phi_{\mathrm{iid}}(n, x^n, y^n)$ depends on (x^n, y^n) only through their quadratic distortion $l = d(x^n, y^n)$, it follows that

$$\Phi_{\mathrm{iid}}(n, l) \sim \frac{\exp\{-nR_{\mathrm{iid}}(l, P_Z, D_2)\}}{s^*(l, P_Z, D_2)\sqrt{\kappa(s^*(l, P_Z, D_2), l, P_Z)}}. \tag{5.63}$$

Finally, note that for any $(\dagger, \ddagger) \in \{\mathrm{sp}, \mathrm{iid}\}^2$, the ensemble JEP in (5.11) satisfies

$$\begin{aligned}&\mathrm{P}_{\mathrm{e},n}^{\dagger,\ddagger}(D_1, D_2|M_1, M_2))\\&\le \Pr\{d(X^n, \hat{X}_1^n) > \lambda D_1 \text{ or } d(X^n, \hat{X}_2^n) > D_2\}\end{aligned} \tag{5.64}$$

$$= \Pr\{d(X^n, \hat{X}_1^n) > \lambda D_1\} + \Pr\{d(X^n, \hat{X}_1^n) \leq \lambda D_1,\ d(X^n, \hat{X}_2^n) > D_2\}. \tag{5.65}$$

The rest of the proof upper bounds (5.65).

5.3.2 *Spherical Codebooks for Both Encoders*

Firstly, consider the case where both encoders use spherical codebooks where $(\dagger, \ddagger) = \{\text{sp}, \text{sp}\}$. Note that the first term in (5.65) is similar to the P2P mismatched lossy source coding problem studied in [7, Section IV. B], except that (i) the distortion level is replaced from D to λD_1, and (ii) the power of codebook is changed from $P_Y = \sigma^2 - D_1$ to $P_Y = \sigma^2 - \lambda D_1$. Choose M_1 such that

$$\log M_1 = \frac{n}{2}\log\frac{\sigma^2}{\lambda D_1} + \sqrt{n\mathrm{V}(\sigma^2, \zeta)}\mathrm{Q}^{-1}(\varepsilon) + O(\log n). \tag{5.66}$$

Similar to [7, cf. (43)–(58)], it follows that

$$\Pr\{d(X^n, \hat{X}_1^n) > \lambda D_1\} \leq \varepsilon + O\left(\frac{1}{\sqrt{n}}\right). \tag{5.67}$$

Thus, the rest of the proof focus on the analysis of the second term in (5.65). For ease of notation, let $\mathbf{Y}$ denotes the random codebooks $(Y^n(1), \ldots, Y^n(M_1))$ of the first encoder and let $\mathbf{y}$ be a particular realization. Recall that given the source sequence x^n and a particular realization of the codebook $\mathbf{y}$, the compressed index of the encoder f_1 is $f_1(x^n)$ and the reproduced source sequence of the first encoder is

$$\hat{x}_1^n = y^n(f_1(x^n)) = \min_{i\in[M_1]} d(x^n, y^n(i)). \tag{5.68}$$

It follows that

$$\begin{aligned}
&\Pr\{d(X^n, \hat{X}_1^n) \leq \lambda D_1,\ d(X^n, \hat{X}_2^n) > D_2\} \\
&= \int_{\substack{x^n,\mathbf{y}:\\ d(x^n,\hat{x}_1^n)\leq\lambda D_1}} P_X^n(x^n) \prod_{i\in[M_1]} f_{\text{sp}}(y^n(i)|\mathbf{0}^n, P_Y) \\
&\quad\times \Pr\{\forall\ j \in [M_2] :\ d(x^n, Z^n(f_1(x^n), j)) > D_2\}\mathrm{d}x^n\mathrm{d}\mathbf{y} \qquad (5.69)\\
&= \int_{\substack{x^n,\mathbf{y}:\\ d(x^n,\hat{x}_1^n)\leq\lambda D_1}} P_X^n(x^n) \prod_{i\in[M_1]} f_{\text{sp}}(y^n(i)|\mathbf{0}^n, P_Y)(1 - \Phi_{\text{sp}}(n, x^n, \hat{x}_1^n))^{M_2}\mathrm{d}x^n\mathrm{d}\mathbf{y} \qquad (5.70)
\end{aligned}$$

$$= \int_{x^n,\mathbf{y}} P_X^n(x^n) \prod_{i\in[M_1]} f_{\mathrm{sp}}(y^n(i)|\mathbf{0}^n, P_Y) \mathbb{1}\{d(x^n, \hat{x}_1^n) < (|\beta_1|^+)^2\}$$

$$\times (1 - \Phi_{\mathrm{sp}}(n, x^n, \hat{x}_1^n))^{M_2} \mathrm{d}x^n \mathrm{d}\mathbf{y} + \int_{x^n,\mathbf{y}} P_X^n(x^n) \prod_{i\in[M_1]} f_{\mathrm{sp}}(y^n(i)|\mathbf{0}^n, P_Y)$$

$$\times \mathbb{1}\{d(x^n, \hat{x}_1^n) \in [(|\beta_1|^+)^2, \lambda D_1]\}(1 - \Phi_{\mathrm{sp}}(n, x^n, \hat{x}_1^n))^{M_2} \mathrm{d}x^n \mathrm{d}\mathbf{y} \quad (5.71)$$

$$= \Pr\{d(X^n, \hat{X}_1^n) < (|\beta_1|^+)^2\} + \int_{x^n,\mathbf{y}} P_X^n(x^n) \prod_{i\in[M_1]} f_{\mathrm{sp}}(y^n(i)|\mathbf{0}^n, P_Y)$$

$$\times \mathbb{1}\{d(x^n, \hat{x}_1^n) \in [(|\beta_1|^+)^2, \lambda D_1]\}(1 - \Phi_{\mathrm{sp}}(n, x^n, \hat{x}_1^n))^{M_2} \mathrm{d}x^n \mathrm{d}\mathbf{y}, \quad (5.72)$$

where (5.70) follows since given x^n, $\mathbf{y}$ and thus $\hat{x}_1^n$, the codewords $(Z^n(\hat{x}_1^n, 1), \ldots, Z^n(\hat{x}_1^n, M_2))$ are independent and generated by the same distribution $f_{\mathrm{sp}}(\cdot|\hat{x}_1^n, \lambda D_1 - D_2)$, (5.71) divides the whole section into two parts and the second term is valid since $(|\beta_1|^+)^2 < \lambda D_1$ and (5.72) follows from the results in Lemma 5.1.

Recall that $l = d(x^n, y^n)$ is the quadratic distortion between x^n and y^n. The second term in (5.72) can be further upper bounded by

$$\int_{(|\beta_1|^+)^2}^{\lambda D_1} f_L(l)(1 - \underline{h}(n, l))^{M_2} \mathrm{d}l \quad (5.73)$$

$$\le \int_{(|\beta_1|^+)^2}^{|P_Z - D_2|} f_L(l)\mathrm{d}l + \int_{|P_Z - D_2|}^{\lambda D_1} f_L(l)(1 - \underline{h}(n, l))^{M_2} \mathrm{d}l \quad (5.74)$$

$$\le \int_{(|\beta_1|^+)^2}^{|P_Z - D_2|} f_L(l)\mathrm{d}l + \int_{|P_Z - D_2|}^{\lambda D_1} f_L(l) \exp\{-M_2 \underline{h}(n, l)\}\mathrm{d}l \quad (5.75)$$

$$\le \int_{(|\beta_1|^+)^2}^{|P_Z - D_2|} f_L(l)\mathrm{d}l + \exp\{-M_2 \underline{h}(n, \lambda D_1)\}, \quad (5.76)$$

where (5.73) follows from the result in (5.61), (5.74) follows since $\beta_1^2 \le |P_Z - D_2|$ (cf. (5.48)) and $\underline{h}(n, l) \in [0, 1]$, (5.75) follows from that $(1 - a)^M \le \exp\{-Ma\}$ for any $a \in [0, 1)$, and (5.76) follows since $\underline{h}(n, l)$ is a decreasing function of l if $l \ge |P_Z - D_2|$.

Combining (5.72) and (5.76), it follows that

$$\Pr\{d(X^n, \hat{X}_1^n) \le \lambda D_1,\ d(X^n, \hat{X}_2^n) > D_2\}$$

$$\le \Pr\{d(X^n, \hat{X}_1^n) \le |P_Z - D_2|\} + \exp\{-M_2 \underline{h}(n, \lambda D_1)\}. \quad (5.77)$$

Choose M_2 such that

$$\log M_2 = -\log \underline{h}(n, \lambda D_1) + \log(\log \sqrt{n}) \quad (5.78)$$

$$= \frac{n}{2} \log \frac{\lambda D_1}{D_2} + O(\log n), \tag{5.79}$$

where (5.79) follows from the definition of $\underline{h}(n,l)$ in (5.61) and the fact that $P_Z = \lambda D_1 - D_2$ and $\Gamma\left(\frac{n+2}{2}\right) / \Gamma\left(\frac{n+1}{2}\right) = O(\sqrt{n})$. With the choice of M_2 in (5.78),

$$\exp\{-M_2 \underline{h}(n, \lambda D_1)\} = \frac{1}{\sqrt{n}}. \tag{5.80}$$

To upper bound the first term in (5.77), similarly to [7, Section IV. C], define the sets

$$\mathcal{P} := \{r \in \mathbb{R} : b < r - \sigma^2 \le a_n\}, \tag{5.81}$$

$$\mathcal{Q} := \{r \in \mathbb{R} : r + P_Y - |P_Z - D_2| \ge 0\}, \tag{5.82}$$

where the value of b will be specified later.

Recall that $w = \frac{\|x^n\|^2}{n}$ is the normalized ℓ_2 norm of x^n. Furthermore, the following definition is useful:

$$R_{\rm sp}(w, P, D) := -\frac{1}{2} \log \left(1 - \frac{(w + P - D)^2}{4wP}\right). \tag{5.83}$$

Recall the definition of $\Psi_{\rm sp}(n, w, D)$ in (5.44). It follows from [7, Eq. (63)–(68)] that

$$\begin{aligned} &\Psi_{\rm sp}(n, w, |P_Z - D_2|) \\ &\quad < \frac{1}{\sqrt{\pi}} \frac{\Gamma(\frac{n}{2})}{\Gamma\left(\frac{n-1}{2}\right)} \exp\{-(n-3) R_{\rm sp}(w, P_Y, |P_Z - D_2|)\} \\ &\quad =: \bar{g}(n, w, |P_Z - D_2|), \end{aligned} \tag{5.84}$$

and similar to [7, cf. (69)–(77)], it follows that

$$\begin{aligned} &\Pr\{d(X^n, \hat{X}_1^n) \ge |P_Z - D_2|\} \\ &= \int_0^\infty (1 - \Psi_{\rm sp}(n, w, |P_Z - D_2|))^{M_1} f_W(w) {\rm d}w \qquad (5.85) \\ &\ge \left(1 - \frac{1}{\sqrt{n}}\right) \Pr\Big\{W \in \mathcal{P} \cap \mathcal{Q},\ \log M_1 \le -\log 2 \\ &\qquad - \log \bar{g}(n, \sigma^2 + b, |P_Z - D_2|) - \frac{1}{2} \log n\Big\}. \qquad (5.86) \end{aligned}$$

The next step focus on further lower bounding the probability term in (5.86) and showing that the probability term tends to 1 asymptotically. Note that $R_{\mathrm{sp}}(w, P, D)$ is a decreasing function of D and is an increasing function of w. For $w_1 = \sigma^2$ and λD_1, it can be verified that $R_{\mathrm{sp}}(w_1, P_Y, \lambda D_1) = \frac{1}{2}\log\frac{\sigma^2}{\lambda D_1}$. Recall that $P_Z = \lambda D_1 - D_2$, since $|\lambda D_1 - 2D_2| < \lambda D_1$, if one choose $w_2 \in \mathbb{R}_+$ such that

$$R_{\mathrm{sp}}(w_2, P_Y, |\lambda D_1 - 2D_2|) = \frac{1}{2}\log\frac{\sigma^2}{\lambda D_1}, \tag{5.87}$$

it follows that $w_2 < w_1 = \sigma^2$. Set $w_3 = \frac{w_1+w_2}{2}$. Thus $w_2 < w_3 < \sigma^2$ and

$$R_{\mathrm{sp}}(w_3, P_Y, |\lambda D_1 - 2D_2|) > R_{\mathrm{sp}}(w_2, P_Y, |\lambda D_1 - 2D_2|) \tag{5.88}$$

$$= \frac{1}{2}\log\frac{\sigma^2}{\lambda D_1}. \tag{5.89}$$

Set the value of b as

$$b := \frac{w_2 - \sigma^2}{2}. \tag{5.90}$$

Using the definition of $\bar{g}(n, w)$ in (5.84), it follows that for some $\delta > 0$,

$$-\log\bar{g}(n, \sigma^2 + b, |P_Z - D_2|) = nR_{\mathrm{sp}}(w_3, P_Y, |\lambda D_1 - 2D_2|) + O(\log n) \tag{5.91}$$

$$= \frac{n}{2}\log\frac{\sigma^2}{\lambda D_1} + n\delta + O(\log n), \tag{5.92}$$

where (5.91) follows since $\Gamma\left(\frac{n}{2}\right)/\Gamma\left(\frac{n-1}{2}\right) = O(\sqrt{n})$ and $P_Z = \lambda D_1 - D_2$, and (5.92) follows from the result in (5.89).

Combining (5.66) and (5.92) and invoking the weak law of large numbers, one can conclude that

$$\lim_{n\to\infty}\Pr\left\{\log M_1 \le -\log 2 - \log\bar{g}(n, \sigma^2 + b) - \frac{1}{2}\log n\right\} = 1. \tag{5.93}$$

Note that (5.93) follows since (5.92) implies that the dominant term in the right-hand side is strictly larger than $\frac{n}{2}\log\frac{\sigma^2}{\lambda D_1}$, which is the dominant term of $\log M_1$ (cf. (5.66)). Similar to [7, Lemma 5], using the weak law of large numbers, the Berry-Esseen theorem [14, 15], and the definition of b, it can be concluded that

$$\Pr\{W \in \mathcal{P} \cap \mathcal{Q}\} \ge 1 - O\left(\frac{1}{\sqrt{n}}\right). \tag{5.94}$$

Finally, combining (5.66), (5.72), (5.79), (5.80), and (5.94), the excess-distortion probability can be upper bounded by

$$\lim_{n\to\infty} \Pr\{d(X^n, \hat{X}_1^n) \le \lambda D_1,\ d(X^n, \hat{X}_2^n) > D_2\} = 0. \tag{5.95}$$

The proof of Theorem 5.1 is completed by combining (5.65), (5.67), (5.95), when both encoders use spherical codebooks.

5.3.3 i.i.d. Gaussian Codebooks for Both Encoders

This subsection presents the proof when both encoders use i.i.d. Gaussian codebooks, i.e., $(\dagger,\ddagger) = \{\mathrm{iid}, \mathrm{iid}\}$. Similar to the case when both encoders use spherical codebooks, the first term in (5.65) is similar to the P2P mismatched lossy source coding problem studied in [7, Section IV. D]. Choose M_1 such that

$$\log M_1 = \frac{n}{2}\log\frac{\sigma^2}{\lambda D_1} + \sqrt{n\mathrm{V}(\sigma^2,\zeta)}\mathrm{Q}^{-1}(\varepsilon) + O(\log n). \tag{5.96}$$

Similar to [7, cf. (90)–(103)], it follows that

$$\Pr\{d(X^n, \hat{X}_1^n) > \lambda D_1\} \le \varepsilon + O\left(\frac{1}{\sqrt{n}}\right). \tag{5.97}$$

The second term in (5.65) is upper bounded as follows:

$$\begin{aligned}
&\Pr\{d(X^n, \hat{X}_1^n) \le \lambda D_1,\ d(X^n, \hat{X}_2^n) > D_2\} \\
&= \int_{\substack{x^n,\mathbf{y}:\\ d(x^n,\hat{x}_1^n)\le\lambda D_1}} P_X^n(x^n)\prod_{i\in[M_1]} f_{\mathrm{iid}}(y^n(i)|\mathbf{0}^n, P_Y) \\
&\quad\times \Pr\{\forall\, j\in[M_2]:\ d(x^n, Z^n(f_1(x^n), j)) > D_2\}\mathrm{d}x^n\mathrm{d}\mathbf{y} \qquad (5.98)\\
&= \int_{\substack{x^n,\mathbf{y}:\\ d(x^n,\hat{x}_1^n)\le\lambda D_1}} P_X^n(x^n)\prod_{i\in[M_1]} f_{\mathrm{iid}}(y^n(i)|\mathbf{0}^n, P_Y)(1-\Phi_{\mathrm{iid}}(n, x^n, \hat{x}_1^n))^{M_2}\mathrm{d}x^n\mathrm{d}\mathbf{y} \qquad (5.99)\\
&= \int_0^{\lambda D_1} f_L(l)(1-\Phi_{\mathrm{iid}}(n,l))^{M_2}\mathrm{d}l \qquad (5.100)\\
&\le \int_0^{|P_Z-D_2|} f_L(l)\mathrm{d}l + \int_{|P_Z-D_2|}^{\lambda D_1} f_L(l)(1-\Phi_{\mathrm{iid}}(n,l))^{M_2}\mathrm{d}l \qquad (5.101)\\
&\le \int_0^{|P_Z-D_2|} f_L(l)\mathrm{d}l + \int_{|P_Z-D_2|}^{\lambda D_1} f_L(l)\exp\{-M_2\Phi_{\mathrm{iid}}(n,l)\}\mathrm{d}l \qquad (5.102)
\end{aligned}$$

$$\leq \Pr\{d(X^n, \hat{X}_1^n) \leq |P_Z - D_2|\} + \exp\{-M_2 \Phi_{\text{iid}}(n, \lambda D_1)\}, \tag{5.103}$$

where (5.99) follows since given x^n, $\mathbf{y}$ and thus $\hat{x}_1^n$, the codewords $(Z^n(\hat{x}_1^n, 1), \ldots, Z^n(\hat{x}_1^n, M_2))$ are independent and generated by the same distribution $f_{\text{iid}}(\cdot|\hat{x}_1^n, \lambda D_1 - D_2)$, (5.100) follows from (5.63), (5.101) follows since $\Phi_{\text{iid}}(n, l) \leq 1$, (5.102) follows since $(1-a)^M \leq \exp\{-Ma\}$ for any $a \in [0, 1)$ and (5.103) since $\Phi_{\text{iid}}(n, l)$ is a decreasing function of l if $l \geq |P_Z - D_2|$.

Choose M_2 such that

$$\log M_2 = -\log \Phi_{\text{iid}}(n, \lambda D_1) + \log(\log \sqrt{n}) \tag{5.104}$$

$$= \frac{n}{2} \log \frac{\lambda D_1}{D_2} + O(\log n). \tag{5.105}$$

It follows that

$$\exp\{-M_2 \Phi_{\text{iid}}(n, \lambda D_1)\} = \frac{1}{\sqrt{n}}. \tag{5.106}$$

The analysis of the first term in (5.103) is similar to spherical codebook (cf. (5.81) to (5.94)) except the following two points: i) replace $\bar{g}(n, w)$ with $\Psi_{\text{iid}}(n, w, |P_Z - D_2|)$ and ii) replace $\mathcal{P} \cap \mathcal{Q}$ with $\mathcal{P}$. Hence,

$$\lim_{n \to \infty} \Pr\{d(X^n, \hat{X}_1^n) \leq \lambda D_1,\ d(X^n, \hat{X}_2^n) > D_2\} = 0. \tag{5.107}$$

The proof of Theorem 5.1 when both encoders use i.i.d. Gaussian codebooks is completed by combining (5.65), (5.97), (5.107).

5.3.4 Different Codebooks for Two Encoders

The proof of Theorem 5.1 when different types of codebooks are used in encoders f_1 and f_2 is provided as follows. When $(\dagger, \ddagger) = \{\text{iid}, \text{sp}\}$, to bound the first term in (5.65), the steps are exactly the same as the case $(\dagger, \ddagger) = \{\text{iid}, \text{iid}\}$ until (5.97). To upper bound the second term in (5.65), the steps are exactly the same as (5.69)–(5.95). When $(\dagger, \ddagger) = \{\text{sp}, \text{iid}\}$, the proof is exactly the same as the case $(\dagger, \ddagger) = \{\text{sp}, \text{sp}\}$ until (5.67). To upper bound the second term in (5.65), the proof is exactly the same as (5.98)–(5.107).

5.4 Chapter Summary

This chapter presents the mismatched code and its low-latency performance for a typical multiterminal compression problem: the successive refinement problem. The mismatched code is used to compress an arbitrary memoryless source with known second moment under quadratic distortion. In the code design, one critical parameter λ is introduced to allocate the compression rates of encoders. The low-latency performance of the mismatched code is analyzed under two performance criteria: the joint excess-distortion probability and separate excess-distortion probabilities. When specialized to an Gaussian memoryless source, the coding scheme and the analysis provide an alternative proof with structured code design to existing results. A key take-away message is that any memoryless source under quadratic distortion measure is strongly successively refinable under the mismatched code: there is no loss of optimality when one separates the compression process into two phases and generates an additional coarse reconstruction.

5.5 Exercises

5.1 What are the applications of the successive refinement problem? What is the difference between the matched and mismatched settings?

5.2 What are the connection between the codewords used by encoders f_1 and f_2 in the mismatched successive refinement problem?

5.3 Why is the normalized power of codewords Y^n chosen as $P_Y = \sigma^2 - D_1$? Show that $\sigma^2 - D_1$ is the optimal power.

5.4 What is the role of the parameter λ? Why cannot simply fix $\lambda = 1$?

5.5 Explain why the non-excess-distortion probability $\Phi_{\mathrm{sp}}(n, x^n, y^n)$ depends on (x^n, y^n) only through their quadratic distortion.

5.6 Let $p \in (0, 0.5)$ and $p > D_1 > D_2 > 0$. Consider a binary memoryless source generated from the distribution $P_X = [p, 1-p]$. Calculate the second-order asymptotic bound in Theorem 5.1.

5.7 Prove Lemma 5.1 concerning the lower bound for the non-excess-distortion probability when spherical codebooks are used.

5.8 Consider a GMS with $P_X = \mathcal{N}(0, 1)$ and let $\varepsilon = 0.1$. Plot the rate region with blocklength from $n = 100$ to $n = 500$.

References

1. B. Rimoldi, Successive refinement of information: characterization of the achievable rates. IEEE Trans. Inf. Theory **40**(1), 253–259 (1994)
2. V. Koshelev, Estimation of mean error for a discrete successive-approximation scheme. Probl. Peredachi Inf. **17**(3), 20–33 (1981)
3. W.H. Equitz, T.M. Cover, Successive refinement of information. IEEE Trans. Inf. Theory **37**(2), 269–275 (1991)
4. L. Zhou, V.Y.F. Tan, M. Motani, Second-order and moderate deviation asymptotics for successive refinement. IEEE Trans. Inf. Theory **63**(5), 2896–2921 (2017)
5. A.No, A. Ingber, T. Weissman, Strong successive refinability and rate-distortion-complexity tradeoff. IEEE Trans. Inf. Theory **62**(6), 3618–3635 (2016)
6. A. Lapidoth, On the role of mismatch in rate distortion theory. IEEE Trans. Inf. Theory **43**(1), 38–47 (1997)
7. L. Zhou, V.Y. Tan, M. Motani, Refined asymptotics for rate-distortion using Gaussian codebooks for arbitrary sources. IEEE Trans. Inf. Theory **65**(5), 3145–3159 (2018)
8. L. Bai, Z. Wu, L. Zhou, Achievable second-order asymptotics for successive refinement using Gaussian codebooks, in *IEEE ISIT* (2021), pp. 2888–2893
9. A. Lapidoth, Nearest neighbor decoding for additive non-Gaussian noise channels. IEEE Trans. Inf. Theory **42**(5), 1520–1529 (1996)
10. J. Scarlett, V.Y.F. Tan, G. Durisi, The dispersion of nearest-neighbor decoding for additive non-Gaussian channels. IEEE Trans. Inf. Theory **63**(1), 81–92 (2017)
11. J.-L. Verger-Gaugry, Covering a ball with smaller equal balls in $\mathbb{R}^n$. Discrete Comput. Geom. **33**(1), 143–155 (2005)
12. V. Kostina, S. Verdú, Fixed-length lossy compression in the finite blocklength regime. IEEE Trans. Inf. Theory **58**(6), 3309–3338 (2012)
13. A. Dembo, O. Zeitouni. *Large Deviations Techniques and Applications*, vol. 38 (Springer, Berlin, 2009)
14. A.C. Berry, The accuracy of the Gaussian approximation to the sum of independent variates. Trans. Am. Math. Soc. **49**(1), 122–136 (1941)
15. C.-G. Esseen, *On the Liapounoff Limit of Error in the Theory of Probability* (Almqvist & Wiksell, 1942)

Chapter 6
Multiuser Transmission

This chapter studies the problem of data transmission over multiuser channels with additive arbitrary noise (AAN). For multiuser communication, a trivial transmission scheme is orthogonal multiple access (OMA), where each user transmits at a different resource block, such as frequency or time [1]. However, OMA cannot satisfy the rate, density, and latency requirements of 6G communication [2]. A feasible solution is to adopt non-orthogonal multiple access (NOMA) so that different users share the same resource block for simultaneous transmission. For uplink communication, corresponding models are the multiple access channel (MAC) [3–5] and the random access channel (RAC) [6, 7]. Compared with MAC that assumes the knowledge of the number and identities of active transmitters, RAC is suitable for practical scenarios of massive access [8], where the number of users is large, but the number and identities of active users are unknown. On the other hand, for downlink communication, the corresponding model is the broadcast channel (BC) [9, 10].

The study of MAC was pioneered by Shannon [3]. Subsequently, for the two-user discrete memoryless MAC with two transmitters and one receiver, Ahlswede [4] and Liao [5] characterized the capacity region that consists of rate pairs such that the error probability of either user vanishes as the blocklength increases. In particular, at the receiver side, *successive interference cancellation (SIC)* was used. Subsequently, the same capacity region was proved under joint decoding by El Gamal and Cover [13]. For a two-user MAC with additive white Gaussian noise (AWGN), the capacity region was derived by Wyner [14] and Cover [15]. The low-latency performance of Gaussian two-user MAC was first characterized by MolavianJazi and Laneman [16] for the achievability part and by Scarlett and Tan [20] for both achievability and converse parts of the simpler setting with degraded message sets. Subsequently, the result in [16] was refined by Yavas, Kostina, and Effros [17], who derived the third-order asymptotics in the achievability part. Corresponding low-latency communication result for discrete memoryless MAC was obtained by Yavas, Kostina, and Effros [18]. Recently, RAC [6, 7] was proposed to account for

L. Zhou, L. Bai, *Low-latency Communication*, Textbooks in Telecommunication Engineering, https://doi.org/10.1007/978-3-032-06627-5_6

massive access, where the number of users is large but the number and identities of active users are unknown. When the number of active users is assumed finite, the low-latency performance was characterized by Yavas, Kostina, and Effros [17, 18].

The study of BC was pioneered by Cover [9], who derived bounds on the capacity region when the error probability vanishes as the blocklength tends to infinity. The capacity region of BC remains open in general. For degraded BC with two users, where one of the receivers has better channel quality than the other user, the capacity was characterized by Bergmans [10, 11] and Gallager [12]. BC with AWGN is a typical example of degraded BC. The low-latency performance of degraded discrete memoryless BC was studied by Tan and Kosut [19], who derived an achievability result for second-order asymptotics.

All the above results assumed that the distribution of the channel noise is perfectly known so that codes can be designed and optimized. However, in practice, obtaining the exact distribution of channel noise is highly challenging. Thus, to account for practice, a mismatched coding scheme, which is ignorant of the exact channel noise distribution, is desired. This chapter presents such coding schemes for MAC, RAC, and BC, and characterizes the achievable low-latency performance [22–25]. The contents in this chapter generalize those in Chap. 3 from point-to-point (P2P) communication to multiuser communication.

6.1 Multiple Access Channel

This section presents the mismatched code and its low-latency performance of MAC with additive arbitrary noise. In particular, SIC and joint nearest neighbor (JNN) decoding are analyzed and compared. Although both decoding algorithms achieve the same asymptotic performance, the low-latency performance of JNN decoding is strictly better.

6.1.1 Problem Formulation

Fix three integers $(n, M_1, M_2) \in \mathbb{N}^3$, an alphabet $\mathcal{Z}$, and a distribution $P_Z \in \mathcal{P}(\mathcal{Z})$. As shown in Fig. 6.1, this section studies a two-user MAC with additive arbitrary noise. For each $i \in [2]$, user i aims to transmit a message $W_i \in [M_i]$ via an additive noise channel, where the message is encoded into a channel input X_i^n. The two channel inputs are corrupted by the additive Z^n that is generated i.i.d. from P_Z, yielding the noisy response $Y^n = X_1^n + X_2^n + Z^n$. Using noisy channel outputs Y^n, the decoder ϕ generates message estimates $(\hat{W}_1, \hat{W}_2)$.

Different from the matched case where the noise distribution P_Z is known, this chapter studies the mismatched case where P_Z is unknown. Instead, the statistics of channel noise distributions can be measured. Thus, consistent with mismatched

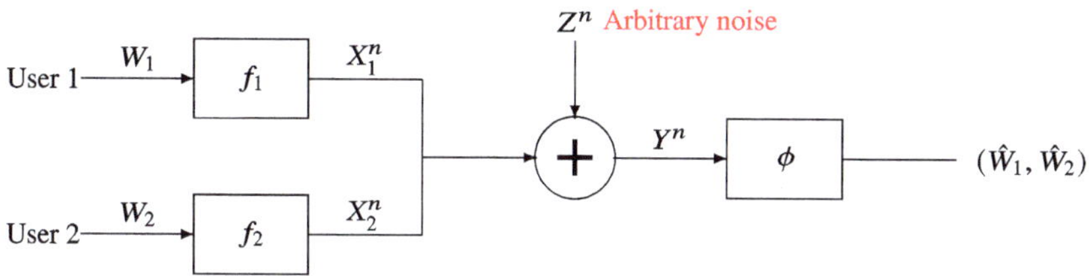

Fig. 6.1 System model for two-user multiple access channel with additive arbitrary noise

communication studies in Chaps. 3 and 4, it is assumed that the second moment of the additive noise distribution is known and the sixth moment is finite, i.e.,

$$\mathbb{E}[Z^2] = 1,\ \zeta := \mathbb{E}[Z^4],\ \mathbb{E}[Z^6] < \infty. \tag{6.1}$$

As shall be demonstrated, the second moment is used in the code design, the fourth moment ζ appears in low-latency performance, while the finiteness of the sixth moment guarantees theoretical derivations using the function version of the Berry-Esseen theorem (cf. Theorem 1.8).

6.1.2 *Coding Scheme*

Fix two positive real numbers $(P_1, P_2) \in \mathbb{R}_+^2$.

When JNN decoding is used, the coding scheme is defined as follows.

Definition 6.1 An (n, M_1, M_2, P_1, P_2)-JNN code for a two-user MAC consists of:

- Random codebooks C_j with M_j codewords $\{X_j^n(1), \ldots, X_j^n(M_j)\}$ subject to a power constraint P_j for each $j \in [2]$,
- Two encoders (f_1, f_2) such that for each $j \in [2]$ and $m_j \in [M_j]$,

$$f_j(m_j) := X_j^n(m_j), \tag{6.2}$$

- A joint nearest neighbor decoder that uses the minimum Euclidean distance decoding, i.e.,

$$\Phi(Y^n) := (\hat{W}_1, \hat{W}_2) = \operatorname*{arg\,min}_{(\bar{w}_1, \bar{w}_2) \in [M_1] \times [M_2]} \left\| Y^n - X_1^n(\bar{w}_1) - X_2^n(\bar{w}_2) \right\|^2. \tag{6.3}$$

SIC is commonly used in practice due to its low complexity. Correspondingly, when SIC decoding is used, the coding scheme is defined as follows.

Definition 6.2 An (n, M_1, M_2, P_1, P_2)-SIC code for a two-user MAC consists of:

- The same codebooks and encoders as in Definition 6.1,
- A successive interference cancellation decoder with two steps. Step 1 applies the nearest neighbor decoder that treats the channel input of one transmitter as noise, i.e.,

$$\hat{W}_1 = \Phi_1(Y^n) := \arg\min_{\bar{w}_1 \in [M_1]} \|Y^n - X_1^n(\bar{w}_1)\|^2. \tag{6.4}$$

  Step 2 subtracts $X_1^n(\hat{W}_1)$ from the received signal to decode $X_2^n(\hat{W}_2)$ through another nearest neighbor decoder, i.e.,

$$\hat{W}_2 = \Phi_2(Y^n) := \arg\min_{\bar{w}_2 \in [M_2]} \left\|Y^n - X_1^n(\hat{W}_1) - X_2^n(\bar{w}_2)\right\|^2. \tag{6.5}$$

Note that in Definition 6.2, a fixed decoding order is assumed. In practice, the decoding order can be determined via the values of the signal-to-noise ratio (SNR) of two users. Usually, the message of the stronger user with a higher SNR value is decoded first so that the overall error probability is minimized.

Inspired by the fact that spherical codebooks achieve better low-latency performance for mismatched channel coding in Chap. 3, both users are assumed to apply spherical codebooks. Specifically, for each $j \in [2]$ and each $t \in [M_j]$, codeword $X_j^n(t)$ is generated from the uniform distribution over a sphere with radius $\sqrt{nP_j}$, i.e.,

$$P_{X_j^n}(x_j^n(t)) := \frac{\delta(\|x_j^n(t)\|^2 - nP_j)}{S_n(\sqrt{nP_j})}, \tag{6.6}$$

where $\delta(\cdot)$ is the Dirac delta function and $S_n(r) = 2\pi^{n/2} r^{n-1}/\Gamma(n/2)$ is the surface area of a radius-r sphere where $\Gamma(\cdot)$ denotes the Gamma function. The spherical codebook is illustrated in Fig. 6.2 when $n = 3$, where each red dot denotes one codeword.

6.1.3 Performance Metric

To evaluate the performance of a code, the following ensemble average error probability is considered:

$$\mathrm{P}_{\mathrm{e}}^n := \Pr\{(\hat{W}_1, \hat{W}_2) \neq (W_1, W_2)\}. \tag{6.7}$$

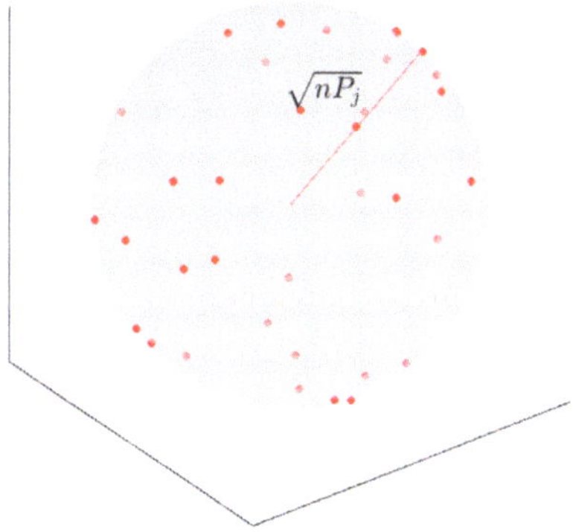

Fig. 6.2 Illustration of a spherical codebook when $n = 3$

The above probability is averaged over the distributions of both random codebooks (C_1, C_2), the messages (W_1, W_2) and the additive noise Z^n.

For multiuser communication, the definition of theoretical benchmarks is significantly different from the P2P case in Chaps. 2, 3, and 4. In particular, since multiple encoders or multiple decoders appear, the theoretical benchmark is characterized by a multi-dimensional region instead of a one-dimensional minimal or maximal rate, analogously to Chap. 5 for multiterminal compression. In the following, detailed definitions of the first- and second-order asymptotics are provided.

Fix any $\dagger \in \{\text{JNN}, \text{SIC}\}$.

Definition 6.3 A rate pair $(R_1, R_2) \in \mathbb{R}_+^2$ is said to be first-order $\dagger$-achievable for the two-user MAC if there exists a sequence of (n, M_1, M_2, P_1, P_2)-$\dagger$ codes such that for each $j \in [2]$,

$$\liminf_{n\to\infty} \frac{1}{n} \log M_j \geq R_j, \tag{6.8}$$

$$\lim_{n\to\infty} \mathrm{P}_\mathrm{e}^n = 0. \tag{6.9}$$

The closure of the set of all first-order achievable rate pairs is called the first-order rate region and denoted by $\mathcal{R}_\dagger$.

Note that $\mathcal{R}_\dagger$ generalizes the theoretical first-order asymptotic transmission rate with vanishing error probability, i.e., $\lim_{\varepsilon\to 0}\lim_{n\to\infty} R^*(n,\varepsilon)$ for channel coding in Theorem 2.3 and mismatched channel coding in Chap. 3. When the noise is AWGN, i.e., $P_Z = \mathcal{N}(\cdot; 0, 1)$, the capacity region of the two-user MAC is

$$\mathcal{R}_{\text{JNN}} = \mathcal{R}_{\text{SIC}} = \{(R_1, R_2) : R_1 \leq C(P_1), R_2 \leq C(P_2), R_1 + R_2 \leq C(P_1 + P_2)\}, \tag{6.10}$$

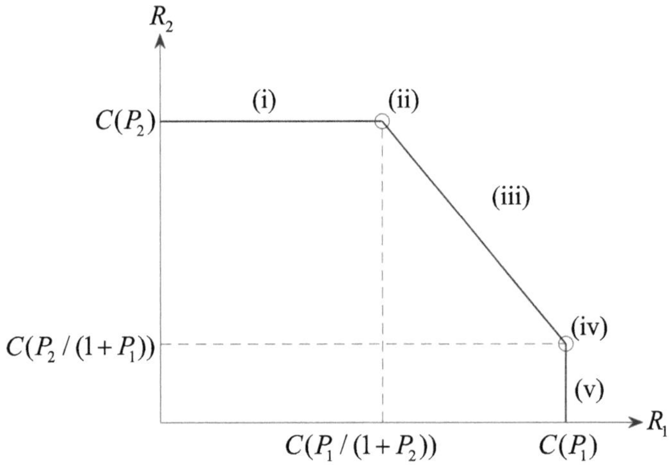

Fig. 6.3 Illustration of the first-order rate region of a two-user Gaussian MAC

where the $C(\cdot)$ is the capacity of an AWGN channel, i.e., for any $P \in \mathbb{R}_+$,

$$C(P) = \frac{1}{2} \log(1 + P). \tag{6.11}$$

The region $\mathcal{R}_{\text{JNN}}$ is illustrated in Fig. 6.3 and divided into several cases, which will be useful for subsequent low-latency performance characterization. Analogously to Chap. 3, the key point of mismatched communication is to use a coding scheme optimal for Gaussian noise for data transmission over an additive arbitrary noise channel. As shown in this chapter, the first-order asymptotics of mismatched MAC equals the corresponding first-order rate region of Gaussian MAC.

Definition 6.4 A pair $(L_1, L_2) \in \mathbb{R}^2$ is said to be second-order $(R_1^*, R_2^*, \varepsilon)$-† achievable for the two-user MAC if there exists a sequence of (n, M_1, M_2, P_1, P_2)-† codes such that for each $j \in [2]$,

$$\limsup_{n\to\infty} \frac{1}{\sqrt{n}}(nR_j^* - \log M_j) \le L_j, \tag{6.12}$$

$$\limsup_{n\to\infty} \mathrm{P}_{\mathrm{e}}^n \le \varepsilon. \tag{6.13}$$

The closure of all second-order $(R_1^*, R_2^*, \varepsilon)$-†-achievable pairs is called the second-order achievable rate region and denoted by $\mathcal{L}_{\dagger}(R_1^*, R_2^*, \varepsilon)$.

Given a rate pair (R_1^*, R_2^*) on the boundary of the first-order rate region $\mathcal{R}_{\dagger}$, the second-order rate pair (L_1, L_2) characterizes the gap to (R_1^*, R_2^*) of the finite blocklength transmission rate. This is because (6.12) implies that $\log M_j \ge nR_j^* -$

$\sqrt{n}L_j$. Thus, the larger the second-order rate region, the larger the finite blocklength transmission rates.

6.1.4 Second-Order Asymptotics

The following definitions are needed. Fix any integer $d \in \mathbb{N}$, positive semi-definite matrix $\mathbf{V}$ of dimension $d \times d$. Recall that $\mathcal{N}(\cdot; \mathbf{0}_d, \mathbf{V})$ denotes the pdf of a d-dimensional Gaussian random vector $S^d = (S_1, \ldots, S_d) \in \mathbb{R}^d$. Given any positive real number $\varepsilon \in (0, 1)$, the complementary cumulative distribution function (ccdf) of a d-dimensional Gaussian random variable is

$$\mathrm{Q_{inv}}(\mathbf{V}, \varepsilon) := \big\{a^d = (a_1, \ldots, a_d) \in \mathbb{R}^d : \Pr\{\mathbf{S}_d \geq a^d\} \geq 1 - \varepsilon\big\}. \tag{6.14}$$

Fix power constraints $(P, P_1, P_2) \in \mathbb{R}_+^3$. Recall that the noise distribution P_Z satisfies the moment constraints in (6.1), especially that the fourth moment is ξ. Define the following dispersion functions:

$$V(P) := \frac{(\xi - 1)P^2 + 4P}{4(1 + P)^2}, \tag{6.15}$$

$$V_{1,2}(P_1, P_2) := \frac{(\xi - 1)P_1 P_2}{4(1 + P_1)(1 + P_2)}, \tag{6.16}$$

$$V_{12}(P_1, P_2) := V(P_1 + P_2) + \frac{P_1 P_2}{(1 + P_1 + P_2)^2}, \tag{6.17}$$

$$V_{i,12}(P_1, P_2) := \frac{(\xi - 1)P_i(P_1 + P_2) + 4P_i}{4(1 + P_i)(1 + P_1 + P_2)}, \quad i \in [2], \tag{6.18}$$

$$V_i(P_1, P_2) := \frac{P_i^2(\xi - 1 + 4P_{3-i}) + 4P_i(1 + P_{3-i})^3}{4(1 + P_{3-i})^2(1 + P_1 + P_2)^2}, \quad i \in [2]. \tag{6.19}$$

Furthermore, define the following dispersion matrices:

$$\mathbf{V}_1(P_1, P_2) := \begin{bmatrix} V(P_2) & V_{2,12}(P_1, P_2) \\ V_{2,12}(P_1, P_2) & V_{12}(P_1, P_2) \end{bmatrix}, \tag{6.20}$$

$$\mathbf{V}_2(P_1, P_2) := \begin{bmatrix} V(P_1) & V_{1,12}(P_1, P_2) \\ V_{1,12}(P_1, P_2) & V_{12}(P_1, P_2) \end{bmatrix}, \tag{6.21}$$

$$\mathbf{V}(P_1, P_2) := \begin{bmatrix} V(P_1) & V_{1,2}(P_1, P_2) & V_{1,12}(P_1, P_2) \\ V_{1,2}(P_1, P_2) & V(P_2) & V_{2,12}(P_1, P_2) \\ V_{1,12}(P_1, P_2) & V_{2,12}(P_1, P_2) & V_{12}(P_1, P_2) \end{bmatrix}. \tag{6.22}$$

The above dispersion functions and matrices are used to characterize achievable second-order asymptotics for MAC.

When JNN decoding is used, the second-order achievable rate region is characterized in the following theorem. Fix $\alpha \in (0, 1)$ and let $\bar{\alpha} := 1 - \alpha$.

Theorem 6.1 *Given boundary rate pairs (R_1^*, R_2^*) on the first-order rate region, the second-order achievable rate region $\mathcal{L}_{\mathrm{JNN}}(R_1^*, R_2^*, \varepsilon)$ for JNN decoding satisfies*

- *Case (i):* $(R_1^*, R_2^*) = (\alpha C(P_1/(1+P_2)), C(P_2))$

$$\mathcal{L}_{\mathrm{JNN}}(R_1^*, R_2^*, \varepsilon) \supseteq \{(L_1, L_2) : L_2 \geq \sqrt{V(P_2)}\mathrm{Q}^{-1}(\varepsilon)\}. \tag{6.23}$$

- *Case (ii):* $(R_1^*, R_2^*) = (C(P_1/(1+P_2)), C(P_2))$

$$\mathcal{L}_{\mathrm{JNN}}(R_1^*, R_2^*, \varepsilon) \supseteq \{(L_1, L_2) : (L_2, L_1 + L_2) \in \mathrm{Q}_{\mathrm{inv}}(\mathbf{V}_1(P_1, P_2), \varepsilon)\}. \tag{6.24}$$

- *Case (iii):* $(R_1^*, R_2^*) = (\alpha C(P_1) + \bar{\alpha} C(P_1/(1+P_2)), \alpha C(P_2/(1+P_1)) + \bar{\alpha} C(P_2))$

$$\mathcal{L}_{\mathrm{JNN}}(R_1^*, R_2^*, \varepsilon) \supseteq \{(L_1, L_2) : L_1 + L_2 \geq \sqrt{V_{12}(P_1, P_2)}\mathrm{Q}^{-1}(\varepsilon)\}. \tag{6.25}$$

- *Case (iv):* $(R_1^*, R_2^*) = (C(P_1), C(P_2/(1+P_1)))$

$$\mathcal{L}_{\mathrm{JNN}}(R_1^*, R_2^*, \varepsilon) \supseteq \{(L_1, L_2) : (L_1, L_1 + L_2) \in \mathrm{Q}_{\mathrm{inv}}(\mathbf{V}_2(P_1, P_2), \varepsilon)\}. \tag{6.26}$$

- *Case (v):* $(R_1^*, R_2^*) = (C(P_1), \alpha C(P_2/(1+P_1)))$

$$\mathcal{L}_{\mathrm{JNN}}(R_1^*, R_2^*, \varepsilon) \supseteq \{(L_1, L_2) : L_1 \geq \sqrt{V(P_1)}\mathrm{Q}^{-1}(\varepsilon)\}. \tag{6.27}$$

The proof of Theorem 6.1 is provided in Sec. 6.4.2.

Following [17], the result in Theorem 6.1 can be written in the following unified form [22].

Corollary 6.1 *For any $\varepsilon \in (0, 1)$, there exists an (n, M_1, M_2, P_1, P_2)-JNN code such that for $\mathrm{P}_{\mathrm{e}}^n \leq \varepsilon$,*

$$\begin{bmatrix} \log M_1 \\ \log M_2 \\ \log M_1 M_2 \end{bmatrix} \in n\mathbf{C}(P_1, P_2) - \sqrt{n}\mathrm{Q}_{\mathrm{inv}}(\mathbf{V}(P_1, P_2), \varepsilon) + O(n^{1/4})\mathbf{1}_3, \tag{6.28}$$

where the capacity vector is

$$\mathbf{C}(P_1, P_2) := [C(P_1), C(P_2), C(P_1 + P_2)]^{\mathrm{T}}. \tag{6.29}$$

Although the above unified form is simple in its expression, the statement in Theorem 6.1 with five explicit cases is easier to follow and enables us to compare the performance of JNN and SIC decoding.

Theorem 6.1 generalizes the result of the AWGN MAC in [16] to the case with additive arbitrary noise. When specialized to the Gaussian MAC, $\xi = 3$ and the above results are consistent with [16, Theorem 1]. Although the more refined third-order asymptotics was characterized [17, Theorem 2], the result was restricted to Gaussian noise, and its proof relies heavily on the properties of Gaussian distributions. In contrast, the result in Theorem 6.1 holds for additive arbitrary noise satisfying the moment constraints, and its proof is based on geometrical analysis, which does not rely on a particular noise distribution.

Given any two real numbers $(a, b) \in \mathbb{R}^2$, define the following set:

$$\mathcal{A}(a, b) := \{(L_1, L_2) \in \mathbb{R}^2 : L_1 \geq a, L_2 \geq b\}. \tag{6.30}$$

When SIC decoding is used, the corresponding achievable second-order asymptotics is as follows.

Theorem 6.2 *Depending on the values of boundary rate pairs (R_1^*, R_2^*) of the first-order achievable rate region $\mathcal{R}_{\mathrm{SIC}}$ in* (6.10), *the second-order achievable rate region $\mathcal{L}_{\mathrm{SIC}}(R_1^*, R_2^*, \varepsilon)$ satisfies*

- *Case (i): $(R_1^*, R_2^*) = (\alpha C(P_1/(1 + P_2)), C(P_2))$*

$$\mathcal{L}_{\mathrm{SIC}}(R_1^*, R_2^*, \varepsilon) \supseteq \bigcup_{\substack{(\varepsilon_1,\varepsilon_2)\in\mathbb{R}_+^2:\\ \varepsilon_1+\varepsilon_2\leq\varepsilon}} \mathcal{A}\Big(\alpha\sqrt{V_1(P_1, P_2)}\mathrm{Q}^{-1}(\varepsilon_1), \sqrt{V(P_2)}\mathrm{Q}^{-1}(\varepsilon_2)\Big). \tag{6.31}$$

- *Case (ii): $(R_1^*, R_2^*) = (C(P_1/(1 + P_2)), C(P_2))$*

$$\mathcal{L}_{\mathrm{SIC}}(R_1^*, R_2^*, \varepsilon) \supseteq \bigcup_{\substack{(\varepsilon_1,\varepsilon_2)\in\mathbb{R}_+^2:\\ \varepsilon_1+\varepsilon_2\leq\varepsilon}} \mathcal{A}\Big(\sqrt{V_1(P_1, P_2)}\mathrm{Q}^{-1}(\varepsilon_1), \sqrt{V(P_2)}\mathrm{Q}^{-1}(\varepsilon_2)\Big). \tag{6.32}$$

- *Case (iii): $(R_1^*, R_2^*) = (\alpha C(P_1)+\bar{\alpha}C(P_1/(1+P_2)), \alpha C(P_2/(1+P_1))+\bar{\alpha}C(P_2))$*

$$\begin{aligned}\mathcal{L}_{\mathrm{SIC}}&(R_1^*, R_2^*, \varepsilon)\\ &\supseteq \bigcup_{\substack{(\varepsilon_1,\varepsilon_2)\in\mathbb{R}_+^2:\\ \varepsilon_1+\varepsilon_2\leq\varepsilon}} \mathcal{A}\Big((\alpha\sqrt{V_1(P_1, P_2)} + \bar{\alpha}\sqrt{V(P_1)})\mathrm{Q}^{-1}(\varepsilon_1),\\ &\qquad (\alpha\sqrt{V(P_2)} + \bar{\alpha}\sqrt{V_2(P_1, P_2)})\mathrm{Q}^{-1}(\varepsilon_2)\Big).\end{aligned} \tag{6.33}$$

- *Case (iv):* $(R_1^*, R_2^*) = (C(P_1), C(P_2/(1+P_1)))$

$$\mathcal{L}_{\text{SIC}}(R_1^*, R_2^*, \varepsilon) \supseteq \bigcup_{\substack{(\varepsilon_1,\varepsilon_2)\in\mathbb{R}_+^2:\\ \varepsilon_1+\varepsilon_2\le\varepsilon}} \mathcal{A}\Big(\sqrt{V(P_1)}\mathrm{Q}^{-1}(\varepsilon_1), \sqrt{V_2(P_1,P_2)}\mathrm{Q}^{-1}(\varepsilon_2)\Big). \tag{6.34}$$

- *Case (v):* $(R_1^*, R_2^*) = (C(P_1), \alpha C(P_2/(1+P_1)))$

$$\mathcal{L}_{\text{SIC}}(R_1^*, R_2^*, \varepsilon) \supseteq \bigcup_{\substack{(\varepsilon_1,\varepsilon_2)\in\mathbb{R}_+^2:\\ \varepsilon_1+\varepsilon_2\le\varepsilon}} \mathcal{A}\Big(\sqrt{V(P_1)}\mathrm{Q}^{-1}(\varepsilon_1), \alpha\sqrt{V_2(P_1,P_2)}\mathrm{Q}^{-1}(\varepsilon_2)\Big). \tag{6.35}$$

For cases (i) and (v), the degenerated cases of $\alpha = 0$ and $\alpha = 1$ are excluded, since these two cases correspond to mismatched P2P channel coding in Chap. 3. The proof of case (ii) is available in Sect. 6.4.3, which combines the proof techniques for mismatched channel coding and mismatched interference channel coding [21]. The proof of all other cases follows due to time sharing or symmetry. Specifically, case (i) is proved by time sharing the results for mismatched channel coding and case (ii), where α denotes the percentage of time that the code for case (ii) is used. Case (iv) is symmetric to case (ii), and case (v) is symmetric to case (i). Case (iii) is proved by time sharing the results of cases (ii) and (iv), where α denotes the percentage of total time where the code for case (iv) is used.

The discussions for mismatched channel coding in Chap. 3 generalize to the MAC case. In a nutshell, when a coding scheme that has guaranteed performance for Gaussian noise is used as a mismatched code, the same first-order asymptotics can be achieved for any additive noise with the same second moment, and the achievable second-order asymptotics depends on the particular noise distribution through the fourth moment. Since the mismatched code is designed without knowledge of the exact noise distribution, its performance might be far away from the optimal code of the matched case. However, the advantage of a mismatched code is its general validity and robustness.

6.1.5 Discussions

This section compares the performance of JNN and SIC decoding for mismatched MAC using Theorems 6.1 and 6.2. Although both JNN and SIC decoding achieve the same first-order rate region for a two-user mismatched MAC, JNN decoding has a strictly larger second-order achievable rate region. Detailed explanation is as follows.

Due to the symmetry of cases (i) and (v), and cases (ii) and (iv), it suffices to consider cases (i), (ii), and (iii). Given $\dagger \in$ (i, ii, iii), let $\mathcal{L}_{\mathrm{JNN},\dagger}$ and $\mathcal{L}_{\mathrm{SIC},\dagger}$ denote inner bounds for the achievable second-order rate region $\mathcal{L}_{\mathrm{JNN}}(R_1^*, R_2^*, \varepsilon)$ and $\mathcal{L}_{\mathrm{SIC}}(R_1^*, R_2^*, \varepsilon)$ in Theorems 6.1 and 6.2, respectively.

First, consider case (i). For JNN decoding, it follows from (6.23) that L_1 can be arbitrarily small. In contrast, for SIC decoding, it follows from (6.31) that L_1 is lower bounded by $\alpha\sqrt{V_1(P_1, P_2)}\mathrm{Q}^{-1}(\varepsilon_1)$. The lower bounds for L_2 in JNN and SIC decoding only differ in the parameter of the ccdf function $\mathrm{Q}^{-1}(\cdot)$. Since $\mathrm{Q}^{-1}(\cdot)$ monotonically decreases in its parameter, it follows from the constraint $\varepsilon \geq \varepsilon_1$ that $\sqrt{V(P_2)}\mathrm{Q}^{-1}(\varepsilon_1) \geq \sqrt{V(P_2)}\mathrm{Q}^{-1}(\varepsilon)$. Thus, for case (i), the second-order achievable rate region of JNN decoding is larger, i.e., $\mathcal{L}_{\mathrm{SIC,\,i}} \subseteq \mathcal{L}_{\mathrm{JNN,\,i}}$.

Now consider case (ii). Recall the definition of the dispersion function $\mathbf{V}_1(P_1, P_2)$ in (6.20) and define the two-dimensional Gaussian random vector $[K_1; K_2] \sim \mathcal{N}(\cdot; \mathbf{0}_2, \mathbf{V}_1(P_1, P_2))$. It follows from (6.24) that the second-order rate region of JNN decoding is equivalently given by

$$\mathcal{L}_{\mathrm{JNN,\,ii}} = \big\{(L_1, L_2) : \Pr\{\{K_1 \geq L_2\} \cup \{K_2 \geq (L_1 + L_2)\}\} \leq \varepsilon\big\}. \tag{6.36}$$

First, consider extreme cases where one of the second-order rates tends to infinity. When $L_1 \to \infty$, $\Pr\{K_2 \leq (L_1 + L_2)\} \to 1$, and

$$\mathcal{L}_{\mathrm{JNN,\,ii}} = \big\{(L_1, L_2) : L_2 \geq \sqrt{V(P_2)}\mathrm{Q}^{-1}(\varepsilon)\big\}. \tag{6.37}$$

The above region matches the second-order achievable rate region $\mathcal{L}_{\mathrm{SIC,\,ii}}$ of SIC decoding in (6.32) when $\varepsilon_2 \to \varepsilon$. However, if $\varepsilon_2 < \varepsilon$, $\mathcal{L}_{\mathrm{SIC,\,ii}} \subseteq \mathcal{L}_{\mathrm{JNN,\,ii}}$ since $\mathrm{Q}^{-1}(\varepsilon_2) > \mathrm{Q}^{-1}(\varepsilon)$. When $L_2 \to \infty$, the probability term in (6.36) does not depend on L_1 anymore, which implies that L_1 can be arbitrarily small in $\mathcal{L}_{\mathrm{JNN,\,ii}}$ for JNN decoding. In contrast, L_1 is strictly lower bounded by $\sqrt{V_1(P_1, P_2)}\mathrm{Q}^{-1}(\varepsilon_1)$ in (6.32) for SIC decoding. Thus, $\mathcal{L}_{\mathrm{SIC,\,ii}} \subsetneq \mathcal{L}_{\mathrm{JNN,\,ii}}$ when either L_1 or L_2 is unbounded.

When the second-order rates L_1 and L_2 are both finite, the comparison is nontrivial. Fix $\varepsilon_1 \in (0, 1)$ such that $\varepsilon_1 < \varepsilon$ and let $L_1' := \sqrt{V_1(P_1, P_2)}\mathrm{Q}^{-1}(\varepsilon_1)$. The set of L_2 such that $(L_1', L_2) \in \mathcal{L}_{\mathrm{JNN,\,ii}}$ is given by

$$\begin{aligned}\mathcal{L}_{\mathrm{JNN,\,ii},L_2} &= \{L_2 : \Pr\{\{K_1 \geq L_2\} \cup \{K_2 - L_1' \geq L_2\}\} \leq \varepsilon\} &(6.38)\\ &= \{L_2 : \Pr\{\max\{K_1, K_2 - L_1'\} \geq L_2\} \leq \varepsilon\}. &(6.39)\end{aligned}$$

Correspondingly, the set of L_2 such that $(L_1', L_2) \in \mathcal{L}_{\mathrm{SIC,ii}}$ satisfies

$$\mathcal{L}_{\mathrm{SIC,\,ii},L_2} = \big\{L_2 :\ \Pr\{K_1 \geq L_2\} \leq \varepsilon_2\big\}. \tag{6.40}$$

The comparison between (6.39) and (6.40) is mathematically challenging. Instead, a numerical example can help. Specifically, in Fig. 6.4, the second-order rate regions of JNN and SIC decoding are plotted when the boundary rate pair is $(R_1^*, R_2^*) =$

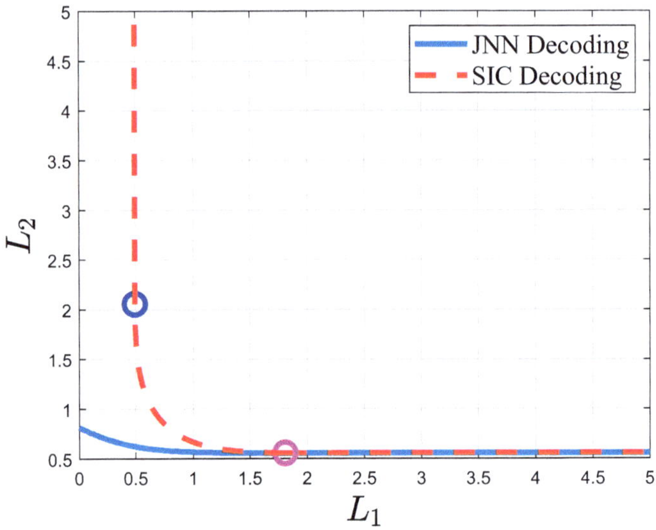

Fig. 6.4 Comparison of second-order rate regions of JNN decoding $\mathcal{L}_{\text{JNN}}$ and SIC decoding $\mathcal{L}_{\text{SIC}}$ for case (ii) when $\varepsilon = 0.2$, $P_1 = 5$, $P_2 = 2$, and $\xi = 3$

$(C(P_1/(1 + P_2)), C(P_2))$, the tolerable error probability is $\varepsilon = 0.2$ the power constraints are $P_1 = 5, P_2 = 2$ and the fourth moment $\xi = 3$. The region above the blue line is the second-order achievable rate region $\mathcal{L}_{\text{JNN, ii}}$ of JNN decoding, and the region above the red dashed line is the second-order achievable rate region $\mathcal{L}_{\text{SIC, ii}}$ of SIC decoding. The pink and deep blue circles denote the cases when $\varepsilon_1 \to 0$ and $\varepsilon_2 \to 0$, respectively. As observed from Fig. 6.4, when $L_1 \to \infty$ or $L_2 \to \infty$, the second-order rates of both JNN and SIC decoding are the same, which is consistent with the theoretical analysis. Furthermore, for a finite L_1, the minimum value of L_2 in $\mathcal{L}_{\text{SIC, ii}}$ for SIC decoding is strictly smaller than the corresponding minimal value of L_2 in $\mathcal{L}_{\text{JNN, ii}}$ for JNN decoding. The above numerical result implies that JNN decoding can be strictly better in terms of low-latency communication performance.

For case (iii), the second-order rate region of JNN in (6.25) can be equivalently written as

$$\mathcal{L}_{\text{JNN,iii}} = \big\{(L_1, L_2) : \Pr\{K_2 \geq L_1 + L_2\} \leq \varepsilon\big\}. \tag{6.41}$$

It follows from (6.36) and (6.41) that $\mathcal{L}_{\text{JNN, ii}} \subseteq \mathcal{L}_{\text{JNN,iii}}$. This is because for any $(L_1, L_2) \in \mathbb{R}^2$, the inequality $\Pr\{K_2 \geq L_1 + L_2\} \leq \Pr\{\{K_1 \geq L_2\} \cup \{K_2 \geq (L_1 + L_2)\}\}$ holds. Analogously, it follows that $\mathcal{L}_{\text{JNN,iv}} \subseteq \mathcal{L}_{\text{JNN,iii}}$. Furthermore, for SIC decoding, $\mathcal{L}_{\text{SIC, iii}}$ is a linear combination of $\mathcal{L}_{\text{SIC, ii}}$ and $\mathcal{L}_{\text{SIC, iv}}$. In Fig. 6.4, it has been shown numerically that $\mathcal{L}_{\text{SIC}} \subseteq \mathcal{L}_{\text{JNN}}$ for case (ii), which is also true for case (iv) due to symmetry. Thus, the same conclusion holds for case (iii) due to the above arguments. To clarify, Fig. 6.5 plots the second-order rate regions of JNN

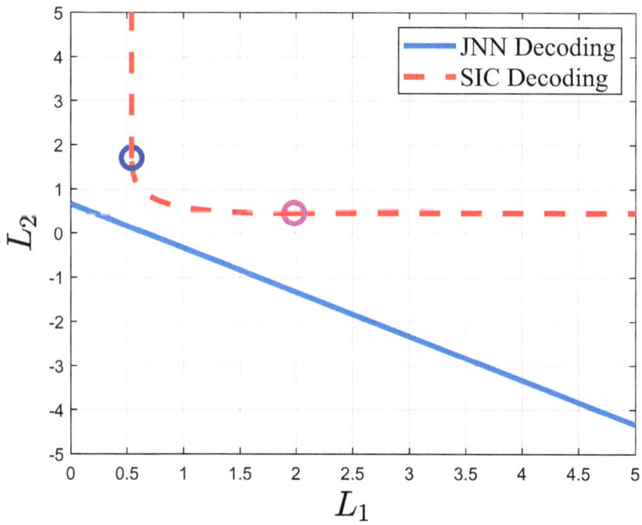

Fig. 6.5 Comparison of second-order rate regions of JNN decoding $\mathcal{L}_{\mathrm{JNN}}$ and SIC decoding $\mathcal{L}_{\mathrm{SIC}}$ for case (iii) when $\varepsilon = 0.2$, $P_1 = 5$, $P_2 = 2$, and $\xi = 3$

and SIC decoding for the same setting as Fig. 6.4 when $\alpha = 0.5$. As observed, the second-order rate region $\mathcal{L}_{\mathrm{SIC,\,iii}}$ of SIC decoding, denoted by the region above the red dashed line, falls strictly inside the second-order rate region $\mathcal{L}_{\mathrm{JNN,\,iii}}$ of JNN decoding, denoted by the region above the blue line.

In summary, the low-latency performance of JNN decoding can be strictly better than SIC decoding, although both decoding algorithms achieve the same asymptotic performance as the blocklength (latency) tends to infinity.

6.2 Random Access Channel

This section presents the mismatched code and its low-latency performance for RAC with additive arbitrary noise. Both SIC and JNN decoding are analyzed. Even asymptotically, JNN decoding is much better than SIC decoding, implying that SIC should not be used in the multiuser uplink communication when the number and identities of active users are unknown.

6.2.1 Problem Formulation

Fix three integers $(K, k, t) \in \mathbb{N}^3$ and a sequence of blocklengths $\mathbf{n} := (n_1, \ldots, n_K) \in \mathbb{N}^K$. As shown in Fig. 6.6, this section studies the random access

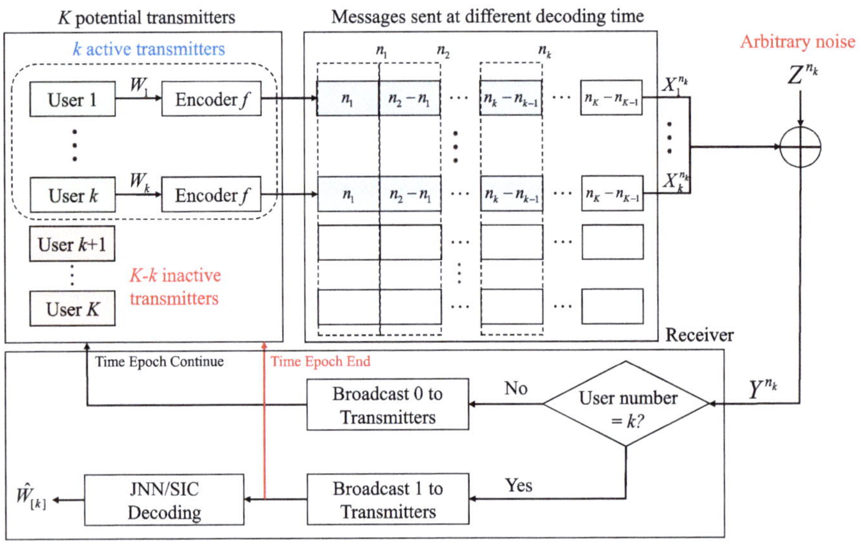

Fig. 6.6 System model for RAC

channel with K transmitters. In each time epoch, k out of all K transmitters are active and the other $K - k$ transmitters are inactive. Only the active users transmit messages. Without loss of generality, assume that the first k users are active. The receiver is assumed to be ignorant of the number k and the identities of active users.

The rateless code [17] is used by the transmitters and the receiver. In particular, all active transmitters use the same codebook and encoder. For each $t \in [K]$, user t generates a codeword of length n_K and divides the codeword into K sub-codewords with lengths $(n_1, n_2 - n_1, \ldots, n_K - n_{K-1})$, respectively. The transmission process is separated into at most K stages, and one-bit feedback signal is sent from the receiver after each stage. In the first stage, each active user transmits the first sub-codeword of length n_1. After receiving a noisy channel output Y^{n_1} that is corrupted by additive arbitrary noise Z^{n_1}, the receiver checks whether the estimated number of active users is one. If yes, a one-bit feedback 1 is sent to all users to terminate the transmission for the current epoch. Otherwise, the receiver broadcasts the other one-bit feedback 0, signaling each active user to continue the transmission process. For each $t \in [2 : K]$, the active users transmitter their t-th sub-codewords of length $n_t - n_{t-1}$. The codewords are corrupted by the additive arbitrary noise $Z^{n_t - n_{t-1}}$, yielding noisy channel output $Y^{n_t - n_{t-1}}$. Using $Y^{n_t - n_{t-1}}$, the receiver first checks whether the estimated number of active users is t. If yes, the receiver terminates the transmission procedure by broadcasting feedback signal 1. Otherwise, the receiver broadcasts 0 to continue the transmission process until the maximal decoding time n_K. When the transmission process is terminated, the receiver uses either a JNN or SIC decoder to estimate the transmitted messages of all active users.

Fix any $k \in [K]$. Assume k active users transmit messages $(W_1, \ldots, W_k) \in [M]^K$. For each $t \in [k]$, the received channel output at time n_t satisfies

$$Y^{n_t} = \sum_{i=1}^{k} X^{n_t}(W_i) + Z^{n_t}, \tag{6.42}$$

where $(X^{n_t}(W_1), \ldots, X^{n_t}(W_k))$ are the first n_t symbols of codewords sent by active users, and Z^{n_t} is the additive noise generated i.i.d. from an unknown distribution P_Z satisfying moment constraints in (6.1).

6.2.2 Coding Scheme

Fix K positive real numbers $(\lambda_1, \ldots, \lambda_K) \in \mathbb{R}_+^K$ and a power constraint $P \in \mathbb{R}_+$. Recall that $\mathbf{n} = (n_1, \ldots, n_K)$ are K fixed blocklengths.

When JNN decoding is used, the mismatched code for RAC is defined as follows.

Definition 6.5 An $(\mathbf{n}, M, P)$-JNN code for a RAC consists of:

- A random codebook C with M codewords $\{X^{n_K}(1), \ldots, X^{n_K}(M)\}$ subject to a maximal power constraint P,
- An encoder f such that for each message $w \in [M]$,

$$f(w) := X^{n_K}(w). \tag{6.43}$$

- A sequence of K decoders $(\Psi_1, \ldots, \Psi_K)$. For each $t \in [K]$, using Y^{n_t}, the decoder Φ_t first checks whether the estimated number of users equals t by verifying the inequality $\rho_t(Y^{n_t}) := \left|\frac{1}{n_t}\|Y^{n_t}\|^2 - (1+tP)\right| \le \lambda_t$. If the inequality is not satisfied, the decoder notifies all active users to continue transmission. Otherwise, the decoder notifies all users to terminate transmission and applies the following JNN decoder:

$$(\hat{W}_1, \ldots, \hat{W}_t) = \Psi_t(Y^{n_t}) := \mathop{\arg\min}_{(\bar{w}_1,\ldots,\bar{w}_t)\in[M]^t} \left\| Y^{n_t} - \sum_{i=1}^{t} X^{n_t}(\bar{w}_i) \right\|^2. \tag{6.44}$$

When SIC decoding is used, the mismatched code for RAC is defined as follows.

Definition 6.6 An $(\mathbf{n}, M, P)$-SIC code is the same as the code in Definition 6.5 except that the JNN decoder in (6.44) is replaced by the following SIC decoder. The first message is decoded using nearest neighbor (NN) decoding and treating interference as noise (TIN):

$$\hat{W}_1 := \mathop{\arg\min}_{\bar{w}\in[M]} \left\| Y^{n_k} - X^{n_k}(\bar{w}) \right\|^2. \tag{6.45}$$

Subsequently, for each $r \in [2:t]$, the message W_r is decoded using interference cancellation with NN and TIN:

$$\hat{W}_r := \arg\min_{\bar{w}\in[M]} \left\| Y^{n_k} - \sum_{i=1}^{r-1} X^{n_k}(\hat{W}_i) - X^{n_k}(\bar{w}) \right\|^2. \tag{6.46}$$

All users in a RAC adopt the same spherical codebook with codewords $\{X^{n_K}(1), \ldots, X^{n_K}(M)\}$. For each $w \in [M]$, the codeword $X^{n_K}(w)$ is generated from the same distribution. For simplicity, let X^{n_K} denote any codeword. The first n_1 symbols X^{n_1} of X^{n_K} are generated from the uniform distribution over the sphere with radius $\sqrt{n_1 P}$, i.e., for any $x^{n_1} \in \mathbb{R}^{n_1}$,

$$P_{X^{n_1}}(x^{n_1}) := \frac{\delta\left(\|x^{n_1}\|^2 - n_1 P\right)}{S_{n_1}(\sqrt{n_1 P})}. \tag{6.47}$$

For each $j \in [2:K]$, the $n_{j-1}+1$ to the n_j symbols $X_{n_{j-1}+1}^{n_j}$ of X^{n_K}, is generated from the uniform distribution over a sphere with radius $\sqrt{(n_j - n_{j-1})P}$, i.e., for any $x_{n_{j-1}+1}^{n_j} \in \mathbb{R}^{n_j - n_{j-1}}$,

$$P_{X_{n_{j-1}+1}^{n_j}}(x_{n_{j-1}+1}^{n_j}) := \frac{\delta(\|x_{n_{j-1}+1}^{n_j}\|^2 - (n_j - n_{j-1})P)}{S_{n_j - n_{j-1}}(\sqrt{(n_j - n_{j-1})P})}. \tag{6.48}$$

The spherical codebook for a RAC is illustrated in Fig. 6.7 when $K = 2$, $n_1 = 2$, and $n_2 = 3$.

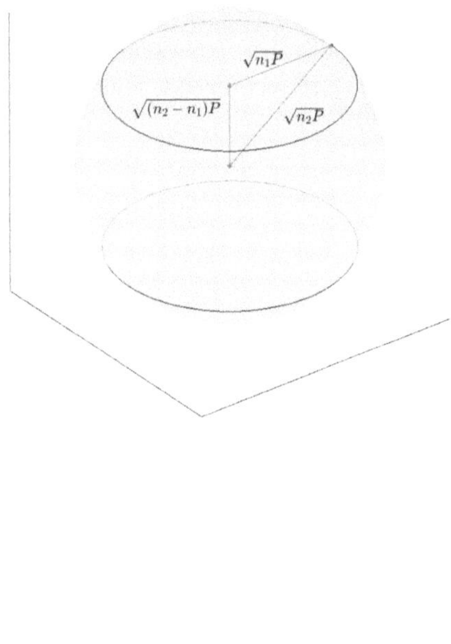

Fig. 6.7 Illustration of the spherical codebook for a RAC when $n_1 = 2$ and $n_2 = 3$ (reproduced from [17, Fig. 2])

6.2.3 Performance Metric

To evaluate the performance of a mismatched RAC code, the generalized error probability is considered. When $k \in [K]$ users are active, who transmit messages $(W_1, \ldots, W_k) \in [M]^k$, the error probability is defined as

$$\mathrm{P}_{\mathrm{e},k}^{n_k} := \Pr\left\{\mathrm{perm}(W_1, \ldots, W_k) \neq \mathrm{perm}(\hat{W}_1, \ldots, \hat{W}_k)\right\}, \tag{6.49}$$

where perm denotes the permutation of messages. Since the identities of active users are unknown, a permutation is necessary since the receiver only aims to decode all messages correctly and does not care which user transmits which message. This is consistent with RAC studies in [17, 18].

Since the same codebook and encoder is used in a RAC, the theoretical benchmark is thus the maximal transmission rate subject to error probability constraints. Specifically, given any K positive real numbers $\boldsymbol{\varepsilon} := (\varepsilon_1, \ldots, \varepsilon_K) \in (0,1)^K$, the finite blocklength benchmark for a mismatched RAC is defined as

$$\begin{aligned} R_{\dagger}^*(\mathbf{n}, \boldsymbol{\varepsilon}) := \sup\Big\{s \in \mathbb{R}_+ :\ &\exists \text{ an } (\mathbf{n}, M, P) - \dagger\text{—code. s.t.} \\ &\forall\, k \in [K], \log M \geq n_k s \text{ and } \mathrm{P}_{\mathrm{e},k}^{n_k} \leq \varepsilon_k\Big\}, \end{aligned} \tag{6.50}$$

for $\dagger \in$ (SIC, JNN). Thus, $R_{\dagger}^*(\mathbf{n}, \boldsymbol{\varepsilon})$ is the maximal transmission rate $\min_{k\in[K]} \frac{\log M}{n_k}$ of any mismatched RAC code such that when the number of active users is any value $k \in [K]$, the error probability is no greater than ε_k. In other words, one RAC code should guarantee the error performance and achieve the rate $R_{\dagger}^*(\mathbf{n}, \boldsymbol{\varepsilon})$ simultaneously for any value k of active user numbers. When $K = 1$, the above definition reduces to the theoretical benchmark for mismatched channel coding in Chap. 3.

It is critical to note that the theoretical benchmark for mismatched RAC is significantly different from that of MAC. This is mainly because different codebooks and encoders are used by different users in a MAC while the same codebook and encoder are used for all active users in a RAC. When different codebooks are used in a MAC, the message rates are thus different, and the theoretical benchmark naturally forms a multidimensional region. In contrast, in a RAC, the same rate is shared by all active users, and thus the theoretical benchmark is a one-dimensional value, analogous to P2P channel coding. The above difference results from the fact that the number and identities of active users in a RAC are unknown to all transmitters and the receiver. Since both are unknown, using different codebooks and encoders is challenging for a RAC.

6.2.4 Second-Order Asymptotics and Discussions

Fix any power constraint $P \in \mathbb{R}_+$. Recall the definition of dispersion function $V(P)$ in (6.15). For each $k \in [K]$ and $r \in [k]$, define the following dispersion functions:

$$V_{\mathrm{cr}}(k,P) := \frac{k(k-1)P^2}{2(1+kP)^2}, \tag{6.51}$$

$$\begin{aligned} V_{\mathrm{rs}}(k,r,P) :=& \frac{1}{4(1+(k-r+1)P)^2(1+(k-r)P)^2} \\ &\times (P^2(\xi-1)+4P(1+(k-r)P)^3 \\ &+4(k-r)P^3+2(k-r)(k-r-1)P^4). \end{aligned} \tag{6.52}$$

Recall that $\mathbf{n} = (n_1,\ldots,n_K) \in \mathbb{N}^K$ are given blocklengths and $\boldsymbol{\varepsilon} = (\varepsilon_1,\ldots,\varepsilon_K) \in (0,1)^K$ are tolerable error probabilities.

Theorem 6.3 *The finite blocklength transmission rate of a mismatched RAC with JNN decoding satisfies*

$$R^*(\mathbf{n},\varepsilon) \geq \min_{k\in[K]} \frac{1}{k}\left(C(kP) - \sqrt{\frac{(V(kP)+V_{\mathrm{cr}}(k,P))}{n_k}}\mathrm{Q}^{-1}(\varepsilon_k) + O\left(\frac{\log n_k}{n_k}\right)\right). \tag{6.53}$$

The proof of Theorem 6.3 is provided in Sect. 6.5.2. In particular, it is proved that there exists an $(\mathbf{n}, M, P)$-JNN code such that for all $k \in [K]$, $\mathrm{P}^n_{\mathrm{e},k} \leq \varepsilon_k$ and

$$\log M \geq \min_{k\in[K]} \frac{1}{k}\left(n_k C(kP) - \sqrt{n_k(V(kP)+V_{\mathrm{cr}}(k,P))}\mathrm{Q}^{-1}(\varepsilon_k) + O(\log n_k)\right). \tag{6.54}$$

Theorem 6.3 lower bounds the second-order achievable rate of a mismatched RAC using a JNN decoder. The result holds for additive arbitrary noise satisfying moment constraints (6.1). When specialized to the Gaussian noise, Theorem 6.3 implies that the coding scheme in Definition 6.5 achieves the same second-order rate as the matched case [17, Theorems 4].

When SIC decoding is used, the corresponding result is as follows.

Theorem 6.4 *The finite blocklength transmission rate of a mismatched RAC with JNN decoding satisfies*

$$R^*(\mathbf{n},\varepsilon) \geq \min_{k\in[K]} \left(C\left(\frac{P}{1+(k-1)P}\right) - \sqrt{\frac{V_{\mathrm{rs}}(k,1,P)}{n_k}}\mathrm{Q}^{-1}(\varepsilon_k) + O\left(\frac{\log n_k}{n_k}\right)\right). \tag{6.55}$$

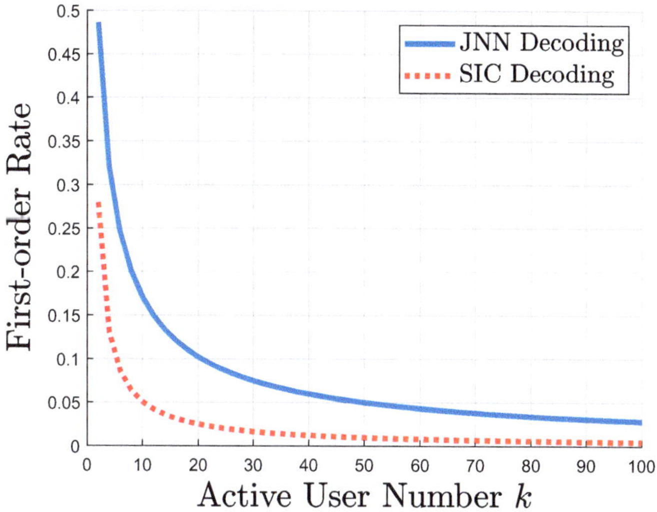

Fig. 6.8 Plot of first-order rates in a RAC for JNN and SIC decoding when $P = 3$

Since all active users adopt the same codebook with M codewords, the first-order achievable rate in (6.55) is the minimum rate to decode the messages of all active users reliably, which corresponds to the decoding of the first message when all the codewords of all other messages are treated as noise. This leads to the signal to interference and noise ratio $\frac{P}{(1+(k-1)P)}$ and thus the first-order rate $C(\frac{P}{(1+(k-1)P)})$. It follows from (6.53) and (6.55) that for mismatched RAC, SIC decoding is strictly inferior to JNN decoding even in the first-order rate. This is because $\frac{C(kP)}{k} > C(\frac{P}{(1+(k-1)P)})$ when $k \geq 2$. Figure 6.8 compares the first-order rates of RAC with JNN and SIC decoding as a function of the active user number k. It is observed that the asymptotic results for both JNN and SIC decoding monotonically decreases as k increases, and JNN is strictly superior to SIC in the first-order rate. Figure 6.9 plots the second-order rates in a RAC for JNN and SIC decoding, which demonstrates the inferiority of SIC decoding for low-latency RAC systems.

6.3 Broadcast Channel

This section presents the mismatched code and its low-latency performance of degraded BC with additive arbitrary noise. In particular, SIC and JNN decoding are analyzed and compared. Analogous to MAC, both decoding algorithms achieve the same asymptotic performance. However, in contrast to MAC, low-latency performance of SIC and JNN decoding is identical.

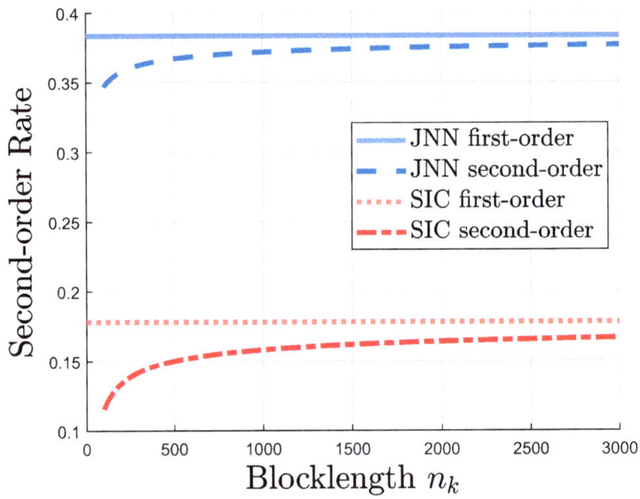

Fig. 6.9 Plot of second-order rates in a RAC for JNN and SIC decoding when $P = 3$, $\xi = 3$, $k = 3$, and $\varepsilon_k = 0.1$

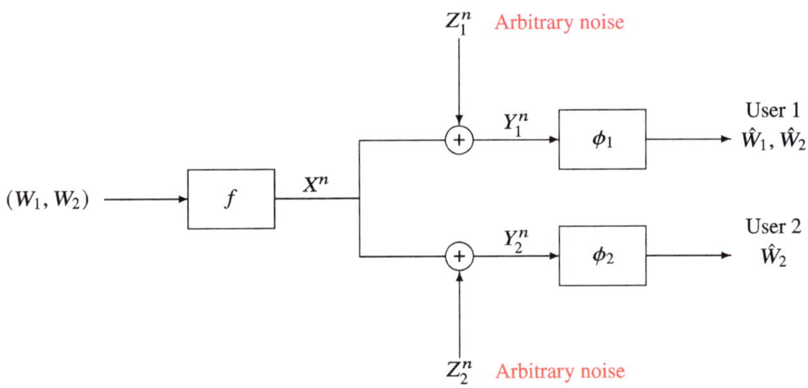

Fig. 6.10 System model for two-user asymmetric broadcast channel with additive arbitrary noise

6.3.1 Problem Formulation

Fix three integers $(n, M_1, M_2) \in \mathbb{N}^3$ and a positive real number $\beta \in (0, 1)$. As shown in Fig. 6.10, two messages $(W_1, W_2) \in [M_1] \times [M_2]$ are transmitted to two users via an encoder f, which maps the messages into a channel input X^n. The channel input is corrupted by two independent additive arbitrary noises (Z_1^n, Z_2^n), yielding noisy channel outputs Y_1^n for user 1 and Y_2^n for user 2, respectively, i.e.,

$$Y_1^n = X^n + Z_1^n, \tag{6.56}$$

$$Y_2^n = X^n + Z_2^n. \tag{6.57}$$

Given Y_1^n, user 1 wishes to decode both messages and the estimates $(\hat{W}_1, \hat{W}_2)$ are obtained via decoder ϕ_1. In contrast, user 2 only wishes to decode one message using Y_2^n and decoder ϕ_2, yielding estimate $\hat{W}_2$.

For each $i \in [2]$, Z_i^n is generated i.i.d. from a distribution P_{Z_i} satisfying the following moment constraints:

$$\mathbb{E}[Z_1^2] = \beta, \zeta_1 := \mathbb{E}[Z_1^4], \mathbb{E}[Z_1^6] < \infty, \tag{6.58}$$

$$\mathbb{E}[Z_2^2] = 1, \zeta_2 := \mathbb{E}[Z_2^4], \mathbb{E}[Z_2^6] < \infty. \tag{6.59}$$

Note that in the above formulation, the noise power of user 1 is assumed to be $\beta < 1$, which models the case where user 1 has better channel equality than user 2. Correspondingly, user 1 is named the strong user, and user 2 is named the weak user.

This section presents a mismatched code to achieve reliable transmission of both messages. Since two users are assumed to decode different messages, the setting is known as the BC with a degraded message set, which is also known as the asymmetric BC (ABC). It is important to note that the above formulation is not restrictive. The assumption of a degraded message set is in fact more general than the assumption that each user decodes only one message. As shown in the proof of this section, for the strong user, decoding the message W_2 for the weak user costs *no* additional rate. Thus, the same transmission rates are also achievable for the non-degraded message set setting.

6.3.2 *Coding Scheme*

This section presents the mismatched code for BC with a degraded message set. The weak user decodes message W_2 using treating interference as noise (TIN) and nearest neighbor (NN) decoding. The strong user applies either joint nearest neighbor (JNN) or successive interference cancellation (SIC) decoding.

Definition 6.7 An (n, M_1, M_2)-JNN code for a two user ABC consists of:

- A set of M_1 codewords $\{V^n(i)\}_{i\in[M_1]}$ known by the encoder f and decoder ϕ_1 of user 1,
- A set of M_2 codewords $\{U^n(i)\}_{i\in[M_2]}$ known by the encoder f and decoders (ϕ_1, ϕ_2) of both users,
- An encoder f that maps each message pair $(w_1, w_2) \in [M_1] \times [M_2]$ into a channel input

$$f(w_1, w_2) := X^n(w_1, w_2) = V^n(w_1) + U^n(w_2). \tag{6.60}$$

- A decoder ϕ_1 for user 1 that applies JNN decoding, i.e.,

$$(\hat{W}_1, \bar{W}_2) := \underset{(\bar{w}_1, \bar{w}_2) \in [M_1] \times [M_2]}{\arg\min} \|Y_1^n - U^n(\bar{w}_2) - V^n(\bar{w}_1)\|^2. \tag{6.61}$$

- A decoder ϕ_2 for user 2 that applies TIN and NN decoding, i.e.,

$$\hat{W}_2 := \underset{\bar{w}_2 \in [M_2]}{\arg\min} \|Y_2^n - U^n(\bar{w}_2)\|^2. \tag{6.62}$$

An (n, M_1, M_2)-SIC code is defined similarly, except that the JNN decoder for user 1 is replaced by the following SIC decoder that operates in two steps. Firstly, TIN and NN decoding are used to decode message W_2:

$$\bar{W}_2 := \underset{\bar{w}_2 \in [M_2]}{\arg\min} \|Y_1^n - U^n(\bar{w}_2)\|^2. \tag{6.63}$$

Subsequently, SIC and NN decoding are used to decode message W_1:

$$\hat{W}_1 := \underset{\bar{w}_1 \in [M_1]}{\arg\min} \|Y_1^n - U^n(\bar{W}_2) - V^n(\bar{w}_1)\|^2. \tag{6.64}$$

Inspired by the point-to-point mismatched channel coding in Chap. 3, spherical codebooks are used. Specifically, given any vector $x^n = (x_1, \ldots, x^n) \in \mathbb{R}^n$ and positive real number $P \in \mathbb{R}_+$, the uniform distribution over the surface of a sphere with radius $\sqrt{nP}$ satisfies

$$f_{\text{sp}}(x^n|P) = \frac{\mathbb{1}(\|x^n\|^2 - nP)}{S_n(\sqrt{nP})}, \tag{6.65}$$

where $S_n(r) = n\pi^{n/2}r^{n-1}/\Gamma(\frac{n+2}{2})$ is the surface area of an n-dimensional sphere with radius r, and $\Gamma(\cdot)$ is the Gamma function. Let $\alpha \in (0, 1)$ and $\bar{\alpha} := 1 - \alpha$. The codewords $\mathbf{V} := \{V^n(i)\}_{i \in [M_1]}$ for user 1 and $\mathbf{U} := \{U^n(i)\}_{i \in [M_2]}$ for user 2 are generated independently as follows. Given any $(v^n.u^n) \in (\mathbb{R}^n)^2$,

$$P_{V^n}(v^n) := f_{\text{sp}}(v^n|\alpha P), \tag{6.66}$$

$$P_{U^n}(u^n) := f_{\text{sp}}(u^n|\bar{\alpha} P). \tag{6.67}$$

Using the above codebook, the transmitter satisfies the maximal power constraint of P when sending any message pair $(w_1, w_2) \in [M_1] \times [M_2]$.

6.3.3 Performance Metric

Two performance metrics are considered: *separate error probabilities (SEP)* and *joint error probability (JEP)*. SEP evaluates the reliability of each user separately:

$$\mathrm{P}^n_{\mathrm{e},1} := \Pr\{\hat{W}_1 \neq W_1 \text{ or } \bar{W}_2 \neq W_2\}, \tag{6.68}$$

$$\mathrm{P}^n_{\mathrm{e},2} := \Pr\{\hat{W}_2 \neq W_2\}, \tag{6.69}$$

while JEP evaluates the reliability of both users simultaneously:

$$\mathrm{P}^n_{\mathrm{e,j}} := \Pr\{\hat{W}_1 \neq W_1 \text{ or } \bar{W}_2 \neq W_2 \text{ or } \hat{W}_2 \neq W_2\}. \tag{6.70}$$

When the strong user only aims to decode message W_1, SEP in (6.68) and JEP (6.70) are replaced by

$$\bar{\mathrm{P}}^n_{\mathrm{e},1} := \Pr\{\hat{W}_1 \neq W_1\}, \tag{6.71}$$

$$\bar{\mathrm{P}}^n_{\mathrm{e},J} := \Pr\{\hat{W}_1 \neq W_1, \hat{W}_2 \neq W_2\}, \tag{6.72}$$

respectively. This corresponds to a symmetric BC, where each user only aims to decode one private message. The results in this section also hold for the symmetric setting. Similarly to MAC, the first- and second-order rate regions for BC characterize the performance of the mismatched code for the asymptotic and low-latency scenarios. To clarify, the definitions are given in the rest of this subsection.

Fix any $\dagger \in \{\text{JNN, SIC}\}$ and $P \in \mathbb{R}_+$. The first-order rate region under SEP is defined as follows.

Definition 6.8 A rate pair $(R_1, R_2) \in \mathbb{R}^2_+$ is said to be $(P, \dagger)$-achievable for mismatched ABC under SEP if there exists a sequence of (n, M_1, M_2)-$\dagger$ codes such that

$$\liminf_{n\to\infty} \frac{1}{n} \log M_i \geq R_i, \tag{6.73}$$

and

$$\max_{i\in[2]} \lim_{n\to\infty} \mathrm{P}^n_{\mathrm{e},i} = 0. \tag{6.74}$$

The convex closure of the set of all $(P, \dagger)$-achievable rate pairs is known as the first-order rate region and denoted by $\mathcal{R}^{\mathrm{SEP}}_{\dagger}(P)$.

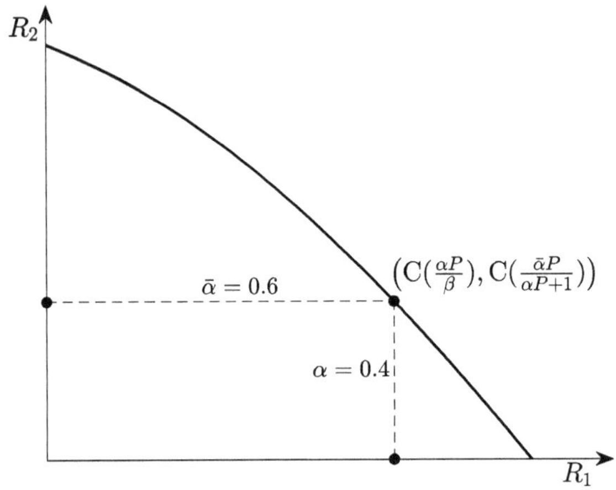

Fig. 6.11 Illustration of the first-order rate region for mismatched BC. Given a particular power allocation vector $(\alpha, \bar{\alpha} = 1 - \alpha)$, the achievable region is characterized by a rectangular. The rate region is obtained by taking the union over all $\alpha \in (0, 1)$

The first-order rate region under JEP is defined similarly except that (6.74) is replaced by

$$\lim_{n\to\infty} \mathrm{P}^n_{\mathrm{e,J}} = 0, \tag{6.75}$$

and the region is denoted by $\mathcal{R}^{\mathrm{JEP}}_{\dagger}(P)$.

A corollary of the results in this section shows that the first-order rate regions of both SIC and JNN are equal and given by the following region $\mathcal{R}_{\mathrm{inner}}(P)$. Recall that $C(P) = \frac{1}{2}\log(1 + P)$ is the capacity of an additive white Gaussian noise channel when the power constraint is P. The rate region $\mathcal{R}_{\mathrm{inner}}(P)$ is defined as follows:

$$\mathcal{R}_{\mathrm{inner}}(P) := \bigcup_{\alpha\in(0,1)} \left\{(R_1, R_2):\ R_1 \le C\Big(\frac{\alpha P}{\beta}\Big), R_2 \le C\Big(\frac{\bar{\alpha}P}{\alpha P + 1}\Big)\right\}. \tag{6.76}$$

When the noise is Gaussian, the rate region $\mathcal{R}_{\mathrm{inner}}(P)$ is optimal [11]. Figure 6.11 illustrates $\mathcal{R}_{\mathrm{inner}}(P)$. Given a particular power allocation vector $(\alpha, \bar{\alpha} = 1 - \alpha)$, the achievable rate region is characterized by a rectangle denoted with boundary denoted by dashed lines. The rate region is obtained by taking the union over all $\alpha \in (0, 1)$. Thus, any value α determines a boundary rate pair $(C(\frac{\alpha P}{\beta}), C(\frac{\bar{\alpha}P}{\alpha P+1}))$.

Analogously to MAC, the second-order rate region for BC is defined for each boundary rate pair on the first-order rate region, characterizing how the finite blocklength rates approach the asymptotic rate pairs. Fix any $\alpha \in (0, 1)$. The second-order rate region under SEP is defined as follows.

Definition 6.9 Given any $(\varepsilon_1, \varepsilon_2) \in (0, 1)^2$, a pair $(L_1, L_2) \in \mathbb{R}^2_+$ is said to be second-order $(\varepsilon_1, \varepsilon_2, \alpha, P, \dagger)$-achievable under SEP if there exists a sequence of

(n, M_1, M_2)-† codes such that

$$\limsup_{n\to\infty} \frac{1}{\sqrt{n}} \left(nC\left(\frac{\alpha P}{\beta}\right) - \log M_1 \right) \le L_1, \tag{6.77}$$

$$\limsup_{n\to\infty} \frac{1}{\sqrt{n}} \left(nC\left(\frac{\bar{\alpha} P}{\alpha P + 1}\right) - \log M_2 \right) \le L_2, \tag{6.78}$$

and for each $i \in [2]$,

$$\lim_{n\to\infty} \mathrm{P}^n_{\mathrm{e},i} \le \varepsilon_i. \tag{6.79}$$

The convex closure of the set of all second-order $(\varepsilon_1, \varepsilon_2, \alpha, P, \dagger)$-achievable rate pairs under SEP is called the second-order rate region and denoted by $\mathcal{L}_{\mathrm{SEP}}(\varepsilon_1, \varepsilon_2, \alpha, P, \dagger)$.

Given any $\varepsilon \in (0, 1)$, the second-order rate region $\mathcal{L}_{\mathrm{JEP}}(\varepsilon, \alpha, P, \dagger)$ under JEP is defined similarly except that the error probability constraint in (6.79) is replaced by

$$\lim_{n\to\infty} \mathrm{P}^n_{\mathrm{e,j}} \le \varepsilon. \tag{6.80}$$

6.3.4 Second-Order Asymptotics and Discussions

6.3.4.1 Preliminaries

The following definitions are needed. Firstly, given any two real numbers $(P_1, P_2) \in \mathbb{R}^2_+$, define two dispersion functions:

$$V_1(P_1, \beta, \zeta) := \frac{P_1^2(\zeta - \beta^2) + 4P_1\beta^3}{4\beta^2(P_1 + \beta)^2}, \tag{6.81}$$

$$V_2(P_1, P_2, \beta, \zeta) := \frac{P_1^2(\zeta - \beta^2 + 4P_2) + 4P_1(P_2 + \beta)^3}{4(P_2 + \beta)^2(P_1 + P_2 + \beta)^2}. \tag{6.82}$$

Recall that (ζ_1, ζ_2) are the fourth moment of the noise distribution in (6.58) and (6.59), respectively. For ease of notation, $V_1(\alpha P)$ and $V_2(\bar{\alpha} P, \alpha P)$ are used to denote $V_1(\alpha P, \beta, \zeta_1)$ and $V_2(\bar{\alpha} P, \alpha P, 1, \zeta_2)$, respectively. Furthermore, define the following dispersion matrix:

$$\mathbf{V}(\alpha, P) := \begin{bmatrix} V_1(\alpha P) & 0 \\ 0 & V_2(\bar{\alpha} P, \alpha P) \end{bmatrix}. \tag{6.83}$$

Recall that $\mathrm{Q}(\cdot)$ is the ccdf of a Gaussian random variable with mean zero and variance one. Finally, recall that in (6.30), given any two real numbers $(a, b) \in \mathbb{R}^2$, the set $\mathcal{A}(a, b)$ is defined as

$$\mathcal{A}(a, b) = \{(L_1, L_2) \in \mathbb{R}^2 : L_1 \geq a, L_2 \geq b\}. \tag{6.84}$$

6.3.4.2 SEP

Fix any power constraint $P \in \mathbb{R}_+$, power allocation parameter $\alpha \in (0, 1)$, and $\dagger \in$ (SIC, JNN). The achievable second-order rate region under SEP is characterized in the following theorem for mismatched ABC.

Theorem 6.5 *Given any $(\varepsilon_1, \varepsilon_2) \in (0, 1)^2$, the second-order rate region under SEP satisfies*

$$\mathcal{L}_{\mathrm{SEP}}(\varepsilon_1, \varepsilon_2, \alpha, P, \dagger) \supseteq \mathcal{A}\Big(\sqrt{V_1(\alpha P)}\mathrm{Q}^{-1}(\varepsilon_1), \sqrt{V_2(\bar{\alpha} P, \alpha P)}\mathrm{Q}^{-1}(\varepsilon_2)\Big). \tag{6.85}$$

The proof of Theorem 6.5 generalizes the proofs of mismatched channel coding [21, Theorem 1] (see also Chap. 3) and interference channel [21, Theorem 2]. Specifically, user 2 aims to decode message W_2 by treating the signal that carries the message W_1 as noise, which is equivalent to the receiver of the interference channel (cf. [21, Section III]). When SIC decoding is used, user 1 firstly decodes W_2 using TIN and NN decoding as user 2, and subsequently applies NN and SIC decoding to decode W_1, which is similar to mismatched channel coding in Chap. 3. In contrast, when JNN decoding is used by user 1, the analysis is significantly different, and a similar analysis to MAC is required.

Theorem 6.5 implies that SIC and JNN decoding achieve the same second-order rate regions. This is in stark contrast with the MAC analysis, where JNN decoding achieves strictly better low-latency performance. The reason that both SIC and JNN decoding for user 1 achieve the same second-order asymptotics is as follows. Since the same coding scheme is used by user 2, the same bound on the message size M_2 is obtained. The difference lies in the bound on M_1 due to user 1. When SIC is used, user 1 can decode W_2 reliability since the same method of TIN and NN decoding as user 2 is used, and user 1 suffers less noise power. Thus, the decoding error event of user 1 results from the event of decoding W_1 incorrectly using SIC and NN decoding, which leads to a bound on M_1. When JNN decoding is used, user 1 searches message pairs $(\bar{w}_1, \bar{w}_2) \in [M_1] \times [M_2]$ in a two-dimensional region simultaneously, which yields bounds on M_1 and M_2. However, since user 1 has better channel quality than user 2, with the constraint on M_2 required by user 2, the decoding error of message W_2 is arbitrarily small and thus the two dimensional search reduces to a one-dimensional search of $\bar{w}_1 \in [M_1]$ with probability of almost one. This way, JNN and SIC decoding lead to roughly the same performance for user 1. Since the computation complexity of JNN and SIC are $O(M_1 M_2)$ and $O(M_1 + M_2)$, respectively, it is preferred to apply SIC.

Theorem 6.5 also holds for symmetric BC, where user 1 decodes message W_1 only. This is mainly because user 1 decodes W_2 for free and thus no decoding error of W_2 occurs for user 1 when user 2 wishes to decode W_2 reliably. When specialized Gaussian noise for both users, it follows that $\zeta_1 = 3\beta^2$ and $\zeta_2 = 3$, and Theorem 6.5 bounds the second-order rate region for Gaussian BC. Such a result complements the discrete case [19, Section IV] and recovers the asymptotic result in [10, 11] as the blocklength n tends to infinity.

6.3.4.3 JEP

Recall the definition of the dispersion matrix $\mathbf{V}(\cdot)$ in (6.83). For any $\varepsilon \in (0, 1)$, recall that $\mathrm{Q_{inv}}(\mathbf{V}, \varepsilon)$ in (6.14) is the ccdf of a Gaussian random vector with zero means and covariance matrix $\mathbf{V}$.

Theorem 6.6 *Given any $\varepsilon \in (0, 1)$, the second-order rate region under JEP satisfies*

$$\mathcal{L}_{\mathrm{JEP}}(\varepsilon, \alpha, P, \dagger) \supseteq \mathrm{Q_{inv}}(\mathbf{V}(\alpha, P), \varepsilon). \tag{6.86}$$

The proof of Theorem 6.6 is analogous to the proof of MAC in Theorems 6.1 and 6.2. Specifically, the random coding union bound [26, Theorem 16] is generalized to BC under both JNN and SIC decoding. Subsequently, the function version Berry-Esseen theorem (cf. Theorem 1.8) and concentration inequalities are used to further upper bound the error probabilities.

6.3.5 *Discussions*

This section compares the second-order rate regions under SEP and JEP. Fix any $\varepsilon \in (0, 1)$. First, consider extreme cases. When $L_1 \to \infty$, the achievable second-order rate region under JEP in (6.86) reduces to

$$\left\{(L_1, L_2) : L_2 \geq \sqrt{V_2(\bar{\alpha}P, \alpha P)}\mathrm{Q}^{-1}(\varepsilon)\right\}. \tag{6.87}$$

Similarly, when $L_2 \to \infty$, the achievable second-order rate region under JEP in (6.86) reduces to

$$\left\{(L_1, L_2) : L_1 \geq \sqrt{V_1(\alpha P)}\mathrm{Q}^{-1}(\varepsilon)\right\}. \tag{6.88}$$

Consider any $(\varepsilon_1, \varepsilon_2) \in (0, 1)^2$ such that $\varepsilon_1 + \varepsilon_2 \leq \varepsilon$. It follows from Theorem 6.5 that when either L_1 or L_2 tends to infinity, the achievable second-order regions under SEP and JEP are the same. Figure 6.12 numerically illustrates this fact for large values of L_1 or L_2.

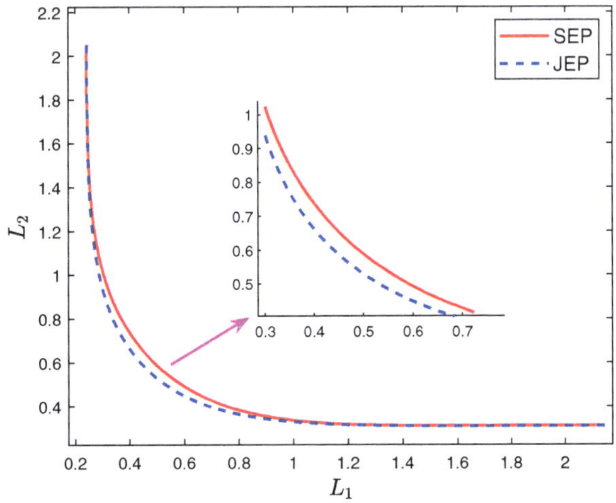

Fig. 6.12 Plot of achievable second-order rate regions under SEP and JEP when $P = 5$, $\alpha = 0.3$, $\beta = 0.6$, $\zeta_1 = 0.3888$, $\zeta_2 = 3$, and $\varepsilon = 0.3$

Next consider non-degenerate cases where (L_1, L_2) are both finite. The dispersion matrix $\mathbf{V}(\alpha, P)$ in (6.83) is diagonal. One might thus believe the second-order rates of users 1 and 2 are independent. However, since L_1 and L_2 are also determined by the joint error probability ε, a natural tradeoff exists for L_1 and L_2. To clarify, Fig. 6.12 plots the achievable second-order rate regions under both SEP and JEP. Under SEP, the sum of error probabilities of users 1 and 2 is constrained $\varepsilon_1 + \varepsilon_2 \leq \varepsilon = 0.3$. It is observed from Fig. 6.12 that the achievable second-order rate regions under SEP and JEP are roughly the same, which implies that both performance criteria lead to the same low-latency performance. Such a result is in contrast with the multiterminal compression problem in Chap. 5.

6.4 Proofs for Multiple Access Channel

6.4.1 Preliminaries

This subsection presents definitions and key lemmas for the proof of Theorems 6.1 and 6.2 concerning low-latency performance of mismatched MAC with JNN and SIC decoding, respectively.

The following definitions are needed. Given any $(x_1^n, x_2^n, y^n) \in (\mathbb{R}^n)^3$, for each $j \in [2]$, define the following mismatched (conditional) mutual information densities

$$\tilde{\imath}(x_1^n; y^n) := nC\left(\frac{P_1}{1+P_2}\right) + \frac{\|y^n\|^2}{2(1+P_1+P_2)} - \frac{\|y^n - x_1^n\|^2}{2(1+P_2)}, \tag{6.89}$$

$$\tilde{\imath}_j^n(x_j^n; y^n|x_{3-j}^n) := nC(P_j) + \frac{\|y^n - x_{3-j}^n\|^2}{2(1+P_j)} - \frac{\|y^n - x_1^n - x_2^n\|^2}{2}, \tag{6.90}$$

$$\tilde{\imath}_{1,2}^n(x_1^n, x_2^n; y^n) := nC(P_1 + P_2) + \frac{\|y^n\|^2}{2(1 + P_1 + P_2)} - \frac{\|y^n - x_1^n - x_2^n\|^2}{2}. \tag{6.91}$$

With the above definitions, the following random coding union (RCU) bound holds for JNN decoding of mismatched MAC.

Lemma 6.1 (RCU Bound for Mismatched MAC with JNN Decoder)
There exists an (n, M_1, M_2, P_1, P_2)-JNN code such that the ensemble average probability satisfies

$$\begin{aligned}
\mathrm{P}_\mathrm{e}^n \leq \mathbb{E}\Big[\min\Big\{1, &(M_1 - 1)\Pr\{\tilde{\imath}_1^n(\bar{X}_1^n; Y^n|X_2^n) \geq \tilde{\imath}_1^n(X_1^n; Y^n|X_2^n)|X_1^n, X_2^n, Y^n\} \\
&+ (M_2 - 1)\Pr\{\tilde{\imath}_2^n(\bar{X}_2^n; Y^n|X_1^n) \geq \tilde{\imath}_2^n(X_2^n; Y^n|X_1^n)|X_1^n, X_2^n, Y^n\} \\
&+ (M_1 - 1)(M_2 - 1)\Pr\{\tilde{\imath}_{1,2}^n(\bar{X}_1^n, \bar{X}_2^n; Y^n) \geq \tilde{\imath}_{1,2}^n(X_1^n, X_2^n; Y^n)|X_1^n, X_2^n, Y^n\}\Big\}\Big],
\end{aligned} \tag{6.92}$$

where $(X_1^n, \bar{X}_1^n, X_2^n, \bar{X}_2^n, Y^n) \sim P_{X_1^n} P_{X_1^n} P_{X_2^n} P_{X_2^n} P_{Y^n|X_1^n X_2^n}$, $P_{X_1^n}$ and $P_{X_2^n}$ are the channel input distribution in (6.6) *and $P_{Y^n|X_1^n X_2^n}$ is induced by the additive noise channel $Y^n = X_1^n + X_2^n + Z^n$.*

Proof Recall the definition of JNN code in Definition 6.1. A decoding error event occurs if one of the following three events occurs:

- $\mathcal{E}_1 : \{\hat{W}_1 \neq W_1, \hat{W}_2 = W_2\}$,
- $\mathcal{E}_2 : \{\hat{W}_1 = W_1, \hat{W}_2 \neq W_2\}$,
- $\mathcal{E}_{1,2} : \{\hat{W}_1 \neq W_1, \hat{W}_2 \neq W_2\}$.

It follows that

$$\mathrm{P}_\mathrm{e}^n = \Pr\left\{\mathcal{E}_1 \cup \mathcal{E}_2 \cup \mathcal{E}_{1,2}\right\} \tag{6.93}$$

$$= \Pr\{\mathcal{E}_1\} + \Pr\{\mathcal{E}_2\} + \Pr\left\{\mathcal{E}_{1,2}\right\}. \tag{6.94}$$

Subsequent analyses bound each term of (6.94). The crux is to use the equivalence between the JNN decoder and the maximum mismatched mutual information density decoder for each error event. Let $(w_1, w_2) \in [M]^2$ be the messages transmitted by the two users. Given that $\mathcal{E}_1$ happens, i.e., the message w_2 is decoded correctly, the JNN decoder in (6.3) satisfies

$$\Phi(Y^n) = \operatorname*{arg\,min}_{\bar{w}_1 \in [M_1]} \|Y^n - X_1^n(\bar{w}_1) - X_2^n(w_2)\|^2 \tag{6.95}$$

$$= \operatorname*{arg\,max}_{\bar{w}_1 \in [M_1]} nC(P_1) + \frac{\|Y^n - X_2^n(w_2)\|^2}{2(1 + P_1)} - \frac{\|Y^n - X_1^n(\bar{w}_1) - X_2^n(w_2)\|^2}{2} \tag{6.96}$$

$$= \arg\max_{\bar{w}_1 \in [M_1]} \tilde{\imath}_1^n(X_1^n(\bar{w}_1); Y^n | X_2^n(w_2)), \tag{6.97}$$

where (6.96) follows since the additive terms are constants when Y^n and w_2 are known. The result in (6.97) implies that under error event $\mathcal{E}_1$, maximizing the mismatched information density $\tilde{\imath}_1^n(X_1^n(\bar{w}_1); Y^n|X_2^n(w_2))$ over $\bar{w}_1 \in [M_1]$ is equivalent to JNN decoding. Analogously, maximizing the mismatched information density $\tilde{\imath}_2^n(X_2^n(\bar{w}_2); Y^n|X_1^n(w_1))$ over $\bar{w}_2 \in [M_2]$ is equivalent to JNN decoding under error event $\mathcal{E}_2$.

Given that error event $\mathcal{E}_{1,2}$ happens, the JNN decoder satisfies

$$\Phi(Y^n) = \arg\min_{(\bar{w}_1, \bar{w}_2) \in [M_1] \times [M_2]} \|Y^n - X_1^n(\bar{w}_1) - X_2^n(\bar{w}_2)\|^2 \tag{6.98}$$

$$= \arg\max_{(\bar{w}_1, \bar{w}_2) \in [M_1] \times [M_2]} nC(P_1 + P_2) + \frac{\|Y^n\|^2}{2(1 + P_1 + P_2)} - \frac{\|Y^n - X_1^n(\bar{w}_1) - X_2^n(\bar{w}_2)\|^2}{2} \tag{6.99}$$

$$= \arg\max_{(\bar{w}_1, \bar{w}_2) \in [M_1] \times [M_2]} \tilde{\imath}_{1,2}^n(X_1^n(\bar{w}_1), X_2^n(\bar{w}_2); Y^n). \tag{6.100}$$

The result in (6.100) implies that under error event $\mathcal{E}_{1,2}$, maximizing the mismatched information density $\tilde{\imath}_{1,2}^n(X_1^n(\bar{w}_1), X_2^n(\bar{w}_2); Y^n)$ over all message pairs $(\bar{w}_1, \bar{w}_2) \in [M_1] \times [M_2]$ is equivalent to JNN decoding.

For ease of notation, given any $(\bar{w}_1, \bar{w}_2) \in [M_1] \times [M_2]$, define the events

$$\mathcal{F}(\bar{w}_1, \bar{w}_2) := \{\|Y^n - X_1^n(\bar{w}_1) - X_2^n(\bar{w}_2)\|^2 \leq \|Y^n - X_1^n(w_1) - X_2^n(w_2)\|^2\}, \tag{6.101}$$

$$\mathcal{G}_1(\bar{w}_1) := \{\tilde{\imath}_1^n(X_1^n(\bar{w}_1); Y^n|X_2^n(w_2)) \geq \tilde{\imath}_1^n(X_1^n(w_1); Y^n|X_2^n(w_2))\}, \tag{6.102}$$

$$\mathcal{G}_2(\bar{w}_2) := \{\tilde{\imath}_2^n(X_2^n(\bar{w}_2); Y^n|X_1^n(w_1)) \geq \tilde{\imath}_2^n(X_2^n(w_2); Y^n|X_1^n(w_1))\}, \tag{6.103}$$

$$\mathcal{G}_{1,2}(\bar{w}_1, \bar{w}_2) := \{\tilde{\imath}_{1,2}^n(X_1^n(\bar{w}_1), X_2^n(\bar{w}_2); Y^n) \geq \tilde{\imath}_{1,2}^n(X_1^n(w_1), X_2^n(w_2); Y^n)\}. \tag{6.104}$$

It follows from (6.94), (6.97), and (6.100) that the conditional error probability given that messages (w_1, w_2) are transmitted satisfies

$$
\begin{aligned}
&\mathrm{P}_{\mathrm{e}}^n(w_1, w_2) \\
&= \Pr\Bigg\{\bigcup_{\substack{\bar{w}_1\in[M_1]:\\ \bar{w}_1\neq w_1}} \mathcal{F}(\bar{w}_1, w_2)\Bigg\} + \Pr\Bigg\{\bigcup_{\substack{\bar{w}_2\in[M_2]:\\ \bar{w}_2\neq w_2}} \mathcal{F}(w_1, \bar{w}_2)\Bigg\} \\
&\quad + \Pr\Bigg\{\bigcup_{\substack{(\bar{w}_1,\bar{w}_2)\in[M_1]\times[M_2]:\\ \bar{w}_1\neq w_1,\bar{w}_2\neq w_2}} \mathcal{F}(\bar{w}_1, \bar{w}_2)\Bigg\} && (6.105) \\
&= \Pr\Bigg\{\bigcup_{\substack{\bar{w}_1\in[M_1]:\\ \bar{w}_1\neq w_1}} \mathcal{G}_1(\bar{w}_1)\Bigg\} + \Pr\Bigg\{\bigcup_{\substack{\bar{w}_2\in[M_2]:\\ \bar{w}_2\neq w_2}} \mathcal{G}_2(\bar{w}_2)\Bigg\} && (6.106) \\
&\quad + \Pr\Bigg\{\bigcup_{\substack{(\bar{w}_1,\bar{w}_2)\in[M_1]\times[M_2]:\\ \bar{w}_1\neq w_1,\bar{w}_2\neq w_2}} \mathcal{G}_{1,2}(\bar{w}_1, \bar{w}_2)\Bigg\} && (6.107) \\
&= \mathbb{E}\Big[(M_1-1)\Pr\{\tilde{\imath}_1^n(\bar{X}_1^n; Y^n|X_2^n) \geq \tilde{\imath}_1^n(X_1^n; Y^n|X_2^n)|X_1^n, X_2^n, Y^n\} \\
&\quad + (M_2-1)\Pr\{\tilde{\imath}_2^n(\bar{X}_2^n; Y^n|X_1^n) \geq \tilde{\imath}_2^n(X_2^n; Y^n|X_1^n)|X_1^n, X_2^n, Y^n\} \\
&\quad + (M_1-1)(M_2-1)\Pr\{\tilde{\imath}_{1,2}^n(\bar{X}_1^n, \bar{X}_2^n; Y^n) \\
&\quad \geq \tilde{\imath}_{1,2}^n(X_1^n, X_2^n; Y^n)|X_1^n, X_2^n, Y^n\}\Big], && (6.108)
\end{aligned}
$$

where (6.105) follows from the definition of JNN decoder and three error events, (6.107) follows due to the equivalence between JNN decoding and maximum mismatched information density decoding, and (6.108) follows from the union bound and the fact given any $(w_1, \bar{w}_1, w_2, \bar{w}_2) \in [M_1]^2 \times [M_2]^2$, the joint distribution of random variables $(X_1^n(w_1), X_1^n(\bar{w}_1), X_2^n(w_2), X_2^n(\bar{w}_2), Y^n)$ is the same and given by $P_{X_1^n} P_{X_1^n} P_{X_2^n} P_{X_2^n} P_{Y^n|X_1^n X_2^n}$.

The proof of Lemma 6.1 by using (6.107) is completed by noting that $\mathrm{P}_{\mathrm{e}}^n(w_1, w_2) \leq 1$ for any $(w_1, w_2) \in [M_1] \times [M_2]$ and thus

$$
\mathrm{P}_{\mathrm{e}}^n = \mathbb{E}[\mathrm{P}_{\mathrm{e}}^n(W_1, W_2)] \leq \mathbb{E}[\min\{1, \mathrm{P}_{\mathrm{e}}^n(W_1, W_2)\}]. \tag{6.109}
$$

□

Analogously, the following RCU bound holds for SIC decoding.

Lemma 6.2 (RCU Bound for Mismatched MAC with SIC Decoder) *There exists an (n, M_1, M_2, P_1, P_2)-SIC code such that the ensemble average probability satisfies*

$$
\begin{aligned}
\mathrm{P}_{\mathrm{e}}^n \leq \mathbb{E}\Big[\min\Big\{1, (M_1-1)\Pr\{\tilde{\imath}^n(\bar{X}_1^n; Y^n) \geq \tilde{\imath}^n(X_1^n; Y^n)|X_1^n, Y^n\}\Big\}\Big] \\
+\mathbb{E}\Big[\min\Big\{1, (M_2-1)\Pr\{\tilde{\imath}_2^n(\bar{X}_2^n; Y^n|X_1^n) \geq \tilde{\imath}_2^n(X_2^n; Y^n|X_1^n)|X_1^n, X_2^n, Y^n\}\Big\}\Big],
\end{aligned}
\tag{6.110}
$$

where $(X_1^n, \bar{X}_1^n, X_2^n, \bar{X}_2^n, Y^n) \sim P_{X_1^n} P_{X_1^n} P_{X_2^n} P_{X_2^n} P_{Y^n|X_1^n X_2^n}$.

Given any (x_1^n, x_2^n, y^n) and any $t \in \mathbb{R}_+$, define the following probability functions

$$g_1(t; y^n, x_2^n) := \Pr\left\{\tilde{\imath}_1^n(\bar{X}_1^n; y^n|x_2^n) \geq t\right\}, \tag{6.111}$$

$$g_2(t; y^n, x_1^n) := \Pr\left\{\tilde{\imath}_2^n(\bar{X}_2^n; y^n|x_1^n) \geq t\right\}, \tag{6.112}$$

$$g_{1,2}(t; y^n) := \Pr\left\{\tilde{\imath}_{1,2}^n(\bar{X}_1^n, \bar{X}_2^n; y^n) \geq t\right\}, \tag{6.113}$$

$$g(t; y^n) := \Pr\{\tilde{\imath}_1^n(\bar{X}_1^n; y^n) \geq t\}, \tag{6.114}$$

where $\bar{X}_j^n$ is distributed according to $P_{X_j^n}$ in (6.6). Each probability term concerns a mismatched (conditional) mutual information density.

For subsequent analyses, define the following constants that depend on (P_1, P_2):

$$\beta_i := \frac{27\sqrt{\pi}(1+P_i)}{\sqrt{8(1+2P_i)}}, \quad i \in [2], \tag{6.115}$$

$$\beta_{1,2} := \frac{9(P_1+P_2)}{2\pi\sqrt{2P_1P_2}}. \tag{6.116}$$

The following lemma upper bounds the above $g(\cdot)$ functions. The proof follows similarly to Lemma 3.2.

Lemma 6.3 *Given any positive real number $t \in \mathbb{R}_+$,*

$$g_1(t; y^n, x_2^n) \leq \beta_1 \exp(-t), \tag{6.117}$$

$$g_2(t; y^n, x_1^n) \leq \beta_2 \exp(-t), \tag{6.118}$$

$$g_{1,2}(t; y^n) \leq \beta_{1,2} \exp(-t), \tag{6.119}$$

$$g(t; y^n) \leq \beta_{1,2} \exp(-t). \tag{6.120}$$

6.4.2 JNN Decoding

This section proves Theorem 6.1. Using the channel law $Y^n = X_1^n + X_n^2 + Z^n$, it follows from (6.89) that

$$\tilde{\imath}_1 := \tilde{\imath}_1^n(X_1^n; Y^n|X_n^2) - nC(P_1) = \frac{\|X_1^n + Z^n\|^2}{2(1+P_1)} - \frac{\|Z^n\|^2}{2} \tag{6.121}$$

$$= \frac{nP_1 + \|Z^n\|^2 + 2\langle X_1^n, Z^n\rangle}{2(1+P_1)} - \frac{\|Z^n\|^2}{2} \tag{6.122}$$

$$= \frac{(n - \|Z^n\|^2)P_1 + 2\langle X_1^n, Z^n\rangle}{2(1+P_1)}. \tag{6.123}$$

Similarly, it follows from (6.91) and (6.90) that

$$\tilde{\imath}_2 := \tilde{\imath}_2^n(X_2^n; Y^n|X_1^n) - nC(P_2) = \frac{(n - \|Z^n\|^2)P_2 + 2\langle X_2^n, Z^n\rangle}{2(1+P_2)}, \tag{6.124}$$

$$\tilde{\imath}_{1,2} := \tilde{\imath}_{1,2}^n(X_1^n, X_2^n; Y^n) - nC(P_1 + P_2) \tag{6.125}$$

$$= \frac{(n - \|Z^n\|^2)(P_1 + P_2) + 2\langle X_1^n + X_2^n, Z^n\rangle + 2\langle X_1^n, X_2^n\rangle}{2(1+P_1+P_2)}. \tag{6.126}$$

Recall that $\beta_1, \beta_2, \beta_{1,2}$ are constants in Lemma 6.120. Define the event

$$\mathcal{S} := \left\{ \begin{bmatrix} \tilde{\imath}_1^n(X_1^n; Y^n|X_2^n) \\ \tilde{\imath}_2^n(X_2^n; Y^n|X_1^n) \\ \tilde{\imath}_{1,2}^n(X_1^n, X_2^n; Y^n) \end{bmatrix} \geq \log \begin{bmatrix} 4M_1\beta_1\sqrt{n} \\ 4M_2\beta_2\sqrt{n} \\ 2M_1M_2\beta_{1,2}\sqrt{n} \end{bmatrix} \right\}. \tag{6.127}$$

For ease of notation, given random variables $(X_1^n, X_2^n, Y^n) \sim P_{X_1^n}P_{X_2^n}P_{Y^n|X_1^nX_2^n}$ in Lemma 6.1, define the following random variables

$$G_1 := g_1(\tilde{\imath}_1(X_1^n; Y^n|X_2^n); Y^n, X_2^n), \tag{6.128}$$

$$G_2 := g_2(\tilde{\imath}_2(X_2^n; Y^n|X_1^n); Y^n, X_1^n), \tag{6.129}$$

$$G_{1,2} := g_{1,2}(\tilde{\imath}_{1,2}^n(X_1^n, X_2^n; Y^n); Y^n). \tag{6.130}$$

It follows from Lemma 6.1 that the ensemble average error probability satisfies

$$\mathrm{P}_\mathrm{e}^n \leq \mathbb{E}[\min\{1, (M_1-1)G_1 + (M_2-1)G_2 + (M_1-1)(M_2-1)G_{1,2}\}] \tag{6.131}$$

$$\leq \mathbb{E}[\min\{1, (M_1 - 1)G_1 + (M_2 - 1)G_2 + (M_1 - 1)(M_2 - 1)G_{1,2}\}\mathbb{1}(\mathcal{S}^c)]$$
$$+ \mathbb{E}[\min\{1, (M_1 - 1)G_1 + (M_2 - 1)G_2 + (M_1 - 1)(M_2 - 1)G_{1,2}\}\mathbb{1}(\mathcal{S})] \tag{6.132}$$

$$\leq \Pr\{\mathcal{S}^c\} + \mathbb{E}[\{(M_1 - 1)G_1 + (M_2 - 1)G_2 + (M_1 - 1)(M_2 - 1)G_{1,2}\}\mathbb{1}(\mathcal{S})] \tag{6.133}$$

$$\leq \Pr\{\mathcal{S}^c\} + M_1\mathbb{E}[G_1\mathbb{1}(\tilde{\imath}_1(X_1^n; Y^n|X_2^n) \geq \log(4M_1\beta_1\sqrt{n}))]$$
$$+ M_2\mathbb{E}[G_2\mathbb{1}(\tilde{\imath}_2(X_2^n; Y^n|X_1^n) \geq \log(4M_2\beta_2\sqrt{n}))]$$
$$+ M_1M_2\mathbb{E}[G_{1,2}\mathbb{1}(\tilde{\imath}_{1,2}(X_1^n, X_2^n; Y^n) \geq \log(2M_1M_2\beta_{1,2}\sqrt{n}))], \tag{6.134}$$

where (6.131) follows since the newly defined random variables G_1 G_2 and $G_{1,2}$ correspond to three terms in right-hand side of the RCU bound in (6.92), (6.134) follows from the union bound and the fact that $\mathcal{S}$ is the intersection of three events.

Using Lemma 6.3, it follows that

$$M_1\mathbb{E}[G_1\mathbb{1}(\tilde{\imath}_1(X_1^n; Y^n|X_2^n) \geq \log(4M_1\beta_1\sqrt{n}))]$$
$$\leq \beta_1 M_1\mathbb{E}\big[\exp\{-\tilde{\imath}_1(X_1^n; Y^n|X_2^n)\}\mathbb{1}\big(\tilde{\imath}_1(X_1^n; Y^n|X_2^n) \geq \log(4M_1\beta_1\sqrt{n})\big)\big] \tag{6.135}$$

$$\leq \beta_1 M_1 \exp(-\log(4M_1\beta_1\sqrt{n})) \tag{6.136}$$

$$= \frac{1}{4\sqrt{n}}. \tag{6.137}$$

Similarly,

$$M_2\mathbb{E}[G_2\mathbb{1}(\tilde{\imath}_2(X_2^n; Y^n|X_1^n) \geq \log(4M_2\beta_2\sqrt{n}))] \leq \frac{1}{4\sqrt{n}}, \tag{6.138}$$

$$M_1M_2\mathbb{E}[G_{1,2}\mathbb{1}(\tilde{\imath}_{1,2}(X_1^n, X_2^n; Y^n) \geq \log(2M_1M_2\beta_{1,2}\sqrt{n}))] \leq \frac{1}{2\sqrt{n}}. \tag{6.139}$$

It remains to bound the probability $\Pr\{\mathcal{S}^C\}$. It follows from the definition of $\mathcal{S}$ in (6.127) that

$$\Pr\{\mathcal{S}^c\} \leq 1 - \Pr\left\{\begin{bmatrix} \tilde{\imath}_1 \\ \tilde{\imath}_2 \\ \tilde{\imath}_{1,2} \end{bmatrix} \geq \log\begin{bmatrix} 4M_1\beta_1\sqrt{n} \\ 4M_2\beta_2\sqrt{n} \\ 2M_1M_2\beta_{1,2}\sqrt{n} \end{bmatrix} - n\mathbf{C}(P_1, P_2)\right\}. \tag{6.140}$$

The rest of this section bounds (6.140) using the function version of the Berry-Esseen theorem (cf. Theorem 1.8). For ease of notation, let

$$\mathbf{i} := [\tilde{\imath}_1; \tilde{\imath}_2; \tilde{\imath}_3], \tag{6.141}$$

$$\boldsymbol{\tau} := \log \begin{bmatrix} 4M_1\beta_1\sqrt{n} \\ 4M_2\beta_2\sqrt{n} \\ 2M_1M_2\beta_{1,2}\sqrt{n} \end{bmatrix} - n\mathbf{C}(P_1, P_2). \tag{6.142}$$

For each $j \in [2]$, let $\tilde{X}_j^n \sim \mathcal{N}(\cdot; \mathbf{0}_n, \mathbf{1}_n)$. It follows from [16, Eq. (57)] that spherical codewords X_1^n and X_2^n can be expressed as functions of i.i.d. random variables. Specifically, for each $(i, j) \subset [n] \times [2]$,

$$X_{ji} = \sqrt{nP_j}\frac{\tilde{X}_{ji}}{\|\tilde{X}_j^n\|}. \tag{6.143}$$

For each $i \in [n]$, define the following random variables:

$$\begin{aligned} A_{1i} &:= 1 - Z_i^2, \quad A_{2i} := \sqrt{P_1}\tilde{X}_{1i}Z_i, \; A_{3i} := \tilde{X}_{1i}^2 - 1, \\ A_{4i} &:= \sqrt{P_2}\tilde{X}_{2i}Z_i, \; A_{5i} := \tilde{X}_{2i}^2 - 1, \quad A_{6i} := \sqrt{P_1P_2}\tilde{X}_{1i}\tilde{X}_{2i}, \end{aligned} \tag{6.144}$$

and let $\mathbf{A}_i := (A_{1i}, \ldots, A_{6i})$. Note that $\mathbf{A}_{[n]} := (\mathbf{A}_1, \ldots, \mathbf{A}_n)$ are zero-mean i.i.d. random vectors with finite third moments. Furthermore, for each $i \in [n]$, the covariance matrix of $\mathbf{A}_i$ is

$$\mathsf{Cov}(\mathbf{A}_i) = \mathrm{diag}[\xi - 1, P_1, 2, P_2, 2, P_1P_2]. \tag{6.145}$$

Given any real numbers $\mathbf{k} := (k_1, \ldots, k_6) \in \mathbb{R}^6$, define three functions:

$$f_1(\mathbf{k}) := \frac{1}{2(1+P_1)}\left(P_1k_1 + \frac{2k_2}{\sqrt{1+k_3}}\right), \tag{6.146}$$

$$f_2(\mathbf{k}) := \frac{1}{2(1+P_2)}\left(P_2k_1 + \frac{2k_4}{\sqrt{1+k_5}}\right), \tag{6.147}$$

$$\begin{aligned} f_3(\mathbf{k}) &:= \frac{1}{2(1+P_1+P_2)} \\ &\times \left((P_1+P_2)k_1 + \frac{2k_2}{\sqrt{1+k_3}} + \frac{2k_4}{\sqrt{1+k_5}} + \frac{2k_6}{\sqrt{(1+k_3)(1+k_5)}}\right), \end{aligned} \tag{6.148}$$

and let $\mathbf{f}(\mathbf{k}) := [f_1(\mathbf{k}), f_2(\mathbf{k}), f_3(\mathbf{k})]^{\mathrm{T}}$ denote the vector of functions. It follows from above definitions that $\mathbf{f}\big(\frac{1}{n}\sum_{i=1}^n \mathbf{A}_i\big) = \frac{1}{n}\mathbf{i}$.

The Jacobian matrix of the function vector $(\cdot)$ around $\mathbb{E}[\mathbf{A}_1] = \mathbf{0}_6$ satisfies

$$\mathbf{J}(\mathbf{A}_1) = \begin{bmatrix} \frac{P_1}{2(1+P_1)} & \frac{1}{1+P_1} & 0 & 0 & 0 & 0 \\ \frac{P_2}{2(1+P_2)} & 0 & 0 & \frac{1}{1+P_2} & 0 & 0 \\ \frac{P_1+P_2}{2(1+P_1+P_2)} & \frac{1}{1+P_1+P_2} & 0 & \frac{1}{1+P_1+P_2} & 0 & \frac{1}{1+P_1+P_2} \end{bmatrix}. \tag{6.149}$$

Recall the definition of dispersion matrix $\mathbf{V}(P_1, P_2)$ in (6.22). It follows that for each $i \in [n]$,

$$\mathbf{J}(\mathbf{A}_i)\,\mathsf{Cov}(\mathbf{A}_i)\mathbf{J}(\mathbf{A}_i)^{\mathrm{T}}$$

$$= \begin{bmatrix} \frac{(\xi-1)P_1^2+4P_1}{4(1+P_1)^2} & \frac{(\xi-1)P_1P_2}{4(1+P_1)(1+P_2)} & \frac{(\xi-1)P_1(P_1+P_2)+4P_1}{4(1+P_1)(1+P_1+P_2)} \\ \frac{(\xi-1)P_1P_2}{4(1+P_1)(1+P_2)} & \frac{(\xi-1)P_2^2+4P_2}{4(1+P_2)^2} & \frac{(\xi-1)P_2(P_1+P_2)+4P_2}{4(1+P_2)(1+P_1+P_2)} \\ \frac{(\xi-1)P_1(P_1+P_2)+4P_1}{4(1+P_1)(1+P_1+P_2)} & \frac{(\xi-1)P_2(P_1+P_2)+4P_2}{4(1+P_2)(1+P_1+P_2)} & \frac{(\xi-1)(P_1+P_2)^2+4(P_1+P_2+P_1P_2)}{4(1+P_1+P_2)^2} \end{bmatrix} \tag{6.150}$$

$$= \mathbf{V}(P_1, P_2). \tag{6.151}$$

Let $\mathbf{S}_3 := [S_1, S_2, S_3]^{\mathrm{T}} \sim \mathcal{N}(\cdot; \mathbf{0}_3, \mathbf{V}(P_1, P_2))$ denote the three-dimensional Gaussian random vector with mean zero and covariance matrix $\mathbf{V}(P_1, P_2)$. It follows from (6.140) and Theorem 1.8 that

$$\Pr\{\mathcal{S}^C\} \le 1 - \Pr\left\{ f\left(\frac{1}{n}\sum_{i=1}^{n}\mathbf{A}_i\right) \ge \frac{\boldsymbol{\tau}}{n} \right\} \tag{6.152}$$

$$\le 1 - \Pr\left\{\mathbf{S}_3 \ge \boldsymbol{\tau}/\sqrt{n}\right\} + O(n^{-1/4}). \tag{6.153}$$

Combining (6.134)(6.137) to (6.139), the error probability of the proposed $(\mathbf{n}, M, P)$-code satisfies

$$\mathrm{P}_{\mathrm{e}}^n \le \Pr\{\mathcal{S}^C\} + O(n^{-1/4}) + \frac{1}{\sqrt{n}} \tag{6.154}$$

$$= 1 - \Pr\left\{\mathbf{S}_3 \ge \boldsymbol{\tau}/\sqrt{n}\right\} + O(n^{-1/4}). \tag{6.155}$$

The rest of the proof analyzes various cases of boundary rate pairs. Fix a rate pair (R_1^*, R_2^*) on the boundary of $\mathcal{R}_{\mathrm{JNN}}$ (cf. (6.10)). To bound the second-order rate region $\mathcal{L}_{\mathrm{JNN}}$ (cf. Definition 6.4), given real numbers $(L_1, L_2) \in \mathbb{R}^2$, let

$$\log M_1 = nR_1^* - \sqrt{n}L_1, \tag{6.156}$$

$$\log M_2 = nR_2^* - \sqrt{n}L_2. \tag{6.157}$$

It follows from (6.142) and (6.153) that

$$\limsup_{n\to\infty} \mathrm{P}_{\mathrm{e}}^n$$

$$\le 1 - \liminf_{n\to\infty} \Pr\{\mathbf{S}_3 \ge \boldsymbol{\tau}/\sqrt{n}\} \tag{6.158}$$

$$= 1 - \liminf_{n\to\infty} \Pr\left\{ \begin{bmatrix} S_1 \\ S_2 \\ S_3 \end{bmatrix} \geq \begin{bmatrix} \sqrt{n}(R_1^* - C(P_1)) - L_1 \\ \sqrt{n}(R_2^* - C(P_2)) - L_2 \\ \sqrt{n}(R_1^* + R_2^* - C(P_1 + P_2)) - (L_1 + L_2) \end{bmatrix} \right\}. \tag{6.159}$$

For subsequent analysis, define the following three events

$$\mathcal{A}_1 := \{S_1 \geq \sqrt{n}(R_1^* - C(P_1)) - L_1\}, \tag{6.160}$$

$$\mathcal{A}_2 := \{S_2 \geq \sqrt{n}(R_2^* - C(P_2)) - L_2\}, \tag{6.161}$$

$$\mathcal{A}_3 := \{S_3 \geq \sqrt{n}(R_1^* + R_2^* - C(P_1 + P_2)) - (L_1 + L_2)\}. \tag{6.162}$$

With the above definition, (6.159) is equivalent to

$$\limsup_{n\to\infty} \mathrm{P}_\mathrm{e}^n \leq 1 - \liminf_{n\to\infty} \Pr\{\mathcal{A}_1 \cap \mathcal{A}_2 \cap \mathcal{A}_3\}. \tag{6.163}$$

Fix $\alpha \in (0, 1)$. First consider case (i) where $(R_1^*, R_2^*) = (\alpha C(P_1/(1 + P_2)), C(P_2))$. It follows that

$$R_1^* < C(P_1), \tag{6.164}$$

$$R_1^* + R_2^* < C(P_1/(1 + P_2)) + C(P_2) = C(P_1 + P_2), \tag{6.165}$$

and thus

$$\lim_{n\to\infty} \sqrt{n}(R_1^* - C(P_1)) - L_1 \to -\infty, \tag{6.166}$$

$$\lim_{n\to\infty} \sqrt{n}(R_1^* + R_2^* - C(P_1 + P_2)) - (L_1 + L_2) \to -\infty. \tag{6.167}$$

As a result,

$$\lim_{n\to\infty} \Pr\{\mathcal{A}_1\} = \lim_{n\to\infty} \Pr\{\mathcal{A}_3\} = 1. \tag{6.168}$$

This way, the right-hand side of (6.163) is further upper bounded as

$$\begin{aligned}
&1 - \liminf_{n\to\infty} \Pr\{\mathcal{A}_1 \cap \mathcal{A}_2 \cap \mathcal{A}_3\} \\
&\leq 1 - \liminf_{n\to\infty} \left(\Pr\{\mathcal{A}_2\} - \Pr\{\mathcal{A}_1^C\} - \Pr\{\mathcal{A}_3^C\} \right) && (6.169) \\
&= 1 - \Pr\{S_2 \geq -L_2\} && (6.170) \\
&= \Pr\{S_2 \geq L_2\}, && (6.171)
\end{aligned}$$

where (6.169) follows since for

$$\Pr\{\mathcal{A}_1 \cap \mathcal{A}_2 \cap \mathcal{A}_3\} = \Pr\left\{\mathcal{A}_2 \cap (\mathcal{A}_1 \cap \mathcal{A}_3)\right\} \tag{6.172}$$
$$\geq \Pr\{\mathcal{A}_2\} - \Pr\{\mathcal{A}_2 \cap (\mathcal{A}_1 \cap \mathcal{A}_3)^C\} \tag{6.173}$$
$$\geq \Pr\{\mathcal{A}_2\} - \Pr\{(\mathcal{A}_1 \cap \mathcal{A}_3)^C\} \tag{6.174}$$
$$= \Pr\{\mathcal{A}_2\} - \Pr\{\mathcal{A}_1^C \cup \mathcal{A}_3^C\} \tag{6.175}$$
$$\geq \Pr\{\mathcal{A}_2\} - \Pr\{\mathcal{A}^C\} - \Pr\{\mathcal{A}_3^C\}, \tag{6.176}$$

(6.170) follows since $R_2^* = C(P_2)$ for case (i), and (6.171) follows since $S_2 \sim \mathcal{N}(0, V(P_2))$ is a zero-mean Gaussian random variable. To make (6.171) less than ε, one requires

$$L_2 \geq \sqrt{V(P_2)}\mathrm{Q}^{-1}(\varepsilon). \tag{6.177}$$

It follows that for case (i), the second-order rate region of mismatched MAC with JNN decoding satisfies

$$\mathcal{L}_{\mathrm{JNN}}(R_1^*, R_2^*, \varepsilon) \supseteq \left\{(L_1, L_2) : L_2 \geq \sqrt{V(P_2)}\mathrm{Q}^{-1}(\varepsilon)\right\}. \tag{6.178}$$

Recall the definition of the dispersion function $\mathbf{V}_1(P_1, P_2)$ in (6.20). Note that $[S_2, S_3]^{\mathrm{T}} \sim \mathcal{N}(\mathbf{0}_2, \mathbf{V}_1(P_1, P_2))$ is a two-dimensional Gaussian random vector, which contains the last two random variables in $\mathbf{S} = [S_1; S_2; S_3]$. Now consider case (ii), where $(R_1^*, R_2^*) = (C(P_1/(1 + P_2)), C(P_2))$. Since $R_1^* < C(P_1)$, (6.166) still holds, which leads to $\lim_{n\to\infty} \Pr\{\mathcal{A}_1\} = 1$. Thus, the right-hand side of (6.163) satisfies

$$\liminf_{n\to\infty}(1 - \Pr\{\mathcal{A}_1 \cap \mathcal{A}_2 \cap \mathcal{A}_3\})$$
$$\leq \liminf_{n\to\infty}(1 - \Pr\{\mathcal{A}_2 \cap \mathcal{A}_3\} - \Pr\{\mathcal{A}_1^C\}) \tag{6.179}$$
$$= 1 - \Pr\{\{S_2 \geq -L_2\} \cap \{S_3 \geq -(L_1 + L_2)\}\} \tag{6.180}$$
$$= 1 - \Pr\{\{S_2 \leq L_2\} \cap \{S_3 \leq L_1 + L_2\}\} \tag{6.181}$$
$$= 1 - \Pr\left\{\begin{bmatrix} S_2 \\ S_3 \end{bmatrix} \leq \begin{bmatrix} L_2 \\ L_1 + L_2 \end{bmatrix}\right\}, \tag{6.182}$$

where (6.179) follows similarly as (6.169), (6.180) follows from the values of boundary rate pairs $(R_1^*, R_2^*) = (C(P_1/(1 + P_2)), C(P_2))$ for case (ii), and (6.181) follows form symmetry since S_2 and S_3 are zero-mean Gaussian random variables. To make (6.182) less than ε, one requires

$$\Pr\left\{\begin{bmatrix} S_2 \\ S_3 \end{bmatrix} \leq \begin{bmatrix} L_2 \\ L_1 + L_2 \end{bmatrix}\right\} \geq 1 - \varepsilon. \tag{6.183}$$

Recall the definition of $\mathrm{Q_{inv}}$ in (6.14). The corresponding second-order rate region of mismatched MAC with JNN decoding for case (ii) satisfies

$$\mathcal{L}_{\mathrm{JNN}} \supseteq \{(L_1, L_2) : (L_2, L_1 + L_2) \in \mathrm{Q_{inv}}(\mathbf{V}_1(P_1, P_2), \varepsilon)\}. \tag{6.184}$$

Next consider case (iii), where $(R_1^*, R_2^*) = (\alpha C(P_1) + \bar{\alpha} C(P_1/(1 + P_2)), \alpha C(P_2/(1 + P_1)) + \bar{\alpha} C(P_2))$ for some $\alpha \in (0, 1)$. Since $R_1^* < C(P_1)$ and $R_2^* < C(P_2)$, it follows that

$$\lim_{n\to\infty} \Pr\{\mathcal{A}_1\} = \lim_{n\to\infty} \Pr\{\mathcal{A}_2\} = 1. \tag{6.185}$$

Analogously to (6.169)-(6.171), the right-hand side of (6.163) satisfies

$$\liminf_{n\to\infty} 1 - \Pr\{\mathcal{A}_1 \cap \mathcal{A}_2 \cap \mathcal{A}_3\} = \Pr\{S_3 \geq L_1 + L_2\}. \tag{6.186}$$

To make (6.186) less than ε, one requires

$$L_1 + L_2 \leq \sqrt{V_{12}(P_1, P_2)}\mathrm{Q}^{-1}(\varepsilon). \tag{6.187}$$

Thus, the second-order rate region of mismatched MAC with JNN decoding for case (iii) satisfies

$$\mathcal{L}_{\mathrm{JNN}}(R_1^*, R_2^*, \varepsilon) \supset \{(L_1, L_2) : L_1 + L_2 \geq \sqrt{V_{12}(P_1, P_2)}\mathrm{Q}^{-1}(\varepsilon)\}. \tag{6.188}$$

The analyses for cases (iv) and (v) are omitted due to their symmetry to cases (ii) and (i), respectively. The proof of Theorem 6.1 is now completed.

6.4.3 SIC Decoding

This section proves Theorem 6.2. As explained below Theorem 6.2, it suffices to consider case (ii) since other cases are either linear combinations (time sharing) of cases (ii) and the P2P case in Chap. 3 or symmetric to cases (ii).

Fix any $\varepsilon \in (0,1)$ and $(\varepsilon_1, \varepsilon_2) \in \mathbb{R}_+^2$ such that $\varepsilon_1 + \varepsilon_2 \leq \varepsilon$. It follows from Lemma 6.2 that the error probability P_e^n of the mismatched code in Definition 6.2 satisfies

$$\begin{aligned}\mathrm{P}_\mathrm{e}^n &\leq \mathbb{E}\Big[\min\Big\{1, (M_1-1)\Pr\{\tilde{\imath}^n(\bar{X}_1^n; Y^n) \geq \tilde{\imath}^n(X_1^n; Y^n)|X_1^n, Y^n\}\Big\}\Big] \\ &\quad + \mathbb{E}\Big[\min\Big\{1, (M_2-1)\Pr\{\tilde{\imath}_2^n(\bar{X}_2^n; Y^n|X_1^n) \geq \tilde{\imath}_2^n(X_2^n; Y^n|X_1^n)|X_1^n, X_2^n, Y^n\}\Big\}\Big].\end{aligned} \tag{6.189}$$

For simplicity, let the first and second terms of the right-hand side of (6.189) be denoted as $\mathrm{P}_{\mathrm{e},1}$ and $\mathrm{P}_{\mathrm{e},2}$, respectively. It follows that

$$\mathrm{P}_{\mathrm{e},1} = \mathbb{E}[\min\{1, M_1 \Pr\{\tilde{\imath}_1(\bar{X}_1^n; Y^n) \geq \tilde{\imath}_1(X_1^n; Y^n)|X_1^n, Y^n\}\}] \tag{6.190}$$

$$\leq \mathbb{E}[\min\{1, M_1 g(\tilde{\imath}_1(X_1^n; Y^n), Y^n)\}] \tag{6.191}$$

$$\leq \Pr\left\{M_1\beta_{1,2}e^{-\tilde{\imath}_1(X_1^n;Y^n)} \geq \frac{1}{\sqrt{n}}\right\} + \frac{1}{\sqrt{n}} \tag{6.192}$$

$$\begin{aligned}&\leq \Pr\left\{\frac{\|X_2^n + Z^n\|^2}{2(1+P_2)} - \frac{\|X_1^n + X_2^n + Z^n\|^2}{2(1+P_1+P_2)}\right. \\ &\qquad\qquad \left.\geq nC\left(\frac{P_1}{1+P_2}\right) - \log M_1 - \log(\beta_{1,2}\sqrt{n})\right\} + \frac{1}{\sqrt{n}},\end{aligned} \tag{6.193}$$

where (6.191) follows from the definition of function g in (6.114), (6.192) follows the result in Lemma 6.3 and the inequality in [21, Eq. (37)], which states that for any real-valued random variable J and integer $n \in \mathbb{N}$,

$$\mathbb{E}[\min\{1, J\}] \leq \Pr\left\{J \geq \frac{1}{\sqrt{n}}\right\} + \frac{1}{\sqrt{n}}, \tag{6.194}$$

(6.193) follows by taking the logarithm at both sides of the inequality and using the definition of mismatched mutual information density of $\tilde{\imath}_1(\cdot)$ in (6.89) and the channel model $Y^n = X_1^n + X_2^n + Z^n$. For ease of notation, define two terms:

$$I := (1+P_1+P_2)\|X_2^n + Z^n\|^2 - (1+P_2)\|X_1^n + X_2^n + Z^n\|^2, \tag{6.195}$$

$$J := 2(1+P_1+P_2)(1+P_2)\left(nC\left(\frac{P_1}{1+P_2}\right) - \log M_1 - \log(\beta_{1,2}\sqrt{n})\right). \tag{6.196}$$

This way, (6.193) is equivalent to

$$\Pr\{\mathcal{E}_1\} \leq \Pr\{I \geq J\} + \frac{1}{\sqrt{n}}. \tag{6.197}$$

For each $(i, j) \in [n] \times [2]$, let $\tilde{X}_{ji}$ be generated from the Gaussian distribution with mean zero and variance one, i.e., $\mathcal{N}(\cdot; 0, 1)$. For subsequent analyses, for each $i \in [n]$, define the following random variables:

$$\begin{aligned}&B_{1i} := Z_i^2 - 1, \quad B_{2i} := \sqrt{P_1}\tilde{X}_{1i}Z_i, \; B_{3i} := \tilde{X}_{1i}^2 - 1,\\&B_{4i} := \sqrt{P_2}\tilde{X}_{2i}Z_i, \; B_{5i} := \tilde{X}_{2i}^2 - 1, \quad B_{6i} := \sqrt{P_1P_2}\tilde{X}_{1i}\tilde{X}_{2i},\end{aligned} \tag{6.198}$$

and let $\mathbf{B}_i := (B_{1i}, \ldots, B_{6i})$ denote the random vector. It follows that $(\mathbf{B}_1, \ldots, \mathbf{B}_n)$ are a sequence of i.i.d. random vectors with finite third moment and for each $i \in [n]$,

$$\mathsf{Cov}(\mathbf{B}_i) = \operatorname{diag}\left[\xi - 1, P_1, 2, P_2, 2, P_1P_2\right]. \tag{6.199}$$

Given any real numbers $\mathbf{b} := (b_1, \ldots, b_6) \in \mathbb{R}^6$, define the function:

$$\begin{aligned}f(\mathbf{b}) :=& P_1b_1 - 2(1+P_2)\frac{b_2}{\sqrt{1+b_3}} + 2P_1\frac{b_4}{\sqrt{1+b_5}}\\&- 2(1+P_2)\frac{b_6}{\sqrt{1+b_3}\sqrt{1+b_5}}.\end{aligned} \tag{6.200}$$

The Jacobian matrix of $f(\cdot)$ at $\mathbb{E}[\mathbf{B}_1] = \mathbf{0}_6$ is

$$\mathbf{J}(\mathbf{B}) = \left[P_1, -2(1+P_2), 0, 2P_1, 0, -2(1+P_2)\right]. \tag{6.201}$$

Thus,

$$\mathbf{J}(\mathbf{B})\,\mathsf{Cov}(\mathbf{B}_1)\mathbf{J}(\mathbf{B})^{\mathrm{T}} = P_1^2(\xi - 1 + 4P_2) + 4P_1(1+P_2)^3 =: \sigma^2 \tag{6.202}$$

Furthermore,

$$\begin{aligned}nf\left(\frac{1}{n}\sum_{i=1}^{n}\mathbf{B}_i\right) &= P_1(\|Z^n\|^2 - n) + 2P_1\langle X_2^n, Z^n\rangle - 2(1+P_2)\langle X_1^n, Z^n\rangle\\&\quad - 2(1+P_1)\langle X_1^n, X_2^n\rangle \qquad (6.203)\\&= I. \qquad (6.204)\end{aligned}$$

It follows from using (6.195) to (6.197) and the function version of the Berry-Esseen theorem (Theorem 1.8) that

$$\begin{aligned}\mathrm{P}_{\mathrm{e},1} &\le \Pr\{I \ge J\} + O\left(\frac{1}{\sqrt{n}}\right) \qquad (6.205)\\&\le \mathrm{Q}\left(\frac{J}{\sqrt{n\sigma}}\right) + O\left(\frac{1}{\sqrt{n}}\right). \qquad (6.206)\end{aligned}$$

To make (6.206) less than ε_1, one requires

$$\frac{J}{\sqrt{n}\sigma} \geq \mathrm{Q}^{-1}\left(\varepsilon_1 + O\left(\frac{1}{\sqrt{n}}\right)\right). \tag{6.207}$$

Applying Taylor Series expansion of $\mathrm{Q}^{-1}(\cdot)$ on ε_1 and using the definition of RHS in (6.196), the maximal value of M_1 such that $\mathrm{P}_{\mathrm{e},1} \leq \varepsilon_1$ satisfies

$$\log M_1 = nC\left(\frac{P_1}{1+P_2}\right) - \sqrt{nV_1(P_1, P_2)}\mathrm{Q}^{-1}(\varepsilon_1) + O(\log n). \tag{6.208}$$

Thus, a lower bound $L_1 \geq \sqrt{nV_1(P_1, P_2)}\mathrm{Q}^{-1}(\varepsilon_1)$ is obtained.

Next consider $\mathrm{P}_{\mathrm{e},2}$. Given that message W_1 is decoded correctly, the remaining error probability satisfies

$$\mathrm{P}_{\mathrm{e},2} = \mathbb{E}[\min\{1, M_2 \Pr\{\tilde{\imath}_2(\bar{X}_2^n; Y^n|X_1^n) \geq \tilde{\imath}_2^n(X_2^n; Y^n|X_1^n)|X_1^n, X_2^n, Y^n\}\}] \tag{6.209}$$

$$\leq \mathbb{E}[\min\{1, M_2 g_2(\tilde{\imath}_2(X_2^n; Y^n|X_1^n), Y^n)\}] \tag{6.210}$$

$$\leq \Pr\left\{M_2\beta_2 e^{-\tilde{\imath}_2(X_2^n;Y^n|X_1^n)} \geq \frac{1}{\sqrt{n}}\right\} + \frac{1}{\sqrt{n}} \tag{6.211}$$

$$\leq \Pr\left\{\frac{\|Z^n\|^2}{2} - \frac{\|X_2^n + Z^n\|^2}{2(1+P_2)} \geq nC(P_2) - \log M_2 - \log(\beta_2\sqrt{n})\right\} + \frac{1}{\sqrt{n}}, \tag{6.212}$$

where (6.210) follows from the definition of $g_2(\cdot)$ in (6.112), (6.211) follows from the result in Lemma 6.3 and the inequality in (6.194), and (6.212) follows from the definition of mismatched mutual information density $\tilde{\imath}_2(\cdot)$ in (6.91).

Note that (6.212) has the same form as the mismatched P2P case in [21, Eq. (42)]. Thus, it follows from [21, Theorem 1] that to ensure $\mathrm{P}_{\mathrm{e},2} \leq \varepsilon_2$, the maximal value of M_2 satisfies

$$\log M_2 = nC(P_2) - \sqrt{nV(P_2)}\mathrm{Q}^{-1}(\varepsilon_2) + O(\log n). \tag{6.213}$$

A lower bound $L_2 \geq \sqrt{nV(P_2)}\mathrm{Q}^{-1}(\varepsilon_2)$ is thus obtained.

The proof for case (ii) of Theorem 6.2 is completed by taking a union over all possible $(\varepsilon_1, \varepsilon_2) \in \mathbb{R}_+^2$ such that $\varepsilon_1 + \varepsilon_2 \leq \varepsilon$.

6.5 Proofs for Random Access Channel

6.5.1 Preliminaries

This section presents preliminaries to prove achievable second-order asymptotics for RAC in Theorems 6.3 and 6.4.

Fix an integer $k \in [K]$. Fix any integers $(t, n_k) \in [k] \times \mathbb{N}$ and channel input-output vectors $(x_{[k]}^{n_k}, y^{n_k}) \in (\mathbb{R}^{n_k})^{k+1}$. The generalized version of mismatched (conditional) mutual information density is defined as

$$\tilde{\imath}_t^{n_k}(x_{[t]}^{n_k}; y^{n_k}|x_{[t+1:k]}^{n_k}) := n_k C(tP) + \frac{\|y^{n_k} - \sum_{i\in[t+1:k]} x_i^{n_k}\|^2}{2(1+tP)} - \frac{\|y^{n_k} - \sum_{i\in[k]} x_i^{n_k}\|^2}{2}, \tag{6.214}$$

$$\bar{\imath}_t^{n_k}(x_t^{n_k}; y^{n_k}|x_{[t-1]}^{n_k}) := n_k C\left(\frac{P}{1+(k-t)P}\right) + \frac{\|y^{n_k} - \sum_{i\in[t-1]} x_i^{n_k}\|^2}{2(1+(k-t+1)P)} - \frac{\|y^{n_k} - \sum_{i\in[t]} x_i^{n_k}\|^2}{2(1+(k-t)P)}, \tag{6.215}$$

where $x_{[t]}^{n_k} = (x_1^{n_k}, \ldots, x_t^{n_k})$ and $x_{[t+1:k]}^{n_k} = (x_{t+1}^{n_k}, \ldots, x_k^{n_k})$. When $t = k$, $x_{[t+1:k]}^{n_k} = 0$ and $\sum_{i\in[t+1:k]} x_i^{n_k} = 0$; when $t = 1$, $x_{[t-1]}^{n_k} = 0$ and $\sum_{i\in[t-1]} x_i^{n_k} = 0$. The above definitions are consistent with the information density in a Gaussian RAC [17, Eq. (10)]. Analogously to (6.114), given any positive real number $s \in \mathbb{R}_+$, define the following probability functions

$$\tilde{g}_t(s; y^{n_k}, x_{[t+1:k]}^{n_k}) := \Pr\left\{\tilde{\imath}_t(\bar{X}_{[t]}^{n_k}; y^{n_k}|x_{[t+1:k]}^{n_k}) \geq s\right\}, \tag{6.216}$$

$$\bar{g}_t(s; y^{n_k}, x_{[t-1]}^{n_k}) := \Pr\left\{\bar{\imath}_t(\bar{X}_t^{n_k}; y^{n_k}|x_{[t-1]}^{n_k}) \geq s\right\}, \tag{6.217}$$

where $\bar{X}_{[t]}^{n_k} = (\bar{X}_1^{n_k}, \ldots, \bar{X}_t^{n_k})$ is a random vector with distributions (6.47) and (6.48). Similarly to Lemma 6.3, the following lemma upper bounds the above functions.

Lemma 6.4 *For each $t \in [k]$, there exists positive real numbers $(\beta_{t,1}, \beta_{t,2}) \in \mathbb{R}_+^2$ that only depend on P such that for any $s \in \mathbb{R}_+$,*

$$\tilde{g}_t(s; y^{n_k}, x_{[t+1:k]}^{n_k}) \leq \beta_{t,1} \exp(-s), \tag{6.218}$$

$$\bar{g}_t(s; y^{n_k}, x_{[t-1]}^{n_k}) \leq \beta_{t,2} \exp(-s). \tag{6.219}$$

6.5.2 JNN Decoding

This section proves Theorem 6.3 for mismatched RAC with JNN decoding. Fix any $k \in [K]$. It follows from the mismatched code in 6.2.2 that when there are k active users among K users, a decoding error event occurs if one of the following events occurs:

- $\mathcal{E}_{\text{rep}}$: two different active users transmit the same message,
- $\mathcal{E}_{\text{time}}$: the receiver stops at a decoding time $n_t < n_k$ for some $t < k$,
- $\mathcal{E}_{\text{msg}}$: the receiver stops at n_t but decodes transmitted messages incorrectly.

Thus, the error probability satisfies

$$\mathrm{P}^n_{\mathrm{e},k} \leq \Pr\{\mathcal{E}_{\text{rep}} \cup \mathcal{E}_{\text{time}} \cup \mathcal{E}_{\text{msg}}\} \tag{6.220}$$

$$\leq \Pr\{\mathcal{E}_{\text{rep}}\} + \Pr\{\mathcal{E}_{\text{time}} \cap \mathcal{E}^c_{\text{rep}}\} + \Pr\{\mathcal{E}_{\text{msg}} \cap \mathcal{E}^c_{\text{rep}}\}. \tag{6.221}$$

The rest of this section bounds each of the three terms in (6.221).

6.5.2.1 Message Repetition and Stopping Time Error Probabilities

First consider the probability $\Pr\{\mathcal{E}_{\text{rep}}\}$ where any two active users transmit the same message. This error event occurs only if the number of active users is greater than one. Since the message of each user is independent and follows the same uniform distribution over the message set $[M]$, it follows that

$$\Pr\{\mathcal{E}_{\text{rep}}\} \leq \sum_{(i,j)\in[K]^2: i\neq j} \Pr\{W_j = W_j\} \tag{6.222}$$

$$\leq \sum_{(i,j)\in[K]^2: i\neq j} \sum_{m\in[M]} \Pr\{W_i = W_j = m\} \tag{6.223}$$

$$= \sum_{(i,j)\in[K]^2: i\neq j} \sum_{m\in[M]} \frac{1}{M^2} \tag{6.224}$$

$$= \sum_{(i,j)\in[K]^2: i\neq j} M \frac{1}{M^2} \tag{6.225}$$

$$= \frac{K(K-1)}{M}. \tag{6.226}$$

Next, consider the probability $\Pr\{\mathcal{E}_{\text{time}} \cap \mathcal{E}^c_{\text{rep}}\}$ where the receiver estimates the number of active users wrongly. It follows from the coding scheme with JNN decoding in Definition 6.5 that

$$\Pr\{\mathcal{E}_{\text{time}} \cap \mathcal{E}^c_{\text{rep}}\} \leq \sum_{t\in[1:k-1]} \Pr\left\{\left|\frac{1}{n_t}\|Y^{n_t}\|^2 - (1+tP)\right| \leq \lambda_t\right\} + \Pr\left\{\left|\frac{1}{n_k}\|Y^{n_k}\|^2 - (1+kP)\right| > \lambda_k\right\}\Bigg\}, \tag{6.227}$$

where the first term bounds the probability that the receiver underestimates the number of active users, while the latter form bounds the probability that the receiver overestimates the number of active users. Using the function version of the Berry-Esseen theorem, it follows that

$$\Pr\left\{\mathcal{E}_{\text{time}} \cap \mathcal{E}^c_{\text{rep}}\right\} \leq O\left(\frac{1}{\sqrt{n_k}}\right). \tag{6.228}$$

6.5.2.2 Decoding Error Probability

Finally, consider the probability $\Pr\{\mathcal{E}_{\text{msg}} \cap \mathcal{E}^c_{\text{rep}}\}$ that the receiver decodes messages of active users wrongly when the number of active users k is estimated correctly. Let $(W_1, \ldots, W_k) \in [M]^k$ be the transmitted messages. Recall that the codebook consists of codewords $(X_1^{n_K}, \ldots, X_M^{n_K})$. For any $k \in [K]$, $t \in [k]$ and $w^t = (w_1, \ldots, w_t) \in [M]^t$, $X^{n_k}(w^t) = (X^{n_k}_{w_1}, \ldots, X^{n_k}_{w_t})$ denotes a collection of t codewords, each with length n_k. Similarly, for any $w^k_{t+1} = (w_{t+1}, \ldots, w_k)$, $X^{n_k}(w^k_{t+1}) = (X^{n_k}_{w_{t+1}}, \ldots, X^{n_k}_{w_k})$.

For each $t \in [k]$, let $\mathcal{E}_t$ denote the error event, where t out of k messages are decoded incorrectly. It follows from the union bound that

$$\Pr\{\mathcal{E}_{\text{msg}} \cap \mathcal{E}^c_{\text{rep}}\} \leq \Pr\left\{\bigcup_{t\in[k]} \mathcal{E}_t \cap \mathcal{E}^c_{\text{rep}}\right\} \tag{6.229}$$

$$\leq \sum_{t\in[k]} \Pr\left\{\mathcal{E}_t \cap \mathcal{E}^c_{\text{rep}}\right\}. \tag{6.230}$$

Fix any $t \in [k]$. Given that the error event $\mathcal{E}_t \cap \mathcal{E}^c_{\text{rep}}$ happens, assume that the first t messages $(W_1, \ldots, W_t)$ out of all messages $(W_1, \ldots, W_k)$ are decoded wrongly. In this case, the JNN decoder operates as follows:

$$\Psi_k(Y^{n_k}) = \operatorname*{arg\,min}_{(\bar{w}_1,\ldots,\bar{w}_t)\in[M]^t} \left\|Y^{n_k} - \sum_{i=1}^{t} X^{n_k}(\bar{w}_i) - \sum_{j=t+1}^{k} X^{n_k}(W_j)\right\|^2 \tag{6.231}$$

$$= \underset{(\bar{w}_1,\dots,\bar{w}_t)\in[M]^t}{\arg\max} \frac{n_k}{2}\log(1+tP) + \frac{\|Y^{n_k} - \sum_{j=t+1}^{k} X^{n_k}(W_j)\|^2}{2(1+tP)} - \frac{\|Y^{n_k} - \sum_{i=1}^{t} X^{n_k}(\bar{w}_i) - \sum_{j=t+1}^{k} X^{n_k}(W_j)\|^2}{2} \tag{6.232}$$

$$= \underset{(\bar{w}_1,\dots,\bar{w}_t)\in[M]^t}{\arg\max} \tilde{\imath}_t^{n_k}(X^{n_k}(\bar{w}^t); Y^{n_k}|X^{n_k}(W_{t+1}^k)), \tag{6.233}$$

where (6.233) follows from the definition of $\tilde{\imath}_t^{n_k}(\cdot)$ in (6.214), $\bar{w}^t = (\bar{w}_1, \dots, \bar{w}_t)$, $X^{n_k}(\bar{w}^t) = (X^{n_k}(\bar{w}_1), \dots, X^{n_k}(\bar{w}_t))$, $W_{t+1}^k = (W_{t+1}, \dots, W_k)$ and $X^{n_k}(W_t^k) = (X^{n_k}(W_{t+1}), \dots, X^{n_k}(W_k))$.

The result in (6.233) implies that under error event $\mathcal{E}_t \cap \mathcal{E}_{\text{rep}}^c$, maximizing the mismatched information density $\tilde{\imath}_s^{n_k}(X^{n_k}(\bar{w}^t); Y^{n_k}|X^{n_k}(W_t^k))$ over $\bar{w}_{[t]} \in [M]^t$ is equivalent to JNN decoding. For ease of notation, given $\bar{w}^t \in [M]^t$, define the event

$$\mathcal{B}(\bar{w}^t) := \left\{\tilde{\imath}_t^{n_k}(X^{n_k}(\bar{w}^t); Y^{n_k}|X^{n_k}(W_{t+1}^k)) \geq \tilde{\imath}_t^{n_k}(X^{n_k}(W^t); Y^{n_k}|X^{n_k}(W_{t+1}^k))\right\}. \tag{6.234}$$

It follows that

$$\Pr\{\mathcal{E}_t \cap \mathcal{E}_{\text{rep}}^c\}$$

$$\leq \binom{k}{t} \Pr\Big\{ \bigcup_{\bar{w}^t\in[M]^t:\ \bar{w}^t\neq W^t} \mathcal{B}(\bar{w}^t)\Big\} \tag{6.235}$$

$$\leq \mathbb{E}\Big[\binom{k}{t} \Pr\Big\{ \bigcup_{\bar{w}^t\in[M]^t:\ \bar{w}^t\neq W^t} \mathcal{B}(\bar{w}^t)|X^{n_k}(W^k), Y^{n_k}\Big\}\Big] \tag{6.236}$$

$$\leq \mathbb{E}\Big[\binom{k}{t}\binom{M-k}{t} \Pr\{\mathcal{B}(\bar{w}^t)|X^{n_k}(W^k), Y^{n_k}\}\Big] \tag{6.237}$$

$$\leq \mathbb{E}\Big[\min\Big\{1, \binom{k}{t}\binom{M-k}{t} \Pr\{\mathcal{B}(\bar{w}^t)|X^{n_k}(W^k), Y^{n_k}\}\Big\}\Big] \tag{6.238}$$

$$\leq \mathbb{E}\Big[\min\Big\{1, \binom{k}{t} M^t \tilde{g}_t(\tilde{\imath}_t(X^{n_k}(W^t); Y^{n_k}|X^{n_k}(W_{t+1}^k)); Y^{n_k}, X^{n_k}(W_{t+1}^k))\Big\}\Big] \tag{6.239}$$

$$\leq \mathbb{E}\Big[\min\Big\{1, \binom{k}{t} M^t \beta_{t,1} \exp(-\tilde{\imath}_t^{n_k}(X^{n_k}(W^t); Y^{n_k}|X^{n_k}(W_{t+1}^k)))\Big\}\Big] \tag{6.240}$$

$$\leq \Pr\Big\{\binom{k}{t} M^t \beta_{t,1} \exp(-\tilde{\imath}_t^{n_k}(X^{n_k}(W^t); Y^{n_k}|X^{n_k}(W_{t+1}^k))) \geq 1/\sqrt{n_k}\Big\} + \frac{1}{\sqrt{n_k}}, \tag{6.241}$$

where (6.235) follows due to symmetry and the fact that there are $\binom{k}{t}$ possibilities of decoding t messages incorrectly among k messages, (6.236) follows from the law of total expectation, (6.237) follows from symmetry of the codebook, (6.238) follows since any probability is no greater than 1, (6.239) follows from the definition of $\tilde{g}$ in (6.216) and the fact that $\binom{M-k}{t} \leq M^t$, (6.240) follows from the result in Lemma 6.3 and the inequality $\frac{\beta_{t,1}\exp(-s)}{\sqrt{n_k}} \leq \beta_{t,1}\exp(-s)$ for any integer n_k, and (6.241) follows from the inequality in (6.194).

The probability term in (6.241) can be further upper bounded as follows:

$$\Pr\left\{\binom{k}{t} M^t \beta_{t,1} \exp(-\tilde{\imath}_t^{n_k}(X^{n_k}(W^t); Y^{n_k}|X^{n_k}(W_{t+1}^k))) \geq \frac{1}{\sqrt{n_k}}\right\}$$

$$\leq \Pr\left\{\tilde{\imath}_t^{n_k}(X^{n_k}(W^t); Y^{n_k}|X^{n_k}(W_{t+1}^k)) \leq \log\left(\binom{k}{t} M^t \beta_{t,1}\sqrt{n_k}\right)\right\} \tag{6.242}$$

$$\leq \Pr\left\{(\|Z^{n_k}\|^2 - n_k)tP - 2\sum_{i=1}^{t}\langle X_i^{n_k}, Z^{n_k}\rangle - 2\sum_{1\leq i<j\leq t}\langle X_i^{n_k}, X_j^{n_k}\rangle \right.$$
$$\left. \geq 2(1+tP)\left(n_k C(tP) - \log\left(\binom{k}{t} M^t \beta_{t,1}\sqrt{n_k}\right)\right)\right\}, \tag{6.243}$$

where (6.242) follows by taking the logarithm at both sides and (6.243) follows from the definition of mismatched mutual information density in (6.214) and the fact that each codeword is generated independently from the same distribution.

Fix any $i \in [n]$. Define the set $\mathcal{D}_t := \{(a,b) \in [t]^2 : a \neq b\}$. For each $l \in [t]$ and $(p,q) \in \mathcal{D}_t$, define the following zero-mean random variables:

$$\begin{aligned} \tilde{A}_{1,i} &:= Z_i^2 - 1 \quad \tilde{A}_{2,i}^{(l)} := \sqrt{P}\tilde{X}_{l,i} Z_i \\ \tilde{A}_{3,i}^{(l)} &:= \tilde{X}_{l,i}^2 - 1 \quad \tilde{A}_{4,i}^{(p,q)} := P\tilde{X}_{p,i}\tilde{X}_{q,i}, \end{aligned} \tag{6.244}$$

and let

$$\tilde{\mathbf{A}}_i := (\tilde{A}_{1,i}, \{\tilde{A}_{2,i}^{(l)}\}_{l\in[t]}, \{\tilde{A}_{3,i}^{(l)}\}_{l\in[t]}, \{\tilde{A}_{4,i}^{(p,q)}\}_{(p,q)\in\mathcal{D}_t}). \tag{6.245}$$

Thus $\tilde{\mathbf{A}}_{[n_k]} = (\tilde{\mathbf{A}}_1, \ldots, \tilde{\mathbf{A}}_{n_k})$ are zero-mean i.i.d. random vectors with finite third moments. The covariance matrix of $\tilde{\mathbf{A}}_i$ is

$$\mathsf{Cov}(\tilde{\mathbf{A}}_1) = \operatorname{diag}\left[\xi - 1, P\cdot \mathbf{1}_t, 2\cdot \mathbf{1}_t, P^2\cdot \mathbf{1}_{t(t-1)}\right]. \tag{6.246}$$

Given any $a_1 \in \mathbb{R}$, $\mathbf{a}_2 = (a_{2,1}, \ldots, a_{2,t}) \in \mathbb{R}^t$, $\mathbf{a}_3 = (a_{3,1}, \ldots, a_{3,t})$ and $\mathbf{a}_4 = (a_4^{p,q})_{(p,q)\in\mathcal{D}_t} \in \mathbb{R}^{t(t-1)}$, define the following function:

$$\tilde{f}(a_1, \mathbf{a}_2, \mathbf{a}_3, \mathbf{a}_4) := tPa_1 - \sum_{l\in[t]}\frac{2a_{2,t}}{\sqrt{1+a_{3,l}}} - \sum_{(p,q)\in\mathcal{D}_t}\frac{2a_4^{(p,q)}}{\sqrt{1+a_{3,p}}\sqrt{1+a_{3,q}}}. \tag{6.247}$$

It follows that

$$n_k \tilde{f}\left(\frac{1}{n_k}\sum_{i\in[n_k]}\tilde{\mathbf{A}}_i\right) = (\|Z^{n_k}\|^2 - n_k)tP - 2\sum_{i\in[t]}\langle X_i^{n_k}, Z^{n_k}\rangle - 2\sum_{(i,j)\in\mathcal{D}_t}\langle X_i^{n_k}, X_j^{n_k}\rangle, \tag{6.248}$$

which is exactly the left hand side of (6.243). The Jacobian of $\tilde{f}$ evaluated at $\mathbf{0}$ is

$$\mathbf{J} = \left[tP, -2\cdot\mathbf{1}_t, \mathbf{0}_t, -2\cdot\mathbf{1}_{t(t-1)}\right]. \tag{6.249}$$

Thus,

$$\mathbf{J}\,\mathsf{Cov}(\tilde{\mathbf{A}}_1)\mathbf{J}^{\mathrm{T}} = tP((\xi+1)tP - 2P + 4) =: \tilde{\sigma}_t^2. \tag{6.250}$$

It follows from the function version Berry-Esseen Theorem (Theorem 1.8) and the results in (6.241) and (6.243) that

$$\Pr\{\mathcal{E}_t \cap \mathcal{E}_{\mathrm{rep}}^c\} \le \mathrm{Q}\left(\frac{2(1+tP)\left(n_k C(tP) - \log\left(\binom{k}{t}M^t\beta_{t,1}\sqrt{n_k}\right)\right)}{\tilde{\sigma}_t\sqrt{n_k}}\right) + O\left(\frac{1}{\sqrt{n_k}}\right). \tag{6.251}$$

6.5.2.3 Final Steps

Fix any $(\varepsilon_{k,1}, \ldots, \varepsilon_{k,k}) \in (0,1)^k$ such that $\sum_{t\in[k]}\varepsilon_{k,t} \le \varepsilon_k$. Choose M such that

$$\log M = \min_{t\in[k]}\frac{1}{t}\left(n_k C(tP) - \sqrt{n_k(V(tP) + V_{\mathrm{cr}}(t,P))}\mathrm{Q}^{-1}(\varepsilon_{k,t}) + O(\log n_k)\right). \tag{6.252}$$

It follows from (6.251) that for each $t \in [k]$,

$$\Pr\{\mathcal{E}_t \cap \mathcal{E}_{\mathrm{rep}}^c\} \le \varepsilon_{k,t} \tag{6.253}$$

Since $\frac{1}{t}C(tP) > \frac{1}{k}C(kP)$ for $t < k$ [17, Lemma 1], when n_k is sufficiently large, the minimum of (6.252) is achieved by $t = k$.

Thus, when n_k is large enough, it follows from (6.221), (6.226), (6.228), (6.252), and (6.253) that

$$\log M = \frac{1}{k}\left(n_k C(kP) - \sqrt{n_k(V(kP) + V_{\mathrm{cr}}(k,P))}\mathrm{Q}^{-1}(\varepsilon_{k,k}) + O(\log n_k)\right), \tag{6.254}$$

and

$$\mathrm{P}_{\mathrm{e},k}^{n_k} \leq \varepsilon_k. \tag{6.255}$$

The proof of Theorem 6.3 is completed by letting $\varepsilon_{k,t} \to 0$ for every $t \in [k-1]$ and $\varepsilon_{k,k} \to \varepsilon_k$.

6.6 Proofs for Broadcast Channel

6.6.1 SEP

6.6.1.1 Preliminaries

Recall that $C(P) = \frac{1}{2}\log(1+P)$. Given any $(u, y_2) \in \mathbb{R}^2$, define the mutual information density

$$\tilde{\imath}(u; y_2) := C\left(\frac{\bar{\alpha}P}{\alpha P+1}\right) + \frac{y_2^2}{2(P+1)} - \frac{(y_2-u)^2}{2(\alpha P+1)}. \tag{6.256}$$

Given any blocklength $n \in \mathbb{N}$ and $(u^n, y^n) \in \mathbb{R}^{2n}$, the accumulated mutual information density is defined as

$$\tilde{\imath}^n(u^n; y_2^n) := \sum_{i=1}^{n} \tilde{\imath}(u_i; y_{2,i}). \tag{6.257}$$

Similarly, for any $(u, v, y_1) \in \mathbb{R}^3$ and $(u^n, v^n, y_1^n) \in \mathbb{R}^{3n}$, define the following mutual information densities and their accumulated versions:

$$\tilde{\imath}(v; y_1|u) := C\left(\frac{\alpha P}{\beta}\right) + \frac{(y_1-u)^2}{2(\alpha P+\beta)} - \frac{(y_1-u-v)^2}{2\beta}, \tag{6.258}$$

$$\tilde{\imath}^n(v^n; y_1^n|u^n) := \sum_{i=1}^{n} \tilde{\imath}(v_i; y_{1,i}|u_i), \tag{6.259}$$

$$\tilde{\imath}(v, u; y_1) := C\left(\frac{P}{\beta}\right) + \frac{y_1^2}{2(P+\beta)} - \frac{(y_1-u-v)^2}{2\beta}, \tag{6.260}$$

$$\tilde{\imath}^n(v^n, u^n; y_1^n) := \sum_{i=1}^{n} \tilde{\imath}(v_i, u_i; y_{1,i}). \tag{6.261}$$

Furthermore, given any non-negative real number $t \in \mathbb{R}_+$ and vectors $(u^n, v^n, y_1^n, y_2^n) \in \mathbb{R}^{4n}$, define the following probability terms.

$$g_1(t; y_1^n, u^n) := \Pr\{\tilde{\imath}^n(\bar{V}^n; y_1^n|u^n) \geq t\}, \tag{6.262}$$

$$g_2(t; y_1^n, v^n) := \Pr\{\tilde{\imath}^n(\bar{U}^n; y_1^n|v^n) \geq t\}, \tag{6.263}$$

$$g_{1,2}(t, y_2^n) := \Pr\{\tilde{\imath}^n(\bar{U}^n, \bar{V}^n; y_1^n) \geq t\}, \tag{6.264}$$

$$g_3(t, y_2^n) := \Pr\{\tilde{\imath}^n(\bar{U}^n; y_2^n) \geq t\}, \tag{6.265}$$

where $\bar{V}^n$ and $\bar{U}^n$ are distributed according to P_{V^n} in (6.66) and P_{U^n} in (6.67), respectively. The above probabilities are all bounded as follows.

Lemma 6.5 *When n is sufficiently large, there exists constants $(K_1, K_2, K_{1,2}, K_3) \in \mathbb{R}_+^4$ such that*

$$g_1(t; y_1^n, u^n) \leq K_1 e^{-t}, \tag{6.266}$$

$$g_2(t; y_1^n, v^n) \leq K_2 e^{-t}, \tag{6.267}$$

$$g_{1,2}(t; y_1^n) \leq K_{1,2} e^{-t}, \tag{6.268}$$

$$g_3(t; y_2^n) \leq K_3 e^{-t}. \tag{6.269}$$

The proof of Lemma 6.5 follows similarly to Lemma 3.2.

Finally, define the following two constants:

$$\Delta_1 := 2\beta(\beta + \alpha P)\Big(C\Big(\frac{\alpha P}{\beta}\Big) - \log M_1 - \log(K_1\sqrt{n})\Big), \tag{6.270}$$

$$\Delta_3 := 2(1 + P)(1 + \alpha P)\Big(C\Big(\frac{\bar{\alpha} P}{\alpha P + 1}\Big) - \log M_2 - \log(K_3\sqrt{n})\Big). \tag{6.271}$$

6.6.1.2 Analysis of User 2

It follows from (6.256) that NN decoding in (6.62) is equivalent to maximizing $\tilde{\imath}^n(U^n(\bar{w}_2); Y_2^n)$ over $\bar{w}_2 \in [M_2]$, i.e.,

$$\phi_2(Y_2^n) = \operatorname*{arg\,min}_{\bar{w}_2 \in [M_2]} \|Y_2^n - U^n(\bar{w}_2)\|^2 \tag{6.272}$$

$$= \operatorname*{arg\,max}_{\bar{w}_2 \in [M_2]} nC\left(\frac{\bar{\alpha} P}{\alpha P + 1}\right) + \frac{\|Y_2^n\|^2}{2(P+1)} - \frac{\|Y_2^n - U^n(\bar{w}_2)\|^2}{2(\alpha P + 1)}. \tag{6.273}$$

It follows from a similar analysis of the random coding union bound [26, Theorem 16] that the error probability $\mathrm{P}_{e,2}^n$ satisfies

$$\mathrm{P}_{e,2}^n \le \mathbb{E}\Big[\min\big\{1, M_2 \Pr\{\tilde{\imath}^n(\bar{U}^n;Y_2^n) \ge \tilde{\imath}^n(U^n;Y_2^n)|U^n,Y_2^n\}\big\}\Big], \tag{6.274}$$

$$\le \mathbb{E}\Big[\min\big\{1, M_2 g_3(\tilde{\imath}^n(U^n;Y_2^n),Y_2^n)\big\}\Big] \tag{6.275}$$

$$\le \mathbb{E}\Big[\min\big\{1, M_2 K_3 \exp(-\tilde{\imath}^n(U^n;Y_2^n))\big\}\Big] \tag{6.276}$$

$$\le \Pr\Big\{M_2 K_3 \exp(-\tilde{\imath}^n(U^n;Y_2^n)) \ge \frac{1}{\sqrt{n}}\Big\} + \frac{1}{\sqrt{n}} \tag{6.277}$$

$$\le \frac{1}{\sqrt{n}} + \Pr\Big\{\log M_2 - \Big(nC\Big(\frac{\bar{\alpha}P}{\alpha P+1}\Big) + \frac{\|Y_2^n\|^2}{2(P+1)} - \frac{\|Y_2^n - U^n\|^2}{2(\alpha P+1)}\Big) \ge -\log(K_3\sqrt{n})\Big\} \tag{6.278}$$

$$= \frac{1}{\sqrt{n}} + \Pr\Big\{\frac{\|V^n+Z_2^n\|^2}{2(\alpha P+1)} - \frac{\|U^n+V^n+Z_2^n\|^2}{2(P+1)} \ge \frac{\Delta_3}{2(\alpha P+1)(P+1)}\Big\}, \tag{6.279}$$

where in (6.274), the random variables satisfy $(\bar{U}^n, U^n, Y_2^n) \sim P_{U^n}P_{U^n}P_{Y_2^n|U^n}$, (6.275) follows from the definition of the probability function $g_3(\cdot)$ in (6.265), (6.276) follows from the result in Lemma 6.5, (6.278) follows from the definition of $\tilde{\imath}^n(u^n;y_2^n)$ in (6.257), and (6.279) follows from the definition of Δ_3 in (6.271) and the channel model $Y^n = U^n + V^n + Z_2^n$.

The rest of the analysis is similar to the analysis of receiver 1 for the mismatched interference channel [21, Section III] with $K = 2$. For ease of notation, let RHS denotes the right-hand side of (6.279). The probability term in the RHS equals to [21, Eq. (110)] except that

1. The interference signal is replaced from $\tilde{X}^n$ to V^n and the corresponding per-symbol average power $\tilde{P}$ is replaced by αP,
2. The transmitted symbol X_1^n is replaced by U^n and its per-symbol average power is replaced from P_1 to $\bar{\alpha}P$,
3. The noise is replaced from Z^n to Z_2^n and the SINR of the channel is replaced from $\bar{P}$ to $\frac{\bar{\alpha}P}{\alpha P+1}$.

It follows [21, Eq. (111-132)] with $K = 2$ that

$$\mathrm{RHS} = \mathrm{Q}\left(\frac{\Delta_3}{\sqrt{nV_2(\bar{\alpha}P,\alpha P)}}\right) + O\left(\frac{1}{\sqrt{n}}\right). \tag{6.280}$$

Choose M_2 such that

$$\log M_2 = nC\Big(\frac{\bar{\alpha}P}{\alpha P+1}\Big) - \sqrt{n}L_2. \tag{6.281}$$

It follows from (6.279) and (6.280) that

$$\mathrm{P}_{\mathrm{e},2}^{n} \leq \mathrm{Q}\left(\frac{\sqrt{n}L_2 - \log(K_2\sqrt{n})}{\sqrt{nV_2(\bar{\alpha}P, \alpha P)}}\right) + O\left(\frac{1}{\sqrt{n}}\right). \tag{6.282}$$

Finally, for any $\varepsilon_2 \in (0,1)$, the desired bound on L_2 in Theorem 6.5 is obtained by equating the right-hand side of (6.282) to ε_2, solving for L_2 and performing a first-order Taylor expansion of $\mathrm{Q}^{-1}(\cdot)$ around ε_2.

6.6.1.3 Analysis of User 1 Under JNN

Recall the definitions of probability functions $g_1(t; y_1^n, u^n)$, $g_2(t; y_1^n, v^n)$, $g_{1,2}(t; y_2^n)$ in (6.262)-(6.264). For ease of notation, define the following random variables:

$$F_1 := g_1(\tilde{\imath}^n(V^n; Y_1^n|U^n); Y_1^n, U^n), \tag{6.283}$$

$$F_2 := g_2(\tilde{\imath}^n(U^n; Y_1^n|V^n); Y_1^n, V^n), \tag{6.284}$$

$$F_{1,2} := g_{1,2}(\tilde{\imath}^n(V^n, U^n; Y_1^n); Y_1^n). \tag{6.285}$$

Note that $(F_1, F_2, F_{1,2})$ are all functions of (U^n, V^n, Y_1^n).

Recall the definition of distributions P_{V^n} in (6.66) and P_{U^n} in (6.67). Recall that $Y_1^n = U^n + V^n + Z_1^n$ with Z_1^n generated i.i.d. from a distribution P_{Z_1} satisfying (6.58). Analogous to the MAC setting in Lemma 6.1, the following RCU bound holds.

Lemma 6.6 *There exist an (n, M_1, M_2)-JNN code such that the ensemble error probability of user 1 satisfies*

$$\mathrm{P}_{\mathrm{e},1}^{n} \leq \mathbb{E}\Big[\min\Big\{1, (M_1-1)F_1 + (M_2-1)F_2 + (M_1-1)(M_2-1)F_{1,2}\Big\}\Big], \tag{6.286}$$

where $(U^n, \bar{U}^n, V^n, \bar{V}^n, Y_1^n) \sim P_{U^n}P_{U^n}P_{V^n}P_{V^n}P_{Y_1^n|U^nV^n}$.

Proof When JNN decoding is used by user 1, the error event can be decomposed into the following three events:

- $\mathcal{E}_{1,1} : \{\hat{W}_1 \neq W_1, \bar{W}_2 = W_2\}$,
- $\mathcal{E}_{1,2} : \{\hat{W}_1 = W_1, \bar{W}_2 \neq W_2\}$,
- $\mathcal{E}_{1,3} : \{\hat{W}_1 \neq W_1, \bar{W}_2 \neq W_2\}$.

Thus, the error probability of user 1 satisfies

$$\mathrm{P}_{\mathrm{e},1}^{n} = \Pr\{\mathcal{E}_{1,1} \cup \mathcal{E}_{1,2} \cup \mathcal{E}_{1,3}\} \tag{6.287}$$

$$* \leq \Pr\{\mathcal{E}_{1,1}\} + \Pr\{\mathcal{E}_{1,2}\} + \Pr\{\mathcal{E}_{1,3}\}. \tag{6.288}$$

Each term in (6.288) is further bounded by using the equivalence between the maximum information density decoding and nearest neighbor decoding, analogously to Lemma 6.1.

Under event $\mathcal{E}_{1,1}$, it follows from the definition of the mutual information density $\tilde{\imath}^n(\cdot)$ in (6.259) that

$$\hat{W}_1 = \arg\min_{\bar{w}_1\in[M_1]} \|Y_1^n - U^n(W_2) - V^n(\bar{w}_1)\|^2 \tag{6.289}$$

$$= \arg\max_{\bar{w}_1\in[M_1]} nC\left(\frac{\alpha P}{\beta}\right) + \frac{\|Y_1^n - U^n(W_2)\|^2}{2(\alpha P+\beta)} - \frac{\|Y_2^n - U^n(W_2) - V^n(\bar{w}_1))\|^2}{2\beta} \tag{6.290}$$

$$= \arg\max_{\bar{w}_1\in[M_1]} \tilde{\imath}^n(V^n(\bar{w}_1); Y_1^n|U^n(W_2)). \tag{6.291}$$

Similarly, under events $\mathcal{E}_{1,2}$ and $\mathcal{E}_{1,3}$, NN decoding is equivalent to maximal mutual information density decoding. The rest of the proof is omitted due to similarity to Lemma 6.1. □

Recall that $(K_1, K_2, K_{1,2})$ are constants in Lemma 6.5. The following definitions are needed for further analyses. Firstly, define three random variables and a random vector:

$$\imath_{s,1} := \frac{\alpha P(n\beta - \|Z_1^n\|^2) + 2\beta\langle V^n, Z_1^n\rangle}{2\beta(\beta+\alpha P)}, \tag{6.292}$$

$$\imath_{s,2} := \frac{\bar{\alpha} P(n\beta - \|Z_1^n\|^2) + 2\beta\langle U^n, Z_1^n\rangle}{2\beta(\beta+\bar{\alpha} P)}, \tag{6.293}$$

$$\imath_{s,3} := \frac{P(n\beta - \|Z_1^n\|^2) + 2\beta(\langle Z_1^n, U^n + V^n\rangle + \langle U^n, V^n\rangle)}{2\beta(\beta+P)}, \tag{6.294}$$

$$\boldsymbol{\imath}_s := [\imath_{s,1}; \imath_{s,2}; \imath_{s,3}]. \tag{6.295}$$

Secondly, define three constant column vectors

$$\mathbf{C}_s := \left[C\left(\frac{\alpha P}{\beta}\right); C\left(\frac{\bar{\alpha} P}{\beta}\right); C\left(\frac{P}{\beta}\right)\right], \tag{6.296}$$

$$\mathbf{M}_s := [\log(\sqrt{n}M_1K_1); \log(\sqrt{n}M_2K_2); \log(\sqrt{n}M_1M_2K_{1,2})], \tag{6.297}$$

$$\boldsymbol{\tau}_s := \mathbf{M}_s - n\mathbf{C}_s. \tag{6.298}$$

Recall the definitions of mutual information densities in (6.259) and (6.261). Define the column random vector

$$\tilde{\boldsymbol{\imath}}_s := \left[\tilde{\imath}^n(V^n; Y_1^n|U^n); \tilde{\imath}^n(U^n; Y_1^n|V^n); \tilde{\imath}^n(V^n, U^n; Y_1^n)\right]. \tag{6.299}$$

Finally, define the following event and shorthand notation

$$\mathcal{S}_{\mathrm{s}} := \left\{\tilde{\boldsymbol{\imath}}_{\mathbf{s}} > \mathbf{M}_{\mathrm{s}}\right\}, \tag{6.300}$$
$$F_{\mathrm{s}} := (M_1 - 1)F_1 + (M_2 - 1)F_2 + (M_1 - 1)(M_2 - 1)F_{1,2}. \tag{6.301}$$

It follows from Lemma 6.6 and the above definitions that

$$\begin{aligned}
\mathrm{P}^n_{\mathrm{e},1} &= \mathbb{E}[\min\{1, F_{\mathrm{s}}\}] && (6.302)\\
&\leq \mathbb{E}[\min\{1, F_{\mathrm{s}}\}\mathbb{1}(\mathcal{S}_{\mathrm{s}}^c)] + \mathbb{E}[\min\left\{1, F_{\mathrm{s}}\}\right\}\mathbb{1}(\mathcal{S}_{\mathrm{s}})] && (6.303)\\
&\leq \Pr\{\mathcal{S}_{\mathrm{s}}^c\} + M_1\mathbb{E}\big[F_1\mathbb{1}(\tilde{\imath}^n(V^n; Y_1^n|U^n) \geq \log(\sqrt{n}M_1K_1))\big]\\
&\quad + M_2\mathbb{E}\big[F_2\mathbb{1}(\tilde{\imath}^n(U^n; Y_1^n|V^n) \geq \log(\sqrt{n}M_2K_2))\big]\\
&\quad + M_1M_2\mathbb{E}\big[F_{1,2}\mathbb{1}(\tilde{\imath}^n(V^n, U^n; Y_1^n) \geq \log(\sqrt{n}M_1M_2K_{1,2}))\big] && (6.304)\\
&\leq \Pr\{\mathcal{S}_{\mathrm{s}}^c\} + O\Big(\frac{1}{\sqrt{n}}\Big) && (6.305)\\
&\leq \Pr\{\boldsymbol{\imath}_{\mathrm{s}} \leq \boldsymbol{\tau}_{\mathrm{s}}\} + O\Big(\frac{1}{\sqrt{n}}\Big). && (6.306)
\end{aligned}$$

where (6.304) follows from the union bound and the definition of $\mathcal{S}_{\mathrm{s}}$ in (6.300), (6.305) follows from the result in Lemma 6.5 and the inequality $\exp\{-a\}\mathbb{1}(a \geq b) \leq \exp\{-b\}$ for any two real numbers $(a, b) \in \mathbb{R}^2$, and (6.306) follows from the definitions of $\boldsymbol{\imath}_{\mathrm{s}}$ in (6.295), $\boldsymbol{\tau}_{\mathrm{s}}$ in (6.298), and $\mathcal{S}_{\mathrm{s}}$ in (6.300), which imply that $\tilde{\boldsymbol{\imath}}_{\mathbf{s}} - \mathbf{M}_{\mathrm{s}} = \boldsymbol{\imath}_{\mathrm{s}} - \boldsymbol{\tau}_{\mathrm{s}}$.

The rest of this section bounds the probability term in (6.306). It follows from the definition of $C(P) = \frac{1}{2}\log(1+P)$ that $C(P)$ increases in its parameter P. Since $(\alpha, \beta) \in (0, 1)^2$ and $\bar{\alpha} = 1 - \alpha$, it follows that

$$C\Big(\frac{\bar{\alpha}P}{\beta}\Big) > C\Big(\frac{\bar{\alpha}P}{\alpha P + \beta}\Big) > C\Big(\frac{\bar{\alpha}P}{\alpha P + 1}\Big), \tag{6.307}$$

Furthermore,

$$\begin{aligned}
C\Big(\frac{P}{\beta}\Big) &= \frac{1}{2}\log\left(\frac{P+\beta}{\beta}\right) && (6.308)\\
&= \frac{1}{2}\log\frac{\beta+\alpha P}{\beta} + \frac{1}{2}\log\frac{P+\beta}{\beta+\alpha P} && (6.309)\\
&= C\Big(\frac{\alpha P}{\beta}\Big) + C\Big(\frac{\bar{\alpha}P}{\alpha P+\beta}\Big), && (6.310)
\end{aligned}$$

Recall that in (6.281), M_2 is chosen such that $\log M_2 = nC(\frac{\bar{\alpha}P}{\alpha P+1}) - \sqrt{n}L_2$, where L_2 was implicitly defined in the end of Sect. 6.6.1.2. Let $L_1 \in \mathbb{R}$ be specified later. Choose M_1 such that

$$\log M_1 = nC\left(\frac{\alpha P}{\beta}\right) - \sqrt{n}L_1. \tag{6.311}$$

Recall that $C(P)$ increases in P, $\beta \in (0, 1)$ and $\bar{\alpha} = 1-\alpha \in (0, 1)$. For sufficiently large n, it follows that

$$\boldsymbol{\tau}_{\mathrm{s}}(2) = nC\Big(\frac{\bar{\alpha}P}{\alpha P+1}\Big) - nC\Big(\frac{\bar{\alpha}P}{\beta}\Big) < 0, \tag{6.312}$$

$$\boldsymbol{\tau}_{\mathrm{s}}(3) = nC\Big(\frac{\bar{\alpha}P}{\alpha P+1}\Big) + nC\Big(\frac{\alpha P}{\beta}\Big) - nC\Big(\frac{P}{\beta}\Big) \tag{6.313}$$

$$= nC\Big(\frac{\bar{\alpha}P}{\alpha P+1}\Big) - \frac{n}{2}\log\frac{P+\beta}{\alpha P+\beta} \tag{6.314}$$

$$= nC\Big(\frac{\bar{\alpha}P}{\alpha P+1}\Big) - nC\Big(\frac{\bar{\alpha}P}{\alpha P+\beta}\Big) < 0. \tag{6.315}$$

It follows from the definition of $\boldsymbol{\imath}_{\mathrm{s}}$ in (6.295) that the random vector has zero means since the average power of noise Z_1^n is β and (Z_1^n, U^n, V^n) are independent. Thus, it follows from the weak law of numbers and the results in (6.312) and (6.315) that as $n \to \infty$,

$$\delta_{\mathrm{s}}^n := \Pr\{\boldsymbol{\imath}_{\mathrm{s}}(2,3) \le \boldsymbol{\tau}_{\mathrm{s}}(2,3)\} \to 0. \tag{6.316}$$

Furthermore, it follows from the definitions of $\boldsymbol{\imath}_{\mathrm{s}}$ in (6.295) and $\boldsymbol{\tau}_{\mathrm{s}}$ in (6.298) that

$$\Pr\{\boldsymbol{\imath}_{\mathrm{s}}(1) \le \boldsymbol{\tau}_{\mathrm{s}}(1)\}$$

$$= \Pr\left\{\frac{\alpha P(n\beta - \|Z_1^n\|^2) + 2\beta\langle V^n, Z_1^n\rangle}{2\beta(\beta+\alpha P)} \le \log(M_1K_1) - nC\Big(\frac{\alpha P}{\beta}\Big)\right\} \tag{6.317}$$

$$= \Pr\left\{\alpha P\|Z_1^n\|^2 - n\alpha\beta P - 2\beta\langle V^n, Z_1^n\rangle \ge \Delta_1\right\}, \tag{6.318}$$

where (6.318) follows from the definition of Δ_1 in (6.270).

The probability in (6.318) is further upper bounded using the function version Berry-Esseen theorem in Theorem 1.8. Let $\tilde{V}^n = (\tilde{V}_1, \ldots, \tilde{V}_n)$ be a sequence of i.i.d. random variables generated from the Gaussian distribution with mean zero and variance one. It follows that the spherical codeword $V^n = \frac{\sqrt{n\alpha P}\tilde{V}^n}{\|\tilde{V}^n\|}$. Fix any $i \in [n]$. Define the random variables

$$D_{1,i} := Z_{1,i}^2 - \beta, \tag{6.319}$$

$$D_{2,i} := -\sqrt{\alpha P}\tilde{V}_i Z_{1,i}, \tag{6.320}$$

$$D_{3,i} := 1 - \tilde{V}_i^2, \tag{6.321}$$

and let $\mathbf{D}_i := [D_{1,i}; D_{2,i}; D_{3,i}]$ be the random column vector. Note that $(\mathbf{D}_1, \ldots, \mathbf{D}_n)$ are i.i.d. random vectors with mean $\mathbf{0}_3$ and finite third moments.

The covariance matrix of $\mathbf{D}_i$ is $\mathrm{Cov}(\mathbf{D}_i) = \mathrm{diag}[\beta^2 - \zeta_1, -\alpha\beta P, -2]$. Given any three real numbers $(d_1, d_2, d_3) \in \mathbb{R}^3$, define the function

$$h(d_1, d_2, d_3) := \alpha P d_1 + \frac{2\beta d_2}{\sqrt{1 + d_3}}. \tag{6.322}$$

It follows that

$$nh\left(\frac{1}{n}\sum_{i=1}^{n}\mathbf{D}_i\right) = \alpha P \|Z_1^n\|^2 - n\beta\alpha P - 2\beta\langle V^n, Z_1^n\rangle \tag{6.323}$$

and Jacobian matrix of the function $h(\cdot)$ around $\mathbb{E}[\mathbf{D}_i] = \mathbf{0}_3$ is $\mathbf{J} = [\alpha P, 2\beta, 0]$. Furthermore,

$$\mathbf{J}\mathrm{Cov}(\mathbf{D}_i)\mathbf{J}^{\mathrm{T}} = 4\beta^2(\alpha P + \beta)^2 V_1(\alpha P) =: \sigma_{\mathrm{s}}. \tag{6.324}$$

It follows from (6.318) and (6.323) that

$$\Pr\{\boldsymbol{\imath}_{\mathrm{s}}(1) \geq \boldsymbol{\tau}_{\mathrm{s}}(1)\} = \Pr\left\{nh\left(\frac{1}{n}\sum_{i=1}^{n}\mathbf{D}_i\right) \geq \Delta_1\right\} \tag{6.325}$$

$$\leq \mathrm{Q}\left(\frac{\Delta_1}{\sqrt{n\sigma_{\mathrm{s}}}}\right) + O\left(\frac{1}{\sqrt{n}}\right), \tag{6.326}$$

$$= \mathrm{Q}\left(\frac{\sqrt{n}L_1 - \log(K_1\sqrt{n})}{\sqrt{nV_1(\alpha P)}}\right) + O\left(\frac{1}{\sqrt{n}}\right), \tag{6.327}$$

where (6.327) follows from the choice of M_1 such that $\log M_1 = nC\left(\frac{\alpha P}{\beta}\right) - \sqrt{n}L_1$ and the definition of Δ_1 in (6.270).

Combining (6.306), (6.316), and (6.327) leads to

$$\mathrm{P}_{\mathrm{e},1}^n \leq 1 - \Pr\{\boldsymbol{\imath}_{\mathrm{s}} \geq \boldsymbol{\tau}_{\mathrm{s}}\} + O\left(\frac{1}{\sqrt{n}}\right) \tag{6.328}$$

$$\leq 1 - \Pr\{\boldsymbol{\imath}_{\mathrm{s}}(1) \geq \boldsymbol{\tau}_{\mathrm{s}}(1)\} + \Pr\{\boldsymbol{\imath}_{\mathrm{s}}(2,3) < \boldsymbol{\tau}_{\mathrm{s}}(2,3)\} + O\left(\frac{1}{\sqrt{n}}\right) \tag{6.329}$$

$$\leq \mathrm{Q}\left(\frac{\sqrt{n}L_1 - \log(K_1\sqrt{n})}{\sqrt{nV_1(\alpha P)}}\right) + \delta_{\mathrm{s}}^n + O\left(\frac{1}{\sqrt{n}}\right). \tag{6.330}$$

where (6.329) follows from the inequality $\Pr\{\mathcal{A} \cap \mathcal{B}\} \geq \Pr\{\mathcal{A}\} - \Pr\{\mathcal{B}^c\}$ for any two events $(\mathcal{A}, \mathcal{B})$.

The desired bound on L_1 in Theorem 6.5 is obtained by setting the right-hand side of (6.327) to ε_1, solving L_1 and performing a first-order Taylor expansion of Q^{-1} around ε_1.

6.6.1.4 Analysis of User 1 Under SIC

When SIC decoding is used, the error probability of user 1 satisfies

$$\mathrm{P}_{\mathrm{e},1}^n = \Pr\{\bar{W}_2 \neq W_2 \text{ or } \hat{W}_1 \neq W_1\} \tag{6.331}$$

$$= \Pr\{\bar{W}_2 \neq W_2\} + \Pr\{\hat{W}_1 \neq W_1, \bar{W}_2 = W_2\} \tag{6.332}$$

$$\leq \Pr\{\bar{W}_2 \neq W_2\} + \Pr\{\hat{W}_1 \neq W_1 | \bar{W}_2 = W_2\}. \tag{6.333}$$

Using similar arguments as in Sect. 6.6.1.2 except that replacing the noise Z_2^n by Z_1^n leads to

$$\Pr\{\bar{W}_2 \neq W_2\} \leq \mathrm{Q}\left(\frac{nC\left(\frac{\bar{\alpha}P}{\alpha P+\beta}\right) - \log M_2 - \log(K_2\sqrt{n})}{\sqrt{nV_2(\bar{\alpha}P, \alpha P)}}\right) + O\left(\frac{1}{\sqrt{n}}\right) \tag{6.334}$$

$$\leq O\left(\frac{1}{\sqrt{n}}\right). \tag{6.335}$$

The result in (6.335) is justified as follows. Recall that in Sect. 6.6.1.2, M_2 is chosen such that $\log M_2 = nC(\frac{\bar{\alpha}P}{\alpha P+1}) - \sqrt{n}L_2$, with L_2 determined by the error probability ε_2 of user 2. Since $C(P)$ decreases in P and $\beta \in (0, 1)$, it follows that

$$C\left(\frac{\bar{\alpha}P}{\alpha P + \beta}\right) > C\left(\frac{\bar{\alpha}P}{\alpha P + 1}\right) \tag{6.336}$$

Thus, it follows from Chebyshev's inequality that the first term in (6.334) is in the order of $O(\frac{1}{n})$. Intuitively, (6.335) follows since user 1 has better channel with lesser noise power than user 2, which implies that the message W_2 can be reliably decoded by user 1 if user 2 can decode W_2 reliably.

The rest of this section bounds the second term $\Pr\{\hat{W}_1 \neq W_1 | \bar{W}_2 = W_2\}$ in (6.333). It follows from similar steps from (6.274) to (6.279) that

$$\Pr\{\hat{W}_1 \neq W_1 | \bar{W}_2 = W_2\}$$

$$\leq \frac{1}{\sqrt{n}} + \Pr\left\{(\alpha P + \beta)\|Z_1^n\|^2 - \beta\big(\|Z_1^n\|^2 + \|V^n\|^2 + 2\langle V^n, Z_1^n\rangle\big) \geq \Delta_1\right\} \tag{6.337}$$

$$= \Pr\left\{\alpha P\|Z_1^n\|^2 - n\alpha\beta P - 2\beta\langle V^n, Z_1^n\rangle \geq \Delta_1\right\} + \frac{1}{\sqrt{n}} \tag{6.338}$$

$$\leq \mathrm{Q}\left(\frac{\sqrt{n}L_1 - \log(K_1\sqrt{n})}{\sqrt{nV_1(\alpha P)}}\right) + O\Big(\frac{1}{\sqrt{n}}\Big), \tag{6.339}$$

where (6.338) follows since the codeword V^n satisfies the equal power constraint αP, i.e., $\|V^n\|^2 = n\alpha P$ and (6.339) follows from the same analysis as (6.318).

The proof is completed by combining (6.333), (6.335), and (6.339) and by calculating the desired bound on L_1 by setting the right-hand side of (6.327) to ε_1, solving L_1 and performing a first-order Taylor expansion of $\mathrm{Q}^{-1}(\cdot)$ around ε_1 with K_1 given in Lemma 6.5.

6.6.2 JEP

This section proves Theorem 6.6 with SIC decoding. The corresponding proof under JNN decoding is left as an exercise.

Recall the definition of $g_3(t; y_2^n)$ in (6.265) and F_1 in (6.283). Define the following random variable:

$$F_3 := g_3(\tilde{\imath}^n(U^n; Y_2^n); Y_2^n), \tag{6.340}$$

$$F_3' := g_3(\tilde{\imath}^n(U^n; Y_1^n); Y_1^n). \tag{6.341}$$

Similar to Lemma 6.6 for SEP with JNN decoding, the following RCU bound is obtained under JEP when SIC decoding is used by user 1.

Lemma 6.7 *There exist an (n, M_1, M_2)-SIC code such that the joint ensemble error probability satisfies*

$$\mathrm{P}_{\mathrm{e,j}}^n \leq \mathbb{E}\big[\min\{1, (M_1 - 1)F_1 + (M_2 - 1)F_3' + (M_2 - 1)F_3\}\big], \tag{6.342}$$

where $(U^n, \bar{U}^n, V^n, \bar{V}^n, Y_1^n, Y_2^n) \sim P_{U^n} P_{U^n} P_{V^n} P_{V^n} P_{Y_1^n|U^n V^n} P_{Y_2^n|U^n V^n}$.

Recall the definition of $\imath_{s,1}$ in (6.292). Define the following random variables:

$$\imath_{js,1} := \imath_{s,1}, \tag{6.343}$$

$$\imath_{js,2} := \frac{\bar{\alpha}P(n\beta - \|Z_1^n\|^2 - 2\langle V^n, Z_1^n\rangle)}{2(\beta+P)(\beta+\alpha P)} + \frac{\langle U^n, Z_1^n\rangle + \langle V^n, U^n\rangle}{\beta+P}, \tag{6.344}$$

$$\imath_{js,3} := \frac{\bar{\alpha}P(n\beta - \|Z_2^n\|^2 - 2\langle V^n, Z_2^n\rangle)}{2(1+P)(1+\alpha P)} + \frac{\langle U^n, Z_2^n\rangle + \langle V^n, U^n\rangle}{1+P}, \tag{6.345}$$

and the column vector

$$\boldsymbol{\imath}_{\mathbf{js}} := [\imath_{js,1}; \imath_{js,2}; \imath_{js,3}]. \tag{6.346}$$

Recall that K_1, K_3 are constants in Lemma 6.5. Furthermore, define three column vectors:

$$\mathbf{C}_{\rm js} := \left[C\Big(\frac{\alpha P}{\beta}\Big); C\Big(\frac{\bar{\alpha}P}{\alpha P+\beta}\Big); C\Big(\frac{\bar{\alpha}P}{\alpha P+1}\Big)\right] \tag{6.347}$$

$$\mathbf{M}_{\rm j,s} := \left[\log(\sqrt{n}M_1K_1); \log(\sqrt{n}M_2K_2); \log(\sqrt{n}M_2K_3)\right], \tag{6.348}$$

$$\boldsymbol{\tau}_{\rm js} := \mathbf{M}_{\rm j,s} - n\mathbf{C}_{\rm js}. \tag{6.349}$$

Finally, define the random variable

$$\tilde{\boldsymbol{\imath}}_{\rm js} := \left[\tilde{\imath}^n(V^n; Y_1^n|U^n); \tilde{\imath}^n(U^n; Y_1^n); \tilde{\imath}^n(U^n; Y_2^n)\right], \tag{6.350}$$

and the event

$$\mathcal{S}_{\rm js} := \left\{\tilde{\boldsymbol{\imath}}_{\rm js} \geq \mathbf{M}_{\rm js}\right\}. \tag{6.351}$$

Using Lemma 6.7, similarly to (6.302) to (6.306) except that (i) replacing $\tilde{\boldsymbol{\imath}}_{\mathbf{s}}$ and $\boldsymbol{\imath}_{\rm s}$ by $\tilde{\boldsymbol{\imath}}_{\rm js}$ and $\boldsymbol{\imath}_{\mathbf{js}}$, respectively, and (ii) replacing $\mathbf{C}_{\mathbf{s}}$ $\mathbf{M}_{\mathbf{s}}$, $\boldsymbol{\tau}_{\rm s}$ and $\mathcal{S}_{\mathbf{s}}$ by $\mathbf{C}_{\mathbf{js}}$, $\mathbf{M}_{\rm j,s}$, $\boldsymbol{\tau}_{\rm js}$ and $\mathcal{S}_{\mathbf{js}}$, respectively, it follows that

$$\mathrm{P}_{\rm e,j}^n \leq 1 - \Pr\{\boldsymbol{\imath}_{\mathbf{js}} \geq \boldsymbol{\tau}_{\rm js}\} + O\Big(\frac{1}{\sqrt{n}}\Big). \tag{6.352}$$

Let $L_2 \in \mathbb{R}$ be specified later. Choose M_2 such that

$$\log M_2 = nC\left(\frac{\bar{\alpha}P}{\alpha P+1}\right) - \sqrt{n}L_2. \tag{6.353}$$

It follows from (6.307) that

$$\boldsymbol{\tau}_{\rm js}(2) = nC\Big(\frac{\bar{\alpha}P}{\alpha P+1}\Big) - nC\Big(\frac{\bar{\alpha}P}{\alpha P+\beta}\Big) < 0. \tag{6.354}$$

Note that the average values of the random vector $\boldsymbol{\iota}_{\mathbf{js}}$ is $\mathbf{0}_3$. It follows from the weak law of large numbers that as $n \to \infty$,

$$\delta_{\mathrm{j}}^n := \Pr\{\boldsymbol{\iota}_{\mathbf{js}}(2) \le \boldsymbol{\tau}_{\mathrm{js}}(2)\} \to 0. \tag{6.355}$$

As a result, the result (6.352) is further upper bounded by

$$\mathrm{P}_{\mathrm{e,j}}^n \le 1 - \Pr\{\boldsymbol{\iota}_{\mathbf{js}}(1,3) \ge \boldsymbol{\tau}_{\mathrm{js}}(1,3)\} - \delta_{\mathrm{j}}^n + O\Big(\frac{1}{\sqrt{n}}\Big). \tag{6.356}$$

The rest of the proof lower bounds the probability term in (6.356) using the function version Berry-Esseen theorem (cf. Theorem 1.8). Let $(\tilde{U}^n, \tilde{V}^n)$ be two independent i.i.d. random vectors generated from the Gaussian distribution with mean zero and variance one. It follows that the spherical codewords satisfy $V^n = \sqrt{n\alpha P}\frac{\tilde{V}^n}{\|\tilde{V}^n\|}$ and $U^n = \sqrt{n\bar{\alpha} P}\frac{\tilde{U}^n}{\|\tilde{U}^n\|}$. Furthermore, for each $i \in [n]$, let $\mathbf{D}_i^{\mathrm{j}} := (D_{1,i}^{\mathrm{j}}, \ldots, D_{8,i}^{\mathrm{j}})$ be a random vector, where

$$\begin{aligned}
D_{1,i}^{\mathrm{j}} &:= \beta - Z_{1,i}^2, & D_{2,i}^{\mathrm{j}} &:= \sqrt{\alpha P}\tilde{V}_i Z_{1,i},\\
D_{3,i}^{\mathrm{j}} &:= \tilde{V}_i^2 - 1, & D_{4,i}^{\mathrm{j}} &:= 1 - Z_{2,i}^2,\\
D_{5,i}^{\mathrm{j}} &:= \sqrt{\bar{\alpha} P}\tilde{U}_i Z_{2,i}, & D_{6,i}^{\mathrm{j}} &:= \tilde{U}_i^2 - 1,\\
D_{7,i}^{\mathrm{j}} &:= \sqrt{\bar{\alpha}\alpha P^2}\tilde{V}_i\tilde{U}_i, & D_{8,i}^{\mathrm{j}} &:= \sqrt{\alpha P}\tilde{V}_i Z_{2,i}.
\end{aligned} \tag{6.357}$$

Given any real numbers $\mathbf{d} = (d_1, \ldots, d_8) \in \mathbb{R}^8$, define the function $f_{\mathrm{j}}(\mathbf{d})$ such that

$$\begin{aligned}
f_{\mathrm{j}}(\mathbf{d}) :=& \frac{1}{2\beta(\beta+\alpha P)}\Big(\alpha P d_1 + \frac{2\beta d_2}{\sqrt{1+d_3}}\Big) + \frac{1}{2(1+P)(1+\alpha P)}\\
&\times \Bigg(\bar{\alpha} P\Big(d_4 - \frac{2d_8}{\sqrt{1+d_3}}\Big)\\
&+ 2(1+\alpha P)\Big(\frac{d_5}{\sqrt{1+d_6}} + \frac{d_7}{\sqrt{1+d_6}\sqrt{1+d_3}}\Big)\Bigg).
\end{aligned} \tag{6.358}$$

It follows that

$$f_{\mathrm{j}}\left(\frac{1}{n}\sum_{i=1}^n \mathbf{D}_i^{\mathrm{j}}\right) = \frac{1}{n}\boldsymbol{\iota}_{\mathrm{js}}(1,3). \tag{6.359}$$

Fix any $i \in [n]$. The covariance matrix of the random vector satisfies

$$\mathrm{Cov}(\mathbf{D}_i^{\mathrm{j}}) = \mathrm{diag}[\zeta_1 - \beta^2, \alpha\beta P, 2, \zeta_2 - 1, \bar{\alpha} P, 2, \alpha\bar{\alpha} P^2, \alpha P] =: \mathbf{V}_{\mathrm{j}}. \tag{6.360}$$

The Jacobian matrix of the function f_{j} around $\mathbb{E}[\mathbf{D}_i^{\mathrm{j}}] = \mathbf{0}_8$ is given by

$$\mathbf{J}_{\mathrm{j}} = [k_1 \alpha P, 2k_1 \beta, 0, k_2 \alpha P, 2K_2(1+\alpha P), 0, 2k_2(1+\alpha P), -2\bar{\alpha} P]. \tag{6.361}$$

Thus,

$$\mathbf{J}_{\mathrm{j}} \mathbf{V}_{\mathrm{j}} \mathbf{J}_{\mathrm{j}}^{\mathrm{T}} = \begin{bmatrix} V_1(\alpha P) & 0 \\ 0 & V_2(\bar{\alpha} P, \alpha P) \end{bmatrix} \tag{6.362}$$

$$= \mathbf{V}\big(\alpha, P\big). \tag{6.363}$$

where $V_1(\alpha P)$, $V_2(\bar{\alpha} P, \alpha P)$ and $\mathbf{V}\big(\alpha, P\big)$ were defined around (6.82).

Let $\mathbf{S}_2 = (S_1, S_2) \sim \mathcal{N}\big(\cdot; \mathbf{0}_2, \mathbf{V}(\alpha, P)\big)$ be a two-dimensional Gaussian random vector with zero means and covariance matrix $\mathbf{V}(\alpha, P)$. It follows from the function version Berry-Esseen theorem and (6.359) that

$$\Pr\{\boldsymbol{\imath}_{\mathbf{js}}(1,3) \geq \boldsymbol{\tau}_{\mathrm{js}}(1,3)\} - \Pr\left\{\mathbf{S}_2 \geq \frac{\boldsymbol{\tau}_{\mathrm{js}}(1,3)}{\sqrt{n}}\right\}$$
$$\geq O\Big(n^{-\frac{1}{4}}\Big). \tag{6.364}$$

Combining (6.356) and (6.364), the joint error probability can be finally upper bounded by

$$\mathrm{P}_{\mathrm{e,j}}^n \leq 1 - \Pr\left\{\mathbf{S}_2 \geq \frac{\boldsymbol{\tau}_{\mathrm{js}}(1,3)}{\sqrt{n}}\right\} + O\Big(n^{-\frac{1}{4}}\Big) \tag{6.365}$$

$$= 1 - \Pr\left\{\mathbf{S}_2 \leq -\frac{\boldsymbol{\tau}_{\mathrm{js}}(1,3)}{\sqrt{n}}\right\} + O\Big(n^{-\frac{1}{4}}\Big). \tag{6.366}$$

Recall the definition of $\mathrm{Q_{inv}}(\mathbf{V}, \varepsilon)$ in (6.14). If

$$-\frac{\boldsymbol{\tau}_{\mathrm{js}}(1,3)}{\sqrt{n}} \in \mathrm{Q_{inv}}\left(\mathbf{V}(\alpha, P), \varepsilon - O\Big(n^{-\frac{1}{4}}\Big)\right), \tag{6.367}$$

the error probability satisfies

$$\limsup_{n \to \infty} \mathrm{P}_{\mathrm{e,j}}^n \leq \varepsilon. \tag{6.368}$$

The proof under JEP is completed by applying the Taylor expansion of the ccdf function $\mathrm{Q_{inv}}(\cdot)$ around ε.

6.7 Chapter Summary

This chapter presents mismatched codes and their low-latency performance for three multiuser communication models: MAC, RAC, and BC. MAC and RAC models uplink communication, while the BC model downlinks communication. For each channel, the mismatched code can be used to transmit messages over an additive, arbitrary noise channel under mild moment conditions on the noise distributions. When decoding multiple messages, either SIC or JNN decoding can be used. For MAC, both SIC and JNN decoding achieve the same asymptotic performance, while JNN decoding achieves strictly better low-latency performance. For RAC, JNN decoding achieves strictly better performance for both the asymptotic and low-latency settings. For BC, both SIC and JNN decoding achieve the same asymptotic and low-latency performance. A key takeaway message is that for BC, SIC decoding should be used due to its low complexity, while JNN decoding should be used for MAC and RAC for enhanced low-latency performance.

6.8 Exercises

6.1 What are orthogonal multiple access and non-orthogonal multiple access? For orthogonal multiple access, what are typical examples?

6.2 What is channel coding model for non-orthogonal multiple access?

6.3 What is the difference between a multiple access channel and a random access channel?

6.4 For any $P \in \mathbb{R}_+$, the capacity function $C(P) = \frac{1}{2}\log(+P)$. Prove that $\frac{C(kP)}{k} > C(P/(1+(k-1)P))$ when $k \geq 2$.

6.5 Prove the random coding union bound in Lemma 6.2 for SIC decoding.

6.6 Prove the result in (6.228) using the function version of the Berry-Esseen theorem in Theorem 1.8.

6.7 Prove Theorem 6.4 concerning second-order asymptotics for mismatched RAC with SIC decoding.

6.8 Prove Theorem 6.6 concerning second-order asymptotics for mismatched BC under JEP performance criterion with JNN decoding for the strong user.

References

1. A.W. Scott, R. Frobenius, *Multiple Access Techniques: FDMA, TDMA, and CDMA* (Wiley-IEEE Press, 2008), pp. 413–429

2. K.B. Letaief, W. Chen, Y. Shi, J. Zhang, Y.A. Zhang, The roadmap to 6G: AI empowered wireless networks. IEEE Commun. Mag. **57**(8), 84–90 (2019)
3. C.E. Shannon, Two-way communication channels, in *The Proceedings of the Fourth Berkeley Symposium on Mathematical Statistics and Probability*, vol. 1 (University of California Press, Berkeley, 1961), pp. 611–645
4. R. Ahlswede, Multiway communication channels, in *IEEE ISIT* (Tsahkadsor, Armenia, 1971), pp. 23–51
5. H. Liao, Multiple access channels. Ph.D. Dissertation, Department of Electrical and Computer Engineering, University of Hawaii, 1972
6. Y. Polyanskiy, A perspective on massive random-access, in *IEEE ISIT* (2017), pp. 2523–2527
7. O. Ordentlich, Y. Polyanskiy, Low complexity schemes for the random access Gaussian channel, in *IEEE ISIT* (2017), pp. 2528–2532
8. Y. Wu, X. Gao, S. Zhou, W. Yang, Y. Polyanskiy, G. Caire, Massive access for future wireless communication systems. IEEE Wirel. Commun. **27**(4), 148–156 (2020)
9. T. Cover, Broadcast channels. IEEE Trans. Inf. Theory **18**(1), 2–14 (1972)
10. P. Bergmans, Random coding theorem for broadcast channels with degraded components. IEEE Trans. Inf. Theory **19**(2), 197–207 (1973)
11. P. Bergmans, A simple converse for broadcast channels with additive white Gaussian noise (corresp.). IEEE Trans. Inf. Theory **20**(2), 279–280 (1974)
12. R.G. Gallager, Capacity and coding for degraded broadcast channels. Probl. Inf. Transm. **10**(3), 3–14 (1974)
13. A. El Gamal, T.M. Cover, Multiple user information theory. Proc. IEEE **68**(12), 1466–1483 (1980)
14. A.D. Wyner. Recent results in the Shannon theory. IEEE Trans. Inf. Theory **20**(1), 2–10 (1974)
15. T.M. Cover, Some advances in broadcast channels, in *Advances in Communication Systems*, ed. by A. J. Viterbi, vol. 4 (Academic Press, San Francisco, 1975), pp. 229–260
16. E. MolavianJazi, J. N. Laneman, A second-order achievable rate region for Gaussian multi-access channels via a central limit theorem for functions. IEEE Trans. Inf. Theory **61**(12), 6719–6733 (2015)
17. R.C. Yavas, V. Kostina, M. Effros, Gaussian multiple and random access channels: finite-blocklength analysis. IEEE Trans. Inf. Theory **67**(11), 6983–7009 (2021)
18. R.C. Yavas, V. Kostina, M. Effros, Random access channel coding in the finite blocklength regime. IEEE Trans. Inf. Theory **67**(4), 2115–2140 (2021)
19. V.Y.F. Tan, O. Kosut, On the dispersions of three network information theory problems. IEEE Trans. Inf. Theory **60**(2), 881–903 (2014)
20. J. Scarlett, V.Y.F. Tan, Second-order asymptotics for the Gaussian MAC with degraded message sets. IEEE Trans. Inf. Theory **61**(12), 6700–6718 (2015)
21. J. Scarlett, V.Y.F. Tan, G. Durisi, The dispersion of nearest-neighbor decoding for additive non-Gaussian channels. IEEE Trans. Inf. Theory **63**(1), 81–92 (2017)
22. Y. Wang, L. Zhou, L. Bai, Achievable second-order asymptotics for additive non-Gaussian MAC and RAC, in *IEEE GLOBECOM* (2023), pp. 3627–3632
23. Y. Wang, L. Bai, Z. Wu, L. Zhou, Achievable second-order asymptotics for MAC and RAC with additive non-Gaussian noise (2024). arXiv: 2410.10312
24. Z. Wu, L. Zhou, J. Xu, L. Bai, Second-order asymptotics for asymmetric broadcast channel with non-Gaussian noise in *IEEE ICCC* (2024), pp. 551–556
25. Z. Wu, L. Bai, J. Xu, L. Zhou, M. Motani, The dispersion of broadcast channels with degraded message sets using Gaussian codebooks (2024). arXiv: 2410.17540
26. Y. Polyanskiy, H.V. Poor, S. Verdú, Channel coding rate in the finite blocklength regime. IEEE Trans. Inf. Theory **56**(5), 2307–2359 (2010)

Part II
Applications

Chapter 7
MIMO

This chapter studies data transmission over a point-to-point multiple-input multiple-output (MIMO) channel with quasi-static fading. The fading parameters, better known as channel state information (CSI), characterize how the transmitted signals are distorted in a multiplier manner, which are either available or unavailable at the transmitter/receiver. There are in total four combinations of CSI availability: no CSI at either the transmitter or the receiver (noCSI), CSI at the receiver (CSIR), CSI at the transmitter (CSIT), and CSI at the transmitter and the receiver (CSITR). The assumption of quasi-static fading is consistent with the practical need of low-latency communication. This is because the short blocklength codes enforced by low-latency communication can be transmitted in a time duration smaller than the channel coherence time. Thus, each symbol of the transmitted signal suffers from the same level of fading [1, Page 2631], [2, Section 5.4.1].

Classical Shannon theory studies the maximal transmission rate when the error probability tends to zero as the blocklength tends to infinity. However, Shannon theory cannot be applied to low-latency communication over a quasi-static fading channel. This is because low-latency enforces finite blocklength, and the error probability of data transmission over a quasi-static fading channel is usually bounded away from zero, e.g., Rayleigh, Rician, or Nakagami fading channels. The latter is because the probability of having arbitrary small fading parameter is non-zero for certain fading distributions. Therefore, the usual notion of capacity is zero for data transmission over a quasi-static fading channel. Instead, the outage capacity is considered, which measures the maximal data transmission rate that ensures reliable communication with high probability. Outage capacity is thus the counterpart of Shannon capacity for quasi-static fading channels and is an asymptotic notion that holds with infinite blocklength.

The low-latency performance of data transmission over quasi-static fading channels requires finite blocklength. Thus, the validity of outage capacity remains unclear. For non-fading single antenna channels, the finite blocklength performance of optimal codes was characterized in Sect. 2.3.2 of Chap. 2. This section generalizes

L. Zhou, L. Bai, *Low-latency Communication*, Textbooks in Telecommunication Engineering, https://doi.org/10.1007/978-3-032-06627-5_7

the results in Sect. 2.3.2 to MIMO quasi-static fading channels and presents the results by Yang, Durisi, Koch, and Polyanskiy [3]. A principal finding is that outage capacity, despite being an asymptotic notion, is a sharp approximation for the finite blocklength theoretical benchmarks for low-latency communication under quasi-static fading. This is because the second-order term, which is non-zero for non-fading channels, equals zero for quasi-static fading channels under all four possibilities of CSI availability.

7.1 Problem Formulation

7.1.1 System Model

Fix integers $(r, t, n, M) \in \mathbb{N}^4$. As shown in Fig. 7.1, this chapter studies a quasi-static MIMO fading channel with t transmit and r receive antennas. Let m denote the minimum number of transmit and receive antennas, i.e., $m := \min\{t, r\}$.

Given a message $W \in [M]$, the encoder generates channel input $\mathsf{X} \in \mathbb{C}^{n\times t}$, which is transmitted via n channel uses. The additive noise $\mathbb{Z} \in \mathbb{C}^{n\times r}$ is generated i.i.d. from the complex Gaussian distribution $\mathcal{CN}(0, 1)$. Thus, the corresponding channel output is

$$\mathbb{Y} = \mathsf{X}\mathbb{H} + \mathbb{Z}. \tag{7.1}$$

Based on the CSI availability, the problem is classified into the following four cases, where for each case, the code is defined.

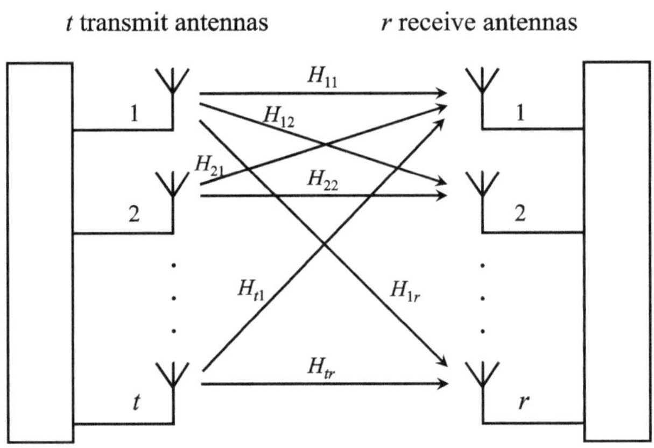

Fig. 7.1 System model for point-to-point quasi-static MIMO fading channel with t transmit and r receive antennas

Definition 7.1 (noCSI) When neither the transmitter nor the receiver is aware of the realizations of the fading matrix $\mathbb{H}$, an $(n, M, \varepsilon)_{\text{no}}$ code consists of:

1. An encoder $f_{\text{no}}\colon [M] \mapsto \mathbb{C}^{n\times t}$ that maps the message $J \in [M]$ to a codeword $\mathsf{X} \in \{\mathsf{C}_1, \cdots, \mathsf{C}_M\}$. The codewords satisfy the power constraint

$$\|\mathsf{C}_i\|_{\mathsf{F}}^2 \le n\rho, \quad i \in [M]; \tag{7.2}$$

2. A decoder $g_{\text{no}}\colon \mathbb{C}^{n\times r} \mapsto [M]$ satisfying a maximum probability of error constraint

$$\max_{1\le j\le M} \Pr\{g_{\text{no}} \ne J|J = j\} \le \varepsilon, \tag{7.3}$$

where $\mathbb{Y}$ is the channel output induced by the transmitted codeword $\mathsf{X} = f_{\text{no}}(j)$ according to (7.1).

Definition 7.2 (CSIR) When only the receiver knows $\mathbb{H}$, an $(n, M, \varepsilon)_{\text{r}}$ code consists of:

1. An encoder $f_{\text{no}}\colon [M] \mapsto \mathbb{C}^{n\times t}$ that maps the message $J \in [M]$ to a codeword $\mathsf{X} \in \{\mathsf{C}_1, \cdots, \mathsf{C}_M\}$. The codewords satisfy the power constraint (7.2);
2. A decoder $g_{\text{r}}\colon \mathbb{C}^{n\times r} \times \mathbb{C}^{t\times r} \mapsto [M]$ satisfying

$$\max_{1\le j\le M} \Pr\{g_{\text{r}}(\mathbb{Y}, \mathbb{H}) \ne J|J = j\} \le \varepsilon. \tag{7.4}$$

Definition 7.3 (CSIT) When only the transmitter knows $\mathbb{H}$, an $(n, M, \varepsilon)_{\text{t}}$ code consists of:

1. An encoder $f_{\text{t}}\colon [M] \times \mathbb{C}^{t\times r} \mapsto \mathbb{C}^{n\times t}$ that maps the message $j \in [M]$ and the channel H to a codeword $\mathsf{X} = f_{\text{t}}(j, \mathsf{H})$ satisfying

$$\|\mathsf{X}\|_{\mathsf{F}}^2 = \|f_{\text{t}}(j, \mathsf{H})\|_{\mathsf{F}}^2 \le n\rho, \quad \forall\, j \in [M],\ \forall\, \mathsf{H} \in \mathbb{C}^{t\times r}. \tag{7.5}$$

2. A decoder $g_{\text{no}}\colon \mathbb{C}^{n\times r} \mapsto [M]$ satisfying (7.3).

Definition 7.4 (CSITR) When both the transmitter and the receiver know $\mathbb{H}$, an $(n, M, \varepsilon)_{\text{tr}}$ code consists of:

1. An encoder $f_{\text{t}}\colon [M] \times \mathbb{C}^{t\times r} \mapsto \mathbb{C}^{n\times t}$ that maps the message $j \in [M]$ and the channel H to a codeword $\mathsf{X} = f_{\text{t}}(j, \mathsf{H})$ satisfying (7.5).
2. A decoder $g_{\text{r}}\colon \mathbb{C}^{n\times r} \times \mathbb{C}^{t\times r} \mapsto [M]$ satisfying (7.4).

Analogously to Chap. 2, the finite blocklength theoretical benchmark is defined as follows with $\dagger \in \{\text{no}, \text{r}, \text{t}, \text{tr}\}$:

$$R_{\dagger}^*(n, \varepsilon) := \sup\left\{\frac{\log M}{n} : \exists\, (n, M, \varepsilon)_{\dagger} \text{ code}\right\}. \tag{7.6}$$

It follows from the assumption of CSI availability that

$$R^*_{\text{no}}(n, \varepsilon) \le R^*_{\text{t}}(n, \varepsilon) \le R^*_{\text{tr}}(n, \varepsilon), \tag{7.7}$$

$$R^*_{\text{no}}(n, \varepsilon) \le R^*_{\text{r}}(n, \varepsilon) \le R^*_{\text{tr}}(n, \varepsilon). \tag{7.8}$$

7.1.2 Outage Probability

This section recalls the notation of outage probability for the special case of $t = r = 1$. Thus, a single-antenna communication system operates over a quasi-static flat-fading channel. Given the fading parameter H, the outage probability with respect to the rate R is defined as

$$F(R) := \Pr\{\log(1 + |H|^2\rho) < R\}, \tag{7.9}$$

where ρ is the signal-to-noise ratio (SNR). For a given $\varepsilon > 0$, the outage capacity (or ε-capacity) C_ε is the supremum of all rates R satisfying $F(R) \le \varepsilon$. For each realization of the fading coefficient $H = h$, the quasi-static fading channel can be viewed as an additive white Gaussian noise (AWGN) channel with channel gain $|h|^2$, for which reliable communication with vanishing error probability is possible if and only if $R < \log(1+|h|^2\rho)$, when the blocklength n is sufficiently large. Thus, the outage probability can be interpreted as the probability that the channel gain H is too small to allow reliable communication with rate R. Corresponding, the outage capacity with error probability ε is defined as

$$G(\varepsilon) := \max\{R \in \mathbb{R}_+ : \ F(R) \le \varepsilon\}. \tag{7.10}$$

The definition of outage probability is straightforward. However, it is itself an asymptotic notion by involving the capacity term $\log(1 + |H|^2\rho)$. Thus, the outage probability is not naturally a performance metric for data transmission over a quasi-static channel. In other words, when $\varepsilon > 0$ is allowed, it remains unclear whether the first-order asymptotic rate $\lim_{n\to\infty} R^*(n, \varepsilon)$ equals the outage capacity $G(\varepsilon)$.

For non-fading channels, it was shown [4–6] that under mild conditions,

$$R^*(n, \varepsilon) = C - \sqrt{\frac{V}{n}}\mathrm{Q}^{-1}(\varepsilon) + O\left(\frac{\log n}{n}\right). \tag{7.11}$$

where C is the channel capacity, V is the channel dispersion (cf. Sect. 2.3.2), and $\mathrm{Q}^{-1}(\cdot)$ is the inverse ccdf function. The approximation in (7.11) implies that to satisfy the error probability constraint $\varepsilon < 0.5$, at a finite blocklength n, one has to suffer a penalty on the transmission rate that scales in the order of $1/\sqrt{n}$ and depends on both the channel dispersion function and the error probability.

For quasi-static fading channels, it follows from [3] that under mild conditions on the fading distribution,

$$R^*(n, \varepsilon) = C_\varepsilon + O\left(\frac{\log n}{n}\right). \tag{7.12}$$

In the above expression, C_ε is roughly the outage capacity $G(\varepsilon)$ in (7.10), with further details specified later. The above result holds for all four cases of CSI availability, and the result differs when CSIT is available or not. Comparing (7.11) and (7.12), it can be observed that for the quasi-static fading case, the $1/\sqrt{n}$ rate penalty is absent. In other words, the second-order term of quasi-static fading channels is zero. This suggests that the maximal achievable rate $R^*(n, \varepsilon)$ converges quickly to C_ε as the blocklength n tends to infinity, indicating that the outage capacity is indeed a valid performance metric for delay-constrained communication over slowly varying fading channels. The rest of this chapter presents and proves the above result.

7.2 Theoretical Results and Discussions

It was noted in [1, Page 2632] that the ε-capacity of quasi-static MIMO fading channel does not depend on whether CSI is available at the receiver. Intuitively, this is true because the channel stays constant during the transmission of a codeword, so it can be accurately estimated at the receiver through the transmission of $\sqrt{n}$ pilot symbols with no rate penalty as the blocklength $n \to \infty$. Thus, this section is divided into two subsections: CSIT and no CSIT. For each case, the finite blocklength and second-order asymptotic bounds are presented and discussed.

7.2.1 CSI Available at the Transmitter

7.2.1.1 Preliminaries

Let $\mathcal{U}_t$ be the set of $t \times t$ positive semidefinite matrices whose trace is upper-bounded by ρ, i.e.,

$$\mathcal{U}_t := \{\mathsf{A} \in \mathbb{C}^{t \times t} : \mathsf{A} \succeq \mathbf{0}, \mathrm{tr}(\mathsf{A}) \leq \rho\}. \tag{7.13}$$

The ε-capacity $C_\varepsilon^{\mathrm{t}}$ is defined as the supremum of all rates R such that the outage probability is no greater than ε, i.e.,

$$C_\varepsilon^{\mathrm{t}} := \sup\{R : F_{\mathrm{t}}(R) \leq \varepsilon\}, \tag{7.14}$$

where the outage probability is defined as

$$F_{\mathrm{t}}(R) := \Pr\left\{ \max_{\mathsf{Q}\in\mathcal{U}_t} \log\det\left(\mathsf{I}_r + \mathbb{H}^{\mathrm{H}}\mathsf{Q}\mathbb{H}\right) < R \right\}. \tag{7.15}$$

Given a particular realization $\mathbb{H} = \mathsf{H}$, the function $\log\det\left(\mathsf{I}_r + \mathsf{H}^{\mathrm{H}}\mathsf{Q}\mathsf{H}\right)$ is maximized by the water-filling power allocation strategy (cf. [7, 8]), which results in

$$\max_{\mathsf{Q}\in\mathcal{U}_t} \log\det\left(\mathsf{I}_r + \mathsf{H}^{\mathrm{H}}\mathsf{Q}\mathsf{H}\right) = \sum_{j\in[m]} \left|\log(\bar{\gamma}\lambda_j)\right|^{+}, \tag{7.16}$$

where the scalars $\lambda_1 \geq \cdots \geq \lambda_m$ denote the m largest eigenvalues of $\mathsf{H}^{\mathrm{H}}\mathsf{H}$, and $\bar{\gamma}$ is the solution of

$$\sum_{j\in[m]} \left|\bar{\gamma} - \frac{1}{\lambda_j}\right|^{+} = \rho. \tag{7.17}$$

Let $\mathcal{A}$ and $\mathcal{B}$ be subspaces in $\mathbb{C}^n$ with $a = \dim(\mathcal{A}) \leq \dim(\mathcal{B}) = b$. The angles $(\theta_1, \ldots, \theta_k)$ are obtained recursively as follows. When $k = 1$, θ_1 is obtained via

$$\cos\theta_1 := \max_{\mathbf{a}\in\mathcal{A},\mathbf{b}\in\mathcal{B}:\|\mathbf{a}\|=\|\mathbf{b}\|=1} |\langle\mathbf{a},\mathbf{b}\rangle|. \tag{7.18}$$

Let a_1 and b_1 be the optimizer of (7.18). Fix any $k \in [2:a]$. The angle θ_k is defined such that

$$\cos\theta_k := \max_{\substack{\mathbf{a}\in\mathcal{A},\mathbf{b}\in\mathcal{B}:\|\mathbf{a}\|=\|\mathbf{b}\|=1,\\ \langle\mathbf{a},\mathsf{a}_i\rangle=\langle\mathbf{b},\mathsf{b}_i\rangle=0,\forall\, i\in[k-1]}} |\langle\mathbf{a},\mathbf{b}\rangle|, \tag{7.19}$$

where $(\mathsf{a}_i, \mathsf{b}_i)$ is the optimizer of (7.19) when $k = i$. Finally, the angle between the subspaces $\mathcal{A}$ and $\mathcal{B}$ is defined as

$$\sin\{\mathcal{A}, \mathcal{B}\} := \prod_{k\in[a]} \sin\theta_k. \tag{7.20}$$

The angle of subspaces is used extensively in this chapter, in the presentation and proof of theoretical results.

7.2.1.2 Finite Blocklength Bounds

This section studies the case where CSI is available at the transmitter but not at the receiver. Firstly, the achievability bound is stated.

Let $\Lambda_1 \geq \cdots \geq \Lambda_m$ be the m largest eigenvalues of $\mathbb{H}\mathbb{H}^{\mathrm{H}}$. For every $\varepsilon \in (0, 1)$ and every $\tau \in (0, \varepsilon)$, let $\gamma_n \in [0, 1]$ be chosen such that

$$\Pr\left\{\sin^2\left\{\mathsf{I}_{n,t}, \sqrt{n}\mathsf{I}_{n,t}\mathrm{diag}\left\{\sqrt{v_1^*\Lambda_1}, \cdots, \sqrt{v_m^*\Lambda_m}, \mathbf{0}_{t-m}\right\} + \mathbb{Z}\right\} \leq \gamma_n\right\} \geq 1 - \varepsilon - \tau, \tag{7.21}$$

where

$$v_j^* = \left|\bar{\gamma} - \frac{1}{\Lambda_j}\right|^+, \quad j \in [r] \tag{7.22}$$

are the water-filling power gains and $\bar{\gamma}$ is defined in (7.17).

Let the Beta distribution [9, Chapter 25] be denoted by Beta$(\cdot, \cdot)$ and let $B^r = (B_1, \ldots, B_r)$ be independent Beta-distributed random variables, where $B_j \sim \mathrm{Beta}(n - t - j + 1, t)$.

Theorem 7.1 *For any* $\varepsilon \in (0, 1)$, *the maximal achievable rate on the quasi-static MIMO fading channel* (7.1) *with CSIT satisfies*

$$R_{\mathrm{t}}^*(n, \varepsilon) \geq \frac{1}{n} \log \frac{\tau}{\Pr\left\{\prod_{j\in[r]} B_j \leq \gamma_n\right\}}. \tag{7.23}$$

The achievability bound is based on a decoder that operates as follows: it first computes the sine of the angle between the subspace spanned by the received matrix $\mathbb{Y}$ and the subspace spanned by each codeword; then, it chooses the first codeword for which the squared sine of the angle is below γ_n. The performance of this decoder is analyzed via the $\kappa\beta$-bound [4, Theorem 25] to a physically degraded channel whose output is span$(\mathbb{Y})$.

Correspondingly, a converse bound is stated as follows. Let $\mathbf{\Lambda} := [\Lambda_1, \cdots, \Lambda_m]^{\mathrm{T}}$. Consider an arbitrary power-allocation function $\mathbf{v} : \mathbb{R}_+^m \mapsto \mathcal{V}_m$, where

$$\mathcal{V}_m := \left\{[p_1, \cdots, p_m] \in \mathbb{R}_+^m : \sum_{j\in[m]} p_j \leq \rho\right\}. \tag{7.24}$$

Let

$$L_n^{\mathrm{tr}}(\mathbf{v}, \mathbf{\Lambda}) := \sum_{i\in[n]} \sum_{j\in[m]} \left(\log\left(1 + \Lambda_j v_j(\mathbf{\Lambda})\right) + 1 - \left|\sqrt{\Lambda_j v_j(\mathbf{\Lambda})} Z_{i,j} - \sqrt{1 + \Lambda_j v_j(\mathbf{\Lambda})}\right|^2\right), \tag{7.25}$$

and

$$S_n^{\mathrm{tr}}(\mathbf{v}, \mathbf{\Lambda}) := \sum_{i \in [n]} \sum_{j \in [m]} \left(\log\left(1 + \Lambda_j v_j(\mathbf{\Lambda})\right) + 1 - \frac{\left|\sqrt{\Lambda_j v_j(\mathbf{\Lambda})} Z_{i,j} - 1\right|^2}{1 + \Lambda_j v_j(\mathbf{\Lambda})} \right), \tag{7.26}$$

where $v_j(\cdot)$ is the j-th coordinate of $\mathbf{v}(\cdot)$, and $\{Z_{i,j}\}_{i \in [n], j \in [m]}$, are i.i.d. $\mathcal{CN}(0, 1)$ distributed random variables. Let

$$c_{\mathrm{tr}}(n) := \left(\frac{(n-1)^n e^{-(n-1)}}{\Gamma(n)} + \frac{\Gamma(n, n-1)}{\Gamma(n)} \right)^m \times \mathbb{E}_{\mathbb{H}} \left[\det(\mathsf{I}_t + \rho \mathbb{H}\mathbb{H}^{\mathrm{H}}) \right], \tag{7.27}$$

and the scalar $\gamma_n(\mathbf{v})$ is the solution of

$$\Pr\left\{ S_n^{\mathrm{tr}}(\mathbf{v}, \mathbf{\Lambda}) \leq n\gamma_n(\mathbf{v}) \right\} = \varepsilon. \tag{7.28}$$

The infimum on the right-hand side of (7.29) is taken over all power allocation functions $\mathbf{v} : \mathbb{R}_+^m \mapsto \mathcal{V}_m$.

The following converse bound assumes that CSITR is available.

Theorem 7.2 *For any n and any $\varepsilon \in (0, 1)$, the maximal achievable rate on the quasi-static MIMO fading channel* (7.1) *with CSITR is upper-bounded by*

$$R_{\mathrm{tr}}^*(n, \varepsilon) \leq \frac{1}{n} \log \frac{c_{\mathrm{tr}}(n)}{\inf_{\mathbf{v}(\cdot)} \Pr\{L_n^{\mathrm{tr}}(\mathbf{v}, \mathbf{\Lambda}) \geq n\gamma_n(\mathbf{v})\}}. \tag{7.29}$$

The converse bound in Theorem 7.2 is proved using the meta-converse bound [4, Theorem 30]. Since CSI is available at both the transmitter and the receiver, the MIMO channel (7.1) can be transformed into a set of parallel quasi-static channels via the singular value decomposition.

7.2.1.3 Second-Order Asymptotics

The following conditions are required. Assume that the fading channel $\mathbb{H}$ satisfies (1) the expectation $\mathbb{E}_{\mathbb{H}}[\det(\mathsf{I}_t + \rho\mathbb{H}\mathbb{H}^{\mathrm{H}})]$ is finite; (2) the joint pdf of the ordered nonzero eigenvalues of $\mathbb{H}^{\mathrm{H}}\mathbb{H}$ exists and is continuously differentiable; (3) $C_\varepsilon^{\mathrm{t}}$ is a point of growth of the outage probability function (7.15), i.e.,

$$F_{\mathrm{t}}'(C_\varepsilon^{\mathrm{t}}) > 0. \tag{7.30}$$

The result below characterizes the ε-dispersion of the quasi-static fading channel (7.1) with CSIT, regardless of whether CSIR is available or not.

Theorem 7.3 *For any* $\varepsilon \in (0, 1)$, *the maximal achievable rate* $R^*(n, \varepsilon)$ *on the channel* (7.1) *with CSIT and CSITR satisfies*

$$\{R_{\mathrm{t}}^*(n, \varepsilon), R_{\mathrm{tr}}^*(n, \varepsilon)\} = C_\varepsilon^{\mathrm{t}} + O\left(\frac{\log n}{n}\right). \tag{7.31}$$

The assumptions on the fading matrix are satisfied by most fading distributions used to model MIMO fading channels, e.g., i.i.d. or correlated Rayleigh, Rician, and Nakagami fading. However, the non-fading MIMO AWGN channel does not meet these assumptions. In fact, the second-order term for MIMO AWGN channel is non-zero [10, Theorem 78].

For the quasi-static MIMO AWGN fading channels, Theorem 7.3 implies that the second-order term equals to zero. Thus, the maximum achievable rate $R^*(n, \varepsilon)$ rapidly converges to the first-order rate C_ε as the blocklength n increases. Such a result validates the adoption of outage capacity for low-latency communication over quasi-static fading channels, although the notation of outage capacity itself is an asymptotic notation.

To prove Theorem 7.3, the achievability part considers CSIT and proves

$$R_{\mathrm{t}}^*(n, \varepsilon) \geq C_\varepsilon^{\mathrm{t}} + O\left(\frac{\log n}{n}\right) \tag{7.32}$$

by judiciously analyzing the asymptotic expansion of the finite blocklength achievability bound in Theorem 7.1. The converse part considers CSITR and proves

$$R_{\mathrm{tr}}^*(n, \varepsilon) \leq C_\varepsilon^{\mathrm{t}} + O\left(\frac{\log n}{n}\right) \tag{7.33}$$

using the finite blocklength converse bound in Theorem 7.2. Finally, the proof is completed by using the relationship of transmission rates in (7.7).

7.2.2 CSI Not Available at the Transmitter

7.2.2.1 Preliminaries

When CSI is not available at the transmitter, the ε-capacity [8, 11] $C_\varepsilon^{\mathrm{no}}$ is the supremum of all rates R such that the corresponding outage probability is no greater than ε, i.e.,

$$C_\varepsilon^{\mathrm{no}} := \sup\{R : F_{\mathrm{no}}(R) \leq \varepsilon\}, \tag{7.34}$$

where

$$F_{\text{no}}(R) := \inf_{\mathsf{Q} \in \mathcal{U}_t} \Pr\left\{\log\det\left(\mathsf{I}_r + \mathbb{H}^{\mathrm{H}}\mathsf{Q}\mathbb{H}\right) < R\right\}, \tag{7.35}$$

is the outage probability for the no-CSIT case. The matrix Q that minimizes the right-hand side of (7.35) is generally unknown, making this case difficult to analyze. The minimization in (7.35) can be restricted to all Q on the boundary of $\mathcal{U}_t$, i.e.,

$$F_{\text{no}}(R) = \inf_{\mathsf{Q} \in \mathcal{U}_t^{\text{e}}} \Pr\left\{\log\det\left(\mathsf{I}_r + \mathbb{H}^{\mathrm{H}}\mathsf{Q}\mathbb{H}\right) < R\right\}, \tag{7.36}$$

where

$$\mathcal{U}_t^{\text{e}} := \{\mathsf{A} \in \mathbb{C}^{t \times t} : \mathsf{A} \succeq \mathbf{0}, \operatorname{tr}(\mathsf{A}) = \rho\}. \tag{7.37}$$

7.2.2.2 Finite Blocklength Bounds

First is the achievability bound, which is derived under noCSI. Assume that for a given $\varepsilon \in (0, 1)$ and any $\tau \in (0, \varepsilon)$, there exists a $\mathsf{Q}^* \in \mathcal{U}_t$ such that

$$F_{\text{no}}(C_\varepsilon^{\text{no}}) = \inf_{\mathsf{Q} \in \mathcal{U}_t} \Pr\left\{\log\det\left(\mathsf{I}_r + \mathbb{H}^{\mathrm{H}}\mathsf{Q}\mathbb{H}\right) \le C_\varepsilon^{\text{no}}\right\} \tag{7.38}$$

$$= \Pr\left\{\log\det\left(\mathsf{I}_r + \mathbb{H}^{\mathrm{H}}\mathsf{Q}^*\mathbb{H}\right) \le C_\varepsilon^{\text{no}}\right\}, \tag{7.39}$$

i.e., the infimum in (7.38) is a minimum. Let $\gamma_n \in [0, 1]$ such that

$$\Pr\left\{\sin^2\{\mathsf{I}_{n,t^*}, \sqrt{n}\mathsf{I}_{n,t^*}\mathsf{U}\mathbb{H} + \mathbb{Z}\} \le \gamma_n\right\} \ge 1 - \varepsilon + \tau \tag{7.40}$$

with $\mathsf{U} \in \mathbb{C}^{t^* \times t}$ satisfying $\mathsf{U}^{\mathrm{H}}\mathsf{U} = \mathsf{Q}^*$. Let $t^* := \operatorname{rank}(\mathsf{Q}^*)$. Recall that $B^r = (B_1, \ldots, B_r)$ are independent Beta-distributed random variables with $B_j \sim \text{Beta}(n - t^* - j + 1, t^*)$ for each $j \in [r]$.

Theorem 7.4 *For any $\varepsilon \in (0, 1)$, the maximal achievable rate on the quasi-static MIMO fading channel* (7.1) *with no-CSI satisfies*

$$R_{\text{no}}^*(n, \varepsilon) \ge \frac{1}{n} \log \frac{\tau}{\Pr\left\{\prod_{j \in [r]} B_j \le \gamma_n\right\}}. \tag{7.41}$$

The proof is identical to the proof of Theorem 7.1, with the only difference that the precoding matrix $\mathsf{P}(\mathbb{H})$ (defined in (7.84) below) is replaced by $\sqrt{n}\mathsf{I}_{n,t^*}\mathsf{U}$.

Next is the finite blocklength converse bound, which is derived under CSIR. Let $\mathcal{U}_t^{\text{e}}$ be as in (7.37). For an arbitrary $\mathsf{Q} \in \mathcal{U}_t^{\text{e}}$, let $\Lambda_1 \ge \cdots \ge \Lambda_m$ be the ordered eigenvalues of $\mathbb{H}^{\mathrm{H}}\mathsf{Q}\mathbb{H}$. Let

$$L_n^{\mathrm{r}}(\mathsf{Q}) := \sum_{i\in[n]}\sum_{j\in[m]}\left(\log\left(1+\Lambda_j\right)+1-\left|\sqrt{\Lambda_j}Z_{i,j}-\sqrt{1+\Lambda_j}\right|^2\right), \tag{7.42}$$

and

$$S_n^{\mathrm{r}}(\mathsf{Q}) := \sum_{i\in[n]}\sum_{j\in[m]}\left(\log\left(1+\Lambda_j\right)+1-\frac{\left|\sqrt{\Lambda_j}Z_{i,j}-1\right|^2}{1+\Lambda_j}\right), \tag{7.43}$$

where $\{Z_{i,j}\}_{i\in[n],j\in[m]}$ are i.i.d. random variables generated from the complex Gaussian distribution $\mathcal{CN}(0, 1)$. Let

$$\begin{aligned} c_{\mathrm{r}}(n) &:= \frac{\pi^{r(r-1)}}{\Gamma_r(n)\Gamma_r(r)}\mathbb{E}\left[\left(1+\rho\|\mathbb{H}\|_{\mathsf{F}}^2\right)^{\lfloor\frac{(r+1)^2}{4}\rfloor}\right] \\ &\times\prod_{i\in[r]}\left[(n+r-2i)^{n+r-2i+1}e^{-(n+r-2i)}+\Gamma(n+r-2i+1,n+r-2i)\right], \end{aligned} \tag{7.44}$$

with $\Gamma_{(\cdot)}(\cdot)$ denoting the complex multivariate Gamma function [12, Eq. (83)], and $\gamma_n(\mathsf{Q})$ is the solution of

$$\Pr\{S_n^{\mathrm{r}}(\mathsf{Q}) \le n\gamma_n(\mathsf{Q})\} = \varepsilon. \tag{7.45}$$

Theorem 7.5 *For any $n \ge r$ and any $\varepsilon \in (0, 1)$, the maximal achievable rate on the quasi-static MIMO fading channel* (7.1) *with CSIR is upper-bounded by*

$$R_{\mathrm{r}}^{*}(n-1,\varepsilon) \le \frac{1}{n-1}\log\frac{c_{\mathrm{r}}(n)}{\inf_{\mathsf{Q}\in\mathcal{U}_t^{\mathrm{e}}}\Pr\{L_n^{\mathrm{r}}(\mathsf{Q})\ge n\gamma_n(\mathsf{Q})\}}. \tag{7.46}$$

The infimum in (7.46) makes the upper bound more difficult to evaluate numerically and to analyze asymptotically up to $O(\log(n)/n)$ terms than the upper bound (7.29) for the CSIT case.

7.2.2.3 Second-Order Asymptotics

To present the result, the following definition of the gradient ∇g of a differentiable function $g : \mathbb{C}^{t\times r} \mapsto \mathbb{R}$ will be needed. Let $\mathsf{L} \in \mathbb{C}^{t\times r}$, write $\nabla g(\mathsf{H}) = \mathsf{L}$ if

$$\frac{d}{dt}g(\mathsf{H}+t\mathsf{A})\bigg|_{t=0} = \mathrm{Re}\Big\{\mathrm{tr}(\mathsf{A}^{\mathrm{H}})\mathsf{L}\Big\},\quad \forall\,\mathsf{A}\in\mathbb{C}^{t\times r}. \tag{7.47}$$

Theorem below establishes the zero-dispersion result for the case of no-CSIT. Because of the analytical intractability of the minimization in the converse bound (7.46), the result of no-CSIT requires more stringent conditions on the fading distribution compared to the CSIT case, and its proof is more involved.

The following conditions are required. Let $f_{\mathbb{H}}$ be the pdf of the fading matrix $\mathbb{H}$. Assume that $\mathbb{H}$ satisfies (1) $f_{\mathbb{H}}$ is a smooth function, i.e., it has derivatives of all orders; (2) there exists a positive constant $a \in \mathbb{R}_+$ such that

$$f_{\mathbb{H}}(\mathsf{H}) \leq a \|\mathsf{H}\|_{\mathsf{F}}^{-2tr - \lfloor \frac{(r+1)^2}{2} \rfloor - 1}, \tag{7.48}$$

$$\|\nabla f_{\mathbb{H}}(\mathsf{H})\|_{\mathsf{F}} \leq a \|\mathsf{H}\|_{\mathsf{F}}^{-2tr-5}. \tag{7.49}$$

(3) The outage probability $F_{\text{no}}(\cdot)$ in (7.35) satisfies

$$\liminf_{\delta \to 0} \frac{F_{\text{no}}(C_\varepsilon^{\text{no}} + \delta) - F_{\text{no}}(C_\varepsilon^{\text{no}})}{\delta} > 0. \tag{7.50}$$

Theorem 7.6 *For any $\varepsilon \in (0, 1)$, the maximal achievable rate $R^*(n, \varepsilon)$ on the quasi-static MIMO fading channel* (7.1) *with no-CSI and CSIR satisfies*

$$\{R^*_{\text{no}}(n, \varepsilon), R^*_{\text{r}}(n, \varepsilon)\} = C_\varepsilon^{\text{no}} + O\left(\frac{\log n}{n}\right). \tag{7.51}$$

The conditions 1–3 for Theorem 7.6 are satisfied by commonly used fading distributions, including Rayleigh, Rician, and Nakagami. Condition 2 requires simply that $f_{\mathbb{H}}$ has a polynomially decaying tail. Condition 3 plays the same role as (7.30) in the CSIT case. The exact counterpart of (7.30) for the no-CSIT case is

$$F'_{\text{no}}(C_\varepsilon^{\text{no}}) > 0. \tag{7.52}$$

Similar to Theorem 7.3, the ε-dispersion is zero for both the CSIR and the no-CSI case. Thus, it has been proven that (under some mild conditions on the fading distribution) the channel dispersion is zero for all four cases of CSI availability. In all these cases, the distribution of the achievable outage capacity is known.

7.2.3 *Numerical Illustrations*

This section illustrates the theoretical results through numerical simulations. Figure 7.2 compares the normal approximation of $R^*_{\text{rt}}(n, \varepsilon)$ with the achievability bound and the converse bound where CSI is available at the transmitter for a quasi-static SIMO fading channel with two receive antennas. The channels between the transmit antenna and each of the two receive antennas are Rician-distributed with K-factor equal to 20 dB. The two channels are assumed to be independent. Also

Fig. 7.2 Achievability and converse bounds for a quasi-static SIMO Rician-fading channel with K-factor equal to 20 dB, two receive antennas, SNR $= -1.55$ dB, and $\varepsilon = 10^{-3}$

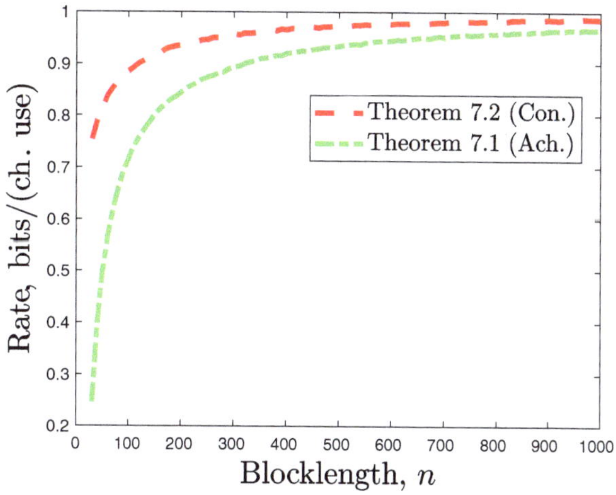

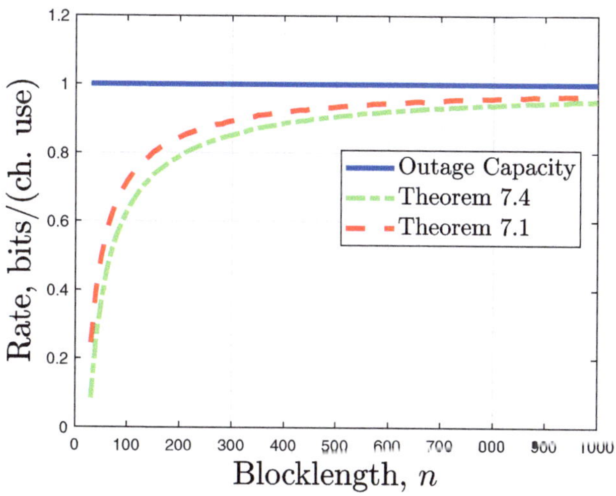

Fig. 7.3 The achievability bounds under both the CSIT and no-CSI scenarios for a quasi-static SIMO Rician-fading channel with K-factor equal to 20 dB, two receive antennas, SNR $= -1.55$ dB, and $\varepsilon = 10^{-3}$

let $\varepsilon = 10^{-3}$ and $\rho = -1.55$ dB so that $C_\varepsilon =$ 1bit/(ch.use). Figure 7.3 illustrates the achievability bounds under both the CSIT and no-CSI scenarios, and also plots the outage capacity C_ε. The channels between the transmit antenna and each of the two receive antennas are also Rician-distributed with K-factor equal to 20 dB. The two channels are assumed to be independent. Also let $\varepsilon = 10^{-3}$ and $\rho = -1.55$ dB so that $C_\varepsilon =$ 1bit/(ch.use). By combining Fig. 7.2, the blocklength $O(\log(n)/n)$ required to achieve 90% of the ε-capacity of the quasi-static fading channel is in the range [120, 320] for the CSIT case and in the range [120, 480] for the no-CSI case. For the AWGN channel, this number is approximately 1420. Figure 7.4 depicts

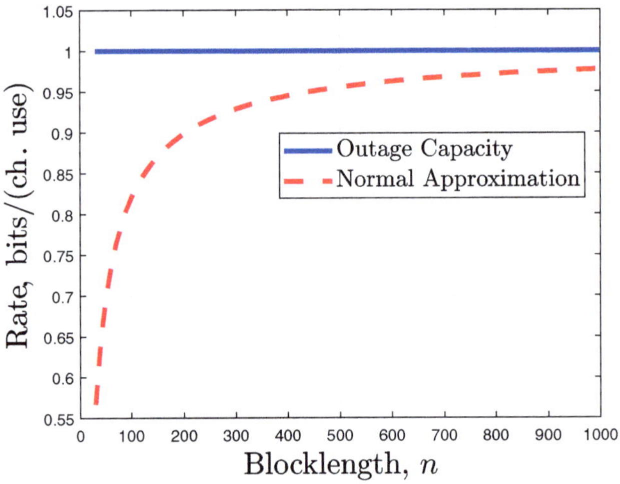

Fig. 7.4 The normal approximation of $R^*_{\mathrm{rt}}(n, \varepsilon)$ for a quasi-static SIMO Rician-fading channel with K-factor equal to 20 dB, two receive antennas, SNR $= -1.55$ dB, and $\varepsilon = 10^{-3}$

the relationship between the normal approximation of $R^*_{\mathrm{rt}}(n, \varepsilon)$ and the outage capacity C_ε. The gap between the normal approximation of $R^*_{\mathrm{rt}}(n, \varepsilon)$ and both the achievability and the converse bounds is less than 0.02 bit/(ch. use) for blocklengths larger than 400. Hence, for the parameters chosen, the prediction (based on zero dispersion) of fast convergence to capacity is validated.

7.3 Proof of CSIT

This section proves the finite blocklength and second-order asymptotic bounds in Theorems 7.1, 7.2, and 7.3.

7.3.1 Preliminaries

The achievability bounds on $R^*(n, \varepsilon)$ applies $\kappa\beta$-bound [4, Theorem 25] to a stochastically degraded channel, whose choice is motivated by geometric considerations. The converse bounds on $R^*(n, \varepsilon)$ use the meta-converse bound [4, Theorem 30].

7.3.1.1 Angle Threshold Decoding

Before establishing the achievability bound, some geometric intuition is provided to guide the choice of the decoder g_{no}. Consider for simplicity a real-valued quasi-static single input single output (SISO) channel, i.e., $t = r = 1$ and thus

$$Y^n = Hx^n + Z^n, \tag{7.53}$$

where Y^n, x^n, and Z^n are n-dimensional vectors, and H is a scalar. As reviewed, the typical error event for the quasi-static fading channel (in the large blocklength regime) is that the instantaneous channel gain H^2 is not large enough to support the desired rate R, i.e., $\frac{1}{2}\log(1 + \rho H^2) < R$ (outage event). For the channel in (7.53), the ε-capacity C_ε, i.e., the largest rate R for which the probability that the channel is in outage is less than ε, is given by

$$C_\varepsilon = \sup\left\{R : \Pr\left\{\frac{1}{2}(1 + \rho H^2) < R\right\} \leq \varepsilon\right\}. \tag{7.54}$$

Roughly speaking, the decoder of a C_ε-achieving code may commit an error only when the channel is in outage.

Pick now an arbitrary codeword x_1^n from the hypersphere $\{x^n \in \mathbb{R}^n : \|x^n\|^2 = n\rho\}$, and let Y^n be the received signal corresponding to x_1^n. Following [13], the angle $\theta(x_1^n, Y^n)$ between x_1^n and Y^n is analyzed as follows. By the law of large numbers, the noise vector Z^n is approximately orthogonal to x_1^n if n is large, i.e.,

$$\frac{\langle x_1^n, Z^n\rangle}{\|x_1^n\|\|Z^n\|} \to 0, \quad n \to \infty. \tag{7.55}$$

Also by the law of large numbers, $\|Z^n\|^2/n \to 1$ as $n \to \infty$. Hence, for a given H and for large n, the angle $\theta(x_1^n, Y^n)$ can be approximated as

$$\theta(x_1^n, Y^n) \approx \arcsin\frac{\|Z^n\|}{\sqrt{H^2\|x_1^n\|^2 + \|Z^n\|^2}} \tag{7.56}$$

$$\approx \arcsin\frac{1}{\rho H^2 + 1}, \tag{7.57}$$

where the first approximation follows by (7.55) and the second approximation follows because $\|Z^n\|^2 \approx n$. As shown in Fig. 7.5, it follows from (7.54) and (7.57) that $\theta(x_1^n, Y^n)$ is larger than $\theta_\varepsilon := \arcsin(e^{-C_\varepsilon})$ in the outage case and smaller than θ_ε otherwise.

This geometric argument suggests the use of a threshold decoder that, for a given received signal Y^n, declares x_i^n to be the transmitted codeword if x_i^n is the only codeword for which $\theta(x_i^n, Y^n) \leq \theta_\varepsilon$. If no codewords or more than one codeword

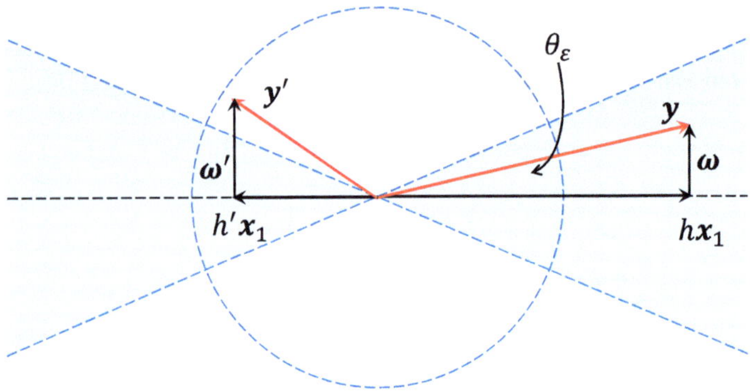

Fig. 7.5 Angle threshold decoding. In the example, the fading realization h' triggers an outage event, h does not

meet this condition, the decoder declares an error. Thresholding angles instead of log-likelihood ratios appears to be a natural approach when CSIR is unavailable. Note that the proposed threshold decoder does neither require CSIR nor knowledge of the fading distribution. As observed, it yields a tight achievability bound at finite blocklength, provided that the threshold θ_ε is chosen appropriately.

The above threshold decoder is generalized to the MIMO case by using (7.18), (7.19), and (7.20).

7.3.1.2 Existing Results

Given two distributions P and Q on a common measurable space $\mathcal{W}$, a randomized test between P and Q is defined as a random transformation $P_{Z|W} : \mathcal{W} \mapsto \{0, 1\}$ where 0 indicates that the test chooses Q. The following performance metric is needed for the test between P and Q:

$$\beta_\alpha(P, Q) := \min \int P_{Z|W}(1|w)Q(dw), \tag{7.58}$$

where the minimum is over all probability distributions $P_{Z|W}$ satisfying

$$\int P_{Z|W}(1|w)P(dw) \geq \alpha. \tag{7.59}$$

Lemma 7.1 (**$\kappa\beta$-Bound**) *Let* A *be the set of all codewords, for any* $0 < \varepsilon < 1$, *there exists an* (M, ε) *code with codewords chosen from* $\mathrm{F} \subset \mathrm{A}$, *satisfying*

$$M \geq \sup_{0<\tau<\varepsilon} \sup_{Q_Y} \frac{\kappa_\tau(\mathrm{F}, Q_Y)}{\sup_{x\in\mathrm{F}} \beta_{1-\varepsilon+\tau}(x, Q_Y)}, \tag{7.60}$$

where $\beta_{(\cdot)}(\cdot,\cdot)$ *is defined in* (7.58) *and*

$$\kappa_\tau(\mathrm{F}, Q_Y) = \inf_{\substack{P_{Z|Y}: \\ \inf_{x\in \mathrm{F}} P_{Z|X}(1|x)\geq \tau}} \sum_{y\in \mathrm{B}} Q_Y(y) P_{Z|Y}(1|y). \tag{7.61}$$

Proof of Lemma 7.1 Construct the codebook sequentially: Step 1. Choose $c_1 \in \mathrm{F}$ arbitrarily. Note that regardless of the choice of c_1, it follows from the definition of $\beta_{(\cdot)}(\cdot,\cdot)$ in (7.58) that

$$\Pr\{Z_{c_1} = 1|X = c_1\} \geq 1 - \varepsilon + \tau. \tag{7.62}$$

Step k. Assume $c_1, \cdots, c_{k-1}$ have been chosen. Choose $c_k \in \mathrm{F}$ so that

$$\begin{aligned}
&\Pr\{Z_{c_1} = 0, \cdots, Z_{c_{k-1}} = 0, Z_{c_k} = 1|X = c_k\} \\
&= \sum_{y\in \mathrm{B}} P_{Y|X}(y|c_k) P_{Z_{c_k}|Y}(1|y) \prod_{i\in[k-1]} P_{Z_{c_i}|Y}(0|y) && (7.63)\\
&> 1 - \varepsilon. && (7.64)
\end{aligned}$$

If step k exists, let M represent the number of steps required to complete the above process; otherwise, let $M = \infty$.

The decoder simply applies the M independent random transformations $P^*_{Z_{c_1}|Y}$, $\cdots, P^*_{Z_{c_M}|Y}$ to the data. If all M outputs are 0, the decoder outputs 0; otherwise, it outputs the smallest i such that $Z_{c_i} = 1$.

It follows from the encoder/decoder construction and (7.58) that the maximal error probability of the code is indeed upper bounded by ε. Let

$$Z^* = \sum_{i\in[M]} Z_{c_i} \tag{7.65}$$

be the binary sum of $(Z_{c_1}, \ldots, Z_{c_M})$. For any $x \in \mathrm{F}$,

$$\begin{aligned}
P_{Z^*|X}(0|x) &= \Pr\{Z^* = 0, Z_x = 1|X = x\} + \Pr\{Z^* = 0, Z_x = 0|X = x\} && (7.66)\\
&\leq 1 - \varepsilon + \Pr\{Z^* = 0, Z_x = 0|X = x\} && (7.67)\\
&\leq 1 - \varepsilon + \Pr\{Z_x = 0|X = x\} && (7.68)\\
&< 1 - \varepsilon + \varepsilon - \tau && (7.69)\\
&= 1 - \tau, && (7.70)
\end{aligned}$$

where (7.67) follows because if $x \in \{c_1, \cdots, c_M\}$, it is impossible that $Z^* = 0$ and $Z_x = 1$ simultaneously, while if $x \in \mathrm{F} - \{c_1, \cdots, c_M\}$, it was not possible to add x to the codebook, and therefore $\Pr\{Z^* = 0, Z_x = 1|X = x\} \leq 1 - \varepsilon$, and (7.69) follows by the construction of Z_x from (7.58). From (7.70), it is concluded that $P_{Z^*|X}$ is such that

$$\inf_{x \in \mathrm{F}} P_{Z^*|X}(1|x) \geq \tau. \tag{7.71}$$

It follows that the decoding error probability is less than $1 - \tau$, implying that the probability of correctly decoding the transmitted codeword as Z^* is at least τ. An upper bound for $\kappa_\tau(\mathrm{F}, Q_Y)$ can be provided as follows:

$$\kappa_\tau(\mathrm{F}, Q_Y) \leq \sum_{y \in \mathrm{B}} Q_Y(y) P_{Z^*|Y}(1|y) \tag{7.72}$$

$$\leq \sum_{y \in \mathrm{B}} Q_Y(y) \sum_{m \in [M]} P_{Z_{cm}|Y}(1|y) \tag{7.73}$$

$$= \sum_{m \in [M]} \beta_{1-\varepsilon+\tau}(c_m, Q_Y) \tag{7.74}$$

$$\leq M \sup_{x \in \mathrm{F}} \beta_{1-\varepsilon+\tau}(x, Q_Y), \tag{7.75}$$

where (7.72) follows because $\kappa_\tau(\mathrm{F}, Q_Y)$ represents the infimum of the decoding error probability for the received signal; the constraint imposed by the infimum can be eliminated by substituting Z^* from (7.71); (7.73) follows from (7.65); and (7.74) follows from the fact that by definition of $P_{Z_{cm}|Y}$, it achieves the minimum in (7.58) for $x = c_m$ and $\alpha = 1 - \varepsilon + \tau$.

This completes the proof of the $\kappa\beta$-bound. □

Lemma 7.2 (Meta-Converse) *For a given code (possibly with a randomized decoder) with codewords belonging to* F*, let*

$$\varepsilon = \text{maximal error probability with } P_{Y|X}, \tag{7.76}$$

$$\varepsilon' = \text{maximal error probability with } Q_{Y|X}. \tag{7.77}$$

Then

$$\inf_{x \in F} \beta_{1-\varepsilon}(P_{Y|X=x}, Q_{Y|X=x}) \leq 1 - \varepsilon'. \tag{7.78}$$

Proof of Lemma 7.2 Consider an (M, ε)-code with codewords $\{c_j \in \mathrm{F}\}_{j=1}^M$ and a randomized decoding rule $P_{Z|Y} : \mathrm{B} \mapsto [M]$. For some j^*

$$\sum_{b \in \mathrm{B}} P_{Z|Y}(j^*|b) Q_{Y|X}(b|j^*) = 1 - \varepsilon', \tag{7.79}$$

and at the same time

$$\sum_{b \in \mathsf{B}} P_{Z|Y}(j^*|b) P_{Y|X}(b|j^*) \geq 1 - \varepsilon, \tag{7.80}$$

where (7.79) represents the case with maximum error probability. Consider the hypothesis test between $P_{Y|X=j^*}$ and $Q_{Y|X=j^*}$ that decides in favor of $P_{Y|X=j^*}$ only when the decoder output is j^*. By (7.80) the probability of correct decision under $P_{Y|X=j^*}$ is at least $1 - \varepsilon$, and therefore

$$1 - \varepsilon' \geq \beta_{1-\varepsilon}(P_{Y|X=j^*}, Q_{Y|X=j^*}) \tag{7.81}$$

$$\geq \inf_{x \in \mathsf{F}} \beta_{1-\varepsilon}(P_{Y|X=x}, Q_{Y|X=x}), \tag{7.82}$$

where (7.81) seeks the minimum of $\beta_{1-\varepsilon}$ for $X = j^*$ over $P_{Z|Y}$, while (7.82) seeks the infimum of $\beta_{1-\varepsilon}$ within the codeword set. This completes the proof of the meta-converse bound. □

7.3.2 *Proof of Finite Blocklength Bounds*

7.3.2.1 Achievability

The achievability bound in Theorem 7.1 can be proven now. In the CSIT scenario, given $\mathbb{H} = \mathsf{H}$, performing a singular value decomposition (SVD) of H:

$$\mathsf{H} = \mathsf{L}\Sigma\mathsf{V}^{\mathrm{H}}, \tag{7.83}$$

where $\mathsf{L} \in \mathbb{C}^{t\times t}$ and $\mathsf{V} \in \mathbb{C}^{r\times r}$ are unitary matrices, and $\Sigma \in \mathbb{C}^{t\times r}$ is a (truncated) diagonal matrix of dimension $t \times r$, whose diagonal elements $\sqrt{\lambda_1}, \cdots, \sqrt{\lambda_m}$, are the ordered singular values of H. It will be convenient to define the following $t \times t$ precoding matrix for each H:

$$\mathsf{P}(\mathsf{H}) := \mathrm{diag}\{\sqrt{nv_1^*}, \cdots, \sqrt{nv_m^*}, \mathbf{0}_{t-m}\}\mathsf{L}^{\mathrm{H}}. \tag{7.84}$$

Precoding employs the water-filling algorithm for power allocation and transforms the signal into a form suitable for channel transmission. Consider a code whose codewords $\{\mathsf{X}_j(\mathbb{H})\}_{j\in[M]}$ have the following structure

$$\mathsf{X}_j(\mathbb{H}) = \Phi_j\mathsf{P}(\mathbb{H}), \quad \Phi_j \in \mathcal{S}_{n,t}, \tag{7.85}$$

where $\mathcal{S}_{n,t} := \{\mathsf{A} \in \mathbb{C}^{n\times t} : \mathsf{A}^{\mathrm{H}}\mathsf{A} = \mathsf{I}_t\}$ denotes the set of all $n \times t$ unitary matrices. Motivated by the geometric considerations reported in Sect. 7.3.1.1, for a given input $\mathsf{X}(\mathbb{H}) = \Phi\mathsf{P}(\mathbb{H})$, a physically degraded version of the channel (7.1) is considered. In this version, the output corresponds to the subspace spanned by the matrix resulting from precoding Φ, transmitting it through the channel and adding noise:

$$\Omega_{\mathbb{Y}} = \text{span}(\Phi \mathsf{P}(\mathbb{H})\mathbb{H} + \mathbb{Z}). \tag{7.86}$$

Note that the rate achievable on (7.86) is a lower bound on the rate achievable on (7.1). Next, the lower bound on this degraded channel is provided by using the $\kappa\beta$-bound [4, Theorem 25].

Following (7.61), the following measure of performance is defined for the composite hypothesis test between an auxiliary output distribution $Q_{\Omega_{\mathbb{Y}}}$ defined on the subspace $\Omega_{\mathbb{Y}}$ and the collection of channel output distributions $\{P_{\Omega_{\mathbb{Y}}|\phi=\Phi}\}_{\Phi \in \mathcal{S}_{n,t}}$:

$$\kappa_\tau(\mathcal{S}_{n,t}, Q_{\Omega_{\mathbb{Y}}}) := \inf \int P_{Z|\Omega_{\mathbb{Y}}}(1|\Omega_{\mathsf{Y}}) Q_{\Omega_{\mathbb{Y}}}(dQ_{\mathsf{Y}}), \tag{7.87}$$

where the infimum is over all probability distributions $P_{Z|\Omega_{\mathbb{Y}}}$ satisfying

$$\int P_{Z|\Omega_{\mathbb{Y}}}(1|\Omega_{\mathsf{Y}}) P_{\Omega_{\mathbb{Y}}|\phi=\Phi}(d\Omega_{\mathsf{Y}}) \geq \tau, \ \ \forall \, \Phi \in \mathcal{S}_{n,t}. \tag{7.88}$$

Using Lemma 7.1 ($\kappa\beta$-bound) to the channel (7.86), for any auxiliary distribution $Q_{\Omega_{\mathbb{Y}}}$:

$$M \geq \frac{\kappa_\tau(\mathcal{S}_{n,t}, Q_{\Omega_{\mathbb{Y}}})}{\sup_{\Phi \in \mathcal{S}_{n,t}} \beta_{1-\varepsilon+\tau}(P_{\Omega_{\mathbb{Y}}|\phi=\Phi}, Q_{\Omega_{\mathbb{Y}}})}, \tag{7.89}$$

where $\beta_{(\cdot)}(\cdot, \cdot)$ is defined in (7.58) and reaches the minimum when

$$\int P_{Z|W}\ (1\ |w)\ P\ (dw) \geq 1 - \varepsilon + \tau. \tag{7.90}$$

Next, the right-hand side of (7.89) is lower-bounded to obtain an expression that can be evaluated numerically.

Fix a $\Phi \in \mathcal{S}_{n,t}$, from geometric analysis, a new variable is defined to represent whether the decoding is correct. Specifically, Z is defined as the indicator function of whether the angle between the input and output is smaller than a given threshold:

$$Z_\Phi(\Omega_{\mathbb{Y}}) = \mathbb{1}\{\sin^2\{\text{span}(\Phi), \Omega_{\mathbb{Y}}\} \leq \gamma_n\}, \tag{7.91}$$

where $\gamma_n \in [0, 1]$ is chosen so that

$$P_{\Omega_{\mathbb{Y}}|\phi=\Phi}[Z_\Phi(\Omega_{\mathbb{Y}}) = 1] \geq 1 - \varepsilon + \tau. \tag{7.92}$$

Furthermore, $Z_{\Phi}(\Omega_{\mathbb{Y}})$ can be viewed as a hypothesis test between $P_{\Omega_{\mathbb{Y}}|\phi=\Phi}$ and $Q_{\Omega_{\mathbb{Y}}}$. Hence, by definition

$$\beta_{1-\varepsilon+\tau}(P_{\Omega_{\mathbb{Y}}|\phi=\Phi}, Q_{\Omega_{\mathbb{Y}}}) \le Q_{\Omega_{\mathbb{Y}}}[Z_{\Phi}(\Omega_{\mathbb{Y}}) = 1], \tag{7.93}$$

for every $\Phi \in \mathcal{S}_{n,t}$. The left-hand side of (7.93) represents the minimum error detection probability for distribution $Q_{\Omega_{\mathbb{Y}}}$, while the right-hand side represents the probability of decoding as $P_{\Omega_{\mathbb{Y}}|\phi=\Phi}$ under distribution $Q_{\Omega_{\mathbb{Y}}}$.

Next, the right-hand side of (7.93) is evaluated by taking the uniform distribution as the auxiliary output distribution, denoted as $Q^{\mathrm{u}}_{\Omega_{\mathbb{Y}}}$. With this choice, the output probabilities for each input are identical, $Q^{\mathrm{u}}_{\Omega_{\mathbb{Y}}}[\sin^2\{\mathrm{span}(\Phi), \Omega_{\mathbb{Y}}\} \le \gamma_n]$ does not depend on $\Phi \in \mathcal{S}_{n,t}$. To simplify calculations, let $\Phi = \mathsf{I}_{n,t}$.

Observe that under $Q^{\mathrm{u}}_{\Omega_{\mathbb{Y}}}$, the squares of the sines of the principle angles between $\mathrm{span}(\mathsf{I}_{n,t})$ and $\Omega_{\mathbb{Y}}$ have the same distribution as the eigenvalues of a complex multivariate Beta-distributed $\mathbb{B} \sim \mathrm{Beta}_r(n-t, t)$ [14, Section 2]. By [15, Corollary 1], the distribution of $\det \mathbb{B}$ coincides with the distribution of $\prod_{i\in[r]} B_i$, where $\{B_i\}_{i\in[r]}$, are independent with $B_i \sim \mathrm{Beta}(n-t-i+1, t)$. Using this result to compute the right-hand side of (7.93) leads to

$$\sup_{\Phi\in\mathcal{S}_{n,t}} \beta_{1-\varepsilon+\tau}(P_{\Omega_{\mathbb{Y}}|\phi=\Phi}, Q_{\Omega_{\mathbb{Y}}}) \le \Pr\left\{\prod_{j\in[r]} B_j \le \gamma_n\right\}, \tag{7.94}$$

where γ_n satisfies

$$\Pr\left\{\sin^2\{\mathsf{I}_{n,t}, \mathsf{I}_{n,t}\mathsf{P}(\mathbb{H})\mathbb{H} + \mathbb{Z}\} \le \gamma_n\right\} \ge 1 - \varepsilon + \tau. \tag{7.95}$$

The left-hand side of (7.95) represents the correct probability of $P_{\Omega_{\mathbb{Y}}|\phi=\Phi}$, which is derived from the definition of $\beta_{(\cdot)}(\cdot,\cdot)$ in (7.58). The following step simplifies the inequality satisfied by γ_n:

$$\Pr\left\{\sin^2\left\{\mathsf{I}_{n,t}, \sqrt{n}\mathsf{I}_{n,t}\mathsf{P}(\mathbb{H})\mathbb{H} + \mathbb{Z}\right\} \le \gamma_n\right\}$$

$$= \Pr\left\{\sin^2\left\{\mathsf{I}_{n,t}, \sqrt{n}\mathsf{I}_{n,t}\mathrm{diag}\left\{\sqrt{v_1^*\Lambda_1}, \cdots, \sqrt{v_m^*\Lambda_m}, \mathbf{0}_{t-m}\right\}\mathbb{V}^{\mathrm{H}} + \mathbb{Z}\right\} \le \gamma_n\right\} \tag{7.96}$$

$$= \Pr\left\{\sin^2\left\{\mathsf{I}_{n,t}, \sqrt{n}\mathsf{I}_{n,t}\mathrm{diag}\left\{\sqrt{v_1^*\Lambda_1}, \cdots, \sqrt{v_m^*\Lambda_m}, \mathbf{0}_{t-m}\right\} + \mathbb{Z}\mathbb{V}\right\} \le \gamma_n\right\} \tag{7.97}$$

$$= \Pr\left\{\sin^2\left\{\mathsf{I}_{n,t}, \sqrt{n}\mathsf{I}_{n,t}\mathrm{diag}\left\{\sqrt{v_1^*\Lambda_1}, \cdots, \sqrt{v_m^*\Lambda_m}, \mathbf{0}_{t-m}\right\} + \mathbb{Z}\right\} \le \gamma_n\right\}, \tag{7.98}$$

where $\mathbb{V}$ contains the right singular vectors of $\mathbb{H}$ (see (7.83)). Here, (7.96) follows from (7.84); (7.97) follows because right-multiplying a matrix A by a unitary matrix does not change the subspace spanned by the columns of A, and hence, it does not change $\sin\{\cdot,\cdot\}$; (7.98) follows because $\mathbb{Z}$ is isotropically distributed, and hence $\mathbb{WV}$ has the same distribution as $\mathbb{Z}$.

At this point, the upper bound on $\beta_{1-\varepsilon+\tau}$ in the $\kappa\beta$-bound is obtained, and it is given by

$$\sup_{\Phi\in\mathcal{S}_{n,t}} \beta_{1-\varepsilon+\tau}(P_{\Omega_{\mathbb{Y}}|\phi=\Phi}, Q_{\Omega_{\mathbb{Y}}}) \le \Pr\left\{\prod_{j\in[r]} B_j \le \gamma_n\right\}, \tag{7.99}$$

where γ_n satisfies

$$\Pr\left\{\sin^2\left\{\mathsf{I}_{n,t}, \sqrt{n}\mathsf{I}_{n,t}\mathrm{diag}\left\{\sqrt{v_1^*\Lambda_1}, \cdots, \sqrt{v_m^*\Lambda_m}, \mathbf{0}_{t-m}\right\} + \mathbb{Z}\right\} \le \gamma_n\right\} \ge 1-\varepsilon+\tau. \tag{7.100}$$

Next, the lower bound of κ_τ needs to be computed. Replace (7.88) with the less stringent constraint that

$$\mathbb{E}_{P_\phi^{\mathrm{u}}}\left[\int P_{Z|\Omega_{\mathbb{Y}}}(1|\Omega_{\mathrm{Y}})P_{\Omega_{\mathbb{Y}}|\phi}(d\Omega_{\mathrm{Y}})\right] \ge \tau, \tag{7.101}$$

where P_ϕ^{u} is the uniform input distribution on $\mathcal{S}_{n,t}$. Since replacing (7.88) by (7.101) enlarges the feasible region of the minimization problem (7.87), an infimum in (7.87) (denoted by $\kappa_\tau^{\mathrm{u}}(\mathcal{S}_{n,t}, Q_{\Omega_{\mathbb{Y}}}^{\mathrm{u}})$) is obtained, which is no larger than $\kappa_\tau(\mathcal{S}_{n,t}, Q_{\Omega_{\mathbb{Y}}}^{\mathrm{u}})$. For all $P_{Z|\Omega_{\mathbb{Y}}}$ that satisfy the relaxed condition:

$$\begin{aligned}
&\int P_{Z|\Omega_{\mathbb{Y}}}(1|\Omega_{\mathrm{Y}})Q_{\Omega_{\mathbb{Y}}}^{\mathrm{u}}(d\Omega_{\mathrm{Y}}) \\
&= \mathbb{E}_{P_\phi^{\mathrm{u}}}\left[\int P_{Z|\Omega_{\mathbb{Y}}}(1|\Omega_{\mathrm{Y}})P_{\Omega_{\mathbb{Y}}|\phi}(d\Omega_{\mathrm{Y}})\right] &(7.102)\\
&\ge \tau. &(7.103)
\end{aligned}$$

Thus, it follows that

$$\kappa_\tau(\mathcal{S}_{n,t}, Q_{\Omega_{\mathbb{Y}}}^{\mathrm{u}}) \ge \kappa_\tau^{\mathrm{u}}(\mathcal{S}_{n,t}, Q_{\Omega_{\mathbb{Y}}}^{\mathrm{u}}) \ge \tau. \tag{7.104}$$

Organize the results to obtain a lower bound for $R_{\mathrm{t}}^*(n,\varepsilon)$. The desired lower bound (7.23) follows by using the inequality (7.94) and (7.104) in (7.89), by taking the logarithm of both sides of (7.89), and by dividing by the blocklength n.

The proof of 7.1 is now completed.

7.3.2.2 Converse

The proof of the converse bound in Theorem 7.2 is presented below. When CSI is available at both the transmitter and the receiver, the MIMO channel (7.1) can be transformed into the following set of m parallel quasi-static channels

$$Y_i^n = x_i^n \sqrt{\Lambda_i} + Z_i^n, \quad i \in [m], \tag{7.105}$$

by performing a singular value decomposition [8, Section 3.1]. Here, $\Lambda_1 \geq \cdots \geq \Lambda_m$ denote the m largest eigenvalues of $\mathbb{H}\mathbb{H}^{\mathrm{H}}$, and $\{Z_i^n\}_{i\in[m]} \sim \mathcal{CN}(\mathbf{0}, \mathsf{I}_n)$ are independent noise vectors. Next, a converse bound for the channel (7.105) is established.

Let $\mathsf{X} = [x_1^n \cdots x_m^n]$ and fix an $(n, M, \varepsilon)_{\mathrm{tr}}$ code. As evident from the definition of the scenario where CSI is available at the transmitter

$$\sum_{i\in[m]} \|x_i^n\|^2 \leq n\rho. \tag{7.106}$$

To simplify the presentation, assume the encoder f_{t} is deterministic, with additional assumptions regarding it. Furthermore, the encoder f_{t} is assumed to act on the pairs $(j, \boldsymbol{\lambda})$ instead of (j, H) in the CSIT definition. The purpose of this approach is to define all channel states on a series of parallel channels, distinguishing them from the overall channel state. Thus, the channel (7.105) and the encoder f_{t} define a random transformation $P_{\mathbb{Y},\boldsymbol{\Lambda}|J}$ from the message set $[M]$ to the space $\mathbb{C}^{n\times m} \times \mathbb{R}_+^m$:

$$P_{\mathbb{Y},\boldsymbol{\Lambda}|J} = P_{\boldsymbol{\Lambda}} P_{\mathbb{Y}|\boldsymbol{\Lambda},J}, \tag{7.107}$$

where $\mathbb{Y} = [Y_1^n, \cdots, Y_m^n]$ and

$$P_{\mathbb{Y}|\boldsymbol{\Lambda}=\boldsymbol{\lambda},J=j} := P_{\mathbb{Y}|\boldsymbol{\Lambda}=\boldsymbol{\lambda},\mathbb{X}=f_{\mathrm{t}}(j,\boldsymbol{\lambda})}. \tag{7.108}$$

$P_{\mathbb{Y},\boldsymbol{\Lambda}|J}$ can be regarded as the channel law associated with

$$J \to \mathbb{Y}, \boldsymbol{\Lambda}. \tag{7.109}$$

To upper-bound $R_{\mathrm{tr}}^*(n, \varepsilon)$, using the meta-converse theorem [4, Theorem 30] on the channel (7.109).

Each codeword X is associated with a power-allocation vector $\tilde{\mathbf{v}}(\mathsf{X})$ whose entries $\tilde{v}_i(\mathsf{X})$ are

$$\tilde{v}_i(\mathsf{X}) := \frac{1}{n}\|x_i^n\|^2, \quad i \in [m]. \tag{7.110}$$

The purpose of defining this power allocation strategy is to distinguish it from the previously used water-filling algorithm, as the new strategy employs an equal allocation approach. Take as auxiliary channel $Q_{\mathbb{Y},\mathbf{\Lambda}|J} = P_{\mathbf{\Lambda}} Q_{\mathbb{Y}|\mathbf{\Lambda},J}$, where

$$Q_{\mathbb{Y}|\mathbf{\Lambda}=\boldsymbol{\lambda},J=j} = \prod_{i\in[m]} Q_{Y_i^n|\mathbf{\Lambda}=\boldsymbol{\lambda},J=j}, \tag{7.111}$$

and

$$Q_{Y_i^n|\mathbf{\Lambda}=\boldsymbol{\lambda},J=j} = \mathcal{CN}\big(\mathbf{0}, \big[1 + (\tilde{v}_i \circ f_{\mathrm{t}}(j,\boldsymbol{\lambda}))\lambda_i\big] \mathsf{I}_n\big). \tag{7.112}$$

Using Lemma 7.2 (meta-converse) to obtain

$$\min_{j\in[M]} \beta_{1-\varepsilon}(P_{\mathbb{Y},\mathbf{\Lambda}|J=j}, Q_{\mathbb{Y},\mathbf{\Lambda}|J=j}) \le 1 - \varepsilon', \tag{7.113}$$

where ε' is the maximal probability of error over $Q_{\mathbb{Y},\mathbf{\Lambda}|J}$. The subsequent proof is divided into two parts: first, evaluate $\beta_{1-\varepsilon}(P_{\mathbb{Y},\mathbf{\Lambda}|J=j}, Q_{\mathbb{Y},\mathbf{\Lambda}|J=j})$, and then relate ε' to $R^*_{\mathrm{tr}}(n, \varepsilon)$ by establishing a converse bound on the auxiliary channel $Q_{\mathbb{Y},\mathbf{\Lambda}|J}$.

Evaluation of $\beta_{1-\varepsilon}(P_{\mathbb{Y},\mathbf{\Lambda}|J=j}, Q_{\mathbb{Y},\mathbf{\Lambda}|J=j})$. Let j^* be the message that achieves the minimum in (7.113), let $f_{\mathrm{t}}^*(\boldsymbol{\lambda}) := f_{\mathrm{t}}(j^*, \boldsymbol{\lambda})$, and let

$$\beta_{1-\varepsilon}(f_{\mathrm{t}}^*) := \beta_{1-\varepsilon}(P_{\mathbb{Y},\mathbf{\Lambda}|J=j^*}, Q_{\mathbb{Y},\mathbf{\Lambda}|J=j^*}). \tag{7.114}$$

Thus, (7.113) can be rewrote as

$$\beta_{1-\varepsilon}(f_{\mathrm{t}}^*) \le 1 - \varepsilon'. \tag{7.115}$$

Now, some mathematical transformations are performed on $\beta_{1-\varepsilon}(f_{\mathrm{t}}^*)$, starting by defining a log-likelihood ratio function to distinguish between these two distributions:

$$r(f_{\mathrm{t}}^*; \mathbb{Y}, \mathbf{\Lambda}) := \log \frac{dP_{\mathbb{Y},\mathbf{\Lambda}|J=j^*}}{dQ_{\mathbb{Y},\mathbf{\Lambda}|J=j^*}}. \tag{7.116}$$

By the Neyman-Pearson lemma [16, Page 300]

$$\beta_{1-\varepsilon}(f_{\mathrm{t}}^*) = Q_{\mathbb{Y},\mathbf{\Lambda}|J=j^*}[r(f_{\mathrm{t}}^*; \mathbb{Y}, \mathbf{\Lambda}) \ge n\gamma_n(f_{\mathrm{t}}^*)], \tag{7.117}$$

where $\gamma_n(f_{\mathrm{t}}^*)$ is the solution of

$$P_{\mathbb{Y},\mathbf{\Lambda}|J=j^*}[r(f_{\mathrm{t}}^*; \mathbb{Y}, \mathbf{\Lambda}) \le n\gamma_n(f_{\mathrm{t}}^*)] = \varepsilon. \tag{7.118}$$

This implies that when the inequality holds, i.e., when the likelihood ratio exceeds the threshold, it represents the error probability of distribution $Q_{\mathbb{Y},\mathbf{\Lambda}|J=j^*}$. Let now $\mathbf{v} := \tilde{\mathbf{v}} \circ f_{\mathrm{t}}^*$, because of the power constraint (7.106), $\mathbf{v}$ is a mapping from $[M]$ to the set $\mathcal{V}_m$, which is defined as

$$\mathcal{V}_m := \left\{ [p_1, \cdots, p_m] \in \mathbb{R}_+^m : \sum_{j\in[m]} p_j \leq \rho \right\}. \tag{7.119}$$

Through algebraic calculations, the distribution functions of the random variable $r(f_{\mathrm{t}}^* \cdot \mathbb{Y}, \mathbf{\Lambda})$ under $Q_{\mathbb{Y},\mathbf{\Lambda}|J=j^*}$ and $P_{\mathbb{Y},\mathbf{\Lambda}|J=j^*}$ are given by

$$L_n^{\mathrm{tr}}(\mathbf{v}, \mathbf{\Lambda}) := \sum_{i\in[n]} \sum_{j\in[m]} \left(\log\left(1 + \Lambda_j v_j(\mathbf{\Lambda})\right) + 1 - \left|\sqrt{\Lambda_j v_j(\mathbf{\Lambda})} Z_{i,j} \right.\right.$$
$$\left.\left. - \sqrt{1 + \Lambda_j v_j(\mathbf{\Lambda})}\right|^2 \right) \tag{7.120}$$

and

$$S_n^{\mathrm{tr}}(\mathbf{v}, \mathbf{\Lambda}) := \sum_{i\in[n]} \sum_{j\in[m]} \left(\log\left(1 + \Lambda_j v_j(\mathbf{\Lambda})\right) + 1 - \frac{\left|\sqrt{\Lambda_j v_j(\mathbf{\Lambda})} Z_{i,j} - 1\right|^2}{1 + \Lambda_j v_j(\mathbf{\Lambda})} \right), \tag{7.121}$$

respectively.

Thus, by reviewing and organizing the above steps, the final expression for calculating $\beta_{1-\varepsilon}(f_{\mathrm{t}}^*)$ is obtained

$$\min_{j\in[M]} \beta_{1-\varepsilon} = \beta_{1-\varepsilon}(P_{\mathbb{Y},\mathbf{\Lambda}|J=j^*}, Q_{\mathbb{Y},\mathbf{\Lambda}|J=j^*}) \tag{7.122}$$
$$= Q_{\mathbb{Y},\mathbf{\Lambda}|J=j^*}[r(f_{\mathrm{t}}^*; \mathbb{Y}, \mathbf{\Lambda}) \geq n\gamma_n(f_{\mathrm{t}}^*)] \tag{7.123}$$
$$= \Pr\{L_n^{\mathrm{tr}}(\mathbf{v}, \mathbf{\Lambda}) \geq n\gamma_n(\mathbf{v})\}, \tag{7.124}$$

where $\gamma_n(f_{\mathrm{t}}^*)$ is the solution of (7.118):

$$P_{\mathbb{Y},\mathbf{\Lambda}|J=j^*}[r(f_{\mathrm{t}}^*; \mathbb{Y}, \mathbf{\Lambda}) \leq n\gamma_n(f_{\mathrm{t}}^*)] = \Pr\{S_n^{\mathrm{tr}}(\mathbf{v}, \mathbf{\Lambda}) \leq n\gamma_n(\mathbf{v})\}. \tag{7.125}$$

At this point, the inequality obtained from the meta-converse lemma can be transformed into an expression for

$$\Pr\{L_n^{\mathrm{rt}}(\mathbf{v}, \mathbf{\Lambda}) \geq n\gamma_n(\mathbf{v})\} \leq 1 - \varepsilon', \tag{7.126}$$

where $\gamma_n(\mathbf{v})$ is the solution of

$$\Pr\{S_n^{\mathrm{tr}}(\mathbf{v}, \mathbf{\Lambda}) \leq n\gamma_n(\mathbf{v})\} = \varepsilon. \tag{7.127}$$

Note that this upper bound depends on the chosen code only through the induced power allocation function $\mathbf{v}$. To conclude, taking the infimum of the left-hand side of (7.126) over all power allocation functions $\mathbf{v}$ to obtain a bound that holds for all $(n, M, \varepsilon)_{\mathrm{tr}}$ codes:

$$\inf_{\mathbf{v}(\cdot)} \Pr\{L_n^{\mathrm{rt}}(\mathbf{v}, \mathbf{\Lambda}) \geq n\gamma_n(\mathbf{v})\} \leq 1 - \varepsilon'. \tag{7.128}$$

Finally, consider the converse on the auxiliary channel, derive the relationship between ε' and $R_{\mathrm{tr}}^*(n, \varepsilon)$. The following lemma serves this purpose.

Lemma 7.3 *For every code with M codewords and blocklength n, the maximum probability of error ε' over the channel $Q_{\mathbb{Y},\mathbf{\Lambda}|J}$ satisfies*

$$1 - \varepsilon' \leq \frac{c_{\mathrm{tr}}(n)}{M}, \tag{7.129}$$

where $c_{\mathrm{tr}}(n)$ is given in (7.27).

Using Lemma 7.3 to obtain

$$\inf_{\mathbf{v}(\cdot)} \Pr\{L_n^{\mathrm{tr}}(\mathbf{v}, \mathbf{\Lambda}) \geq n\gamma_n(\mathbf{v})\} \leq 1 - \varepsilon' \leq \frac{c_{\mathrm{tr}}(n)}{M}. \tag{7.130}$$

Organize the results to obtain an upper bound for $R_{\mathrm{tr}}^*(n, \varepsilon)$. The desired lower bound (7.29) follows by taking the logarithm on both sides of (7.130) and dividing by n.

The proof of 7.2 is now completed.

7.3.3 *Proof of Second-Order Asymptotics*

7.3.3.1 Achievability

To prove (7.32), the achievability bound (7.23) is studied in the large-n limit. The denominator on the right-hand side of (7.23) is first analyzed. Let $\alpha = n - t - r > 0$. Then,

$$\Pr\left\{\prod_{i\in[r]} B_i \leq \gamma_n\right\} = \Pr\left\{\prod_{i\in[r]} B_i^{-\alpha} \geq \gamma_n^{-\alpha}\right\} \tag{7.131}$$

$$\leq \frac{\mathbb{E}\left[\prod_{i\in[r]} B_i^{-\alpha}\right]}{\gamma_n^{-\alpha}} \tag{7.132}$$

$$= \gamma_n^{n-t-r} \prod_{i\in[r]} \mathbb{E}\left[B_i^{-(n-t-r)}\right], \tag{7.133}$$

where (7.132) follows from Markov's inequality, and (7.133) follows because the $B_1, \cdots, B_r$ are independent. Recalling that $B_i \sim \text{Beta}(n-t-i+1, t)$, for every $i \in [r]$

$$\mathbb{E}\left[B_i^{-(n-t-r)}\right] = \frac{\Gamma(n-i+1)}{\Gamma(n-t-i+1)\Gamma(t)} \int_0^1 s^{r-i}(1-s)^{t-1} ds \tag{7.134}$$

$$\leq \frac{\Gamma(n-i+1)}{\Gamma(n-t-i+1)\Gamma(t)} \tag{7.135}$$

$$\leq n^t. \tag{7.136}$$

Substituting (7.136) into (7.133):

$$\Pr\left\{\prod_{i\in[r]} B_i \leq \gamma_n\right\} \leq n^{rt} \gamma_n^{n-t-r}. \tag{7.137}$$

Setting $\tau = 1/n$ and $\gamma_n = \exp(-C_\varepsilon^{\mathrm{t}} + O(1/n))$ in (7.23), and using (7.137):

$$\frac{\log M}{n} \geq C_\varepsilon^{\mathrm{t}} - (1+rt)\frac{\log n}{n} + O\left(\frac{1}{n}\right). \tag{7.138}$$

To conclude the proof, it remains to show that there exists a $\gamma_n = \exp(-C_\varepsilon^{\mathrm{t}} + O(1/n))$ satisfying (7.21). To this end, note that

$$\Pr\left\{\sin^2\left\{\mathsf{I}_{n,t}, \sqrt{n}\mathsf{I}_{n,t}\text{diag}\left\{\sqrt{v_1^*\Lambda_1}, \cdots, \sqrt{v_m^*\Lambda_m}, \mathbf{0}_{t-m}\right\} + \mathbb{Z}\right\} \leq \gamma_n\right\}$$

$$\geq \Pr\left\{\prod_{j\in[m]} \sin^2\left\{\mathbf{e}_j, \sqrt{nv_j^*\Lambda_j}\mathbf{e}_j + Z_j^n\right\} \leq \gamma_n\right\} \tag{7.139}$$

$$= \Pr\left\{\prod_{j\in[m]} \sin^2\left\{\mathbf{e}_1, \sqrt{nv_j^*\Lambda_j}\mathbf{e}_1 + Z_j^n\right\} \leq \gamma_n\right\}. \tag{7.140}$$

Here, (7.139) follows from the product of sines of principal angles, the result in [3, Lemma 13] by letting $\mathbf{e}_j$ and Z_j^n stand for the j-th column of $\mathsf{I}_{n,t}$ and $\mathbb{Z}$, respectively; (7.140) follows by symmetry. Next note that the random variable $\sin^2\{\mathbf{e}_1, \sqrt{nv_j^*\Lambda_j}\mathbf{e}_1 + Z_j^n\}$ has the same distribution as

$$T_j := \frac{\sum_{i\in[2,n]} |W_{i,j}|^2}{\left|\sqrt{nv_j^*\Lambda_j} + W_{1,j}\right|^2 + \sum_{i\in[2,n]} |W_i, j|^2}. \tag{7.141}$$

Thus,

$$\Pr\left\{\prod_{j\in[m]} \sin^2\left\{\mathbf{e}_1, \sqrt{nv_j^*\Lambda_j}\mathbf{e}_1 + Z_j^n\right\} \le \gamma_n\right\} = \Pr\left\{\prod_{j\in[m]} T_j \le \gamma_n\right\}. \tag{7.142}$$

To evaluate the right-hand side of (7.142), it is observed that, by the law of large numbers, the noise term $\frac{1}{n}\sum_{i\in[2,n]}|W_{i,j}|^2$ in (7.141) concentrates around 1 as $n \to \infty$. Hence, it is expected that for all $\gamma > 0$

$$\Pr\left\{\prod_{j\in[m]} T_j \le \gamma\right\} \to \Pr\left\{\prod_{j\in[m]} \frac{1}{v_j^*\Lambda_j + 1} \le \gamma\right\} \text{ as } n \to \infty. \tag{7.143}$$

Next, this statement is made rigorous by showing that, for all γ in a certain neighborhood of $e^{-C_\varepsilon^{\mathrm{t}}}$,

$$\Pr\left\{\prod_{j\in[m]} T_j \le \gamma\right\} \ge \Pr\left\{\prod_{j\in[m]} \frac{1}{v_j^*\Lambda_j + 1} \le \gamma\right\} + O\left(\frac{1}{n}\right), \tag{7.144}$$

where the term $O(1/n)$ is uniform in γ. Since the joint pdf of $\Lambda_1 v_1^*, \cdots, \Lambda_m v_m^*$ is not continuously differentiable, partition $\mathbb{R}_{\ge}^m$ into m non-intersecting subregions $\{\mathcal{W}_j\}_{j\in[m]}$, [7, Eq. (24)]

$$\mathcal{W}_j := \left\{x^n \in \mathbb{R}_{\ge}^m : \frac{1}{x_{j+1}} > \frac{1}{j}\sum_{l\in[j]}\frac{1}{x_l} + \frac{\rho}{j} \ge \frac{1}{x_j}\right\}, \quad j \in [m-1], \tag{7.145}$$

and

$$\mathcal{W}_m := \left\{x^n \in \mathbb{R}_{\ge}^m : \frac{1}{m}\sum_{l\in[m]}\frac{1}{x_l} + \frac{\rho}{m} \ge \frac{1}{x_m}\right\}. \tag{7.146}$$

Note that $\bigcup_{j=1}^m \mathcal{W}_j = \mathbb{R}_{\ge}^m$. For every $\boldsymbol{\lambda} \in \mathcal{W}_j$, the water-filling solution gives exactly j active eigenmodes, i.e.,

$$v_1^*(\boldsymbol{\lambda}) \ge \cdots \ge v_j^*(\boldsymbol{\lambda}) > v_{j+1}^*(\boldsymbol{\lambda}) = \cdots = v_m^*(\boldsymbol{\lambda}) = 0. \tag{7.147}$$

To simplify notation, using $T_j^{(u)}$ to denote the random variable T_j conditioned on the event $\boldsymbol{\Lambda} \in \mathrm{Int}(\mathcal{W}_u)$, $u \in [m]$. Furthermore, let $\boldsymbol{\Lambda}^{(u)}$ and $\tilde{\boldsymbol{\Lambda}}^{(u)}$ denote the random vectors $[\Lambda_1, \cdots, \Lambda_u]^{\mathrm{T}}$ and $[\Lambda_1 v_1^*(\boldsymbol{\Lambda}), \cdots, \Lambda_u v_u^*(\boldsymbol{\Lambda})]^{\mathrm{T}}$, conditioned on the event $\boldsymbol{\Lambda} \in \mathrm{Int}(\mathcal{W}_u)$, respectively. Using these definitions, the left-hand side of (7.144) can be rewritten as

$$\Pr\left\{ \prod_{j\in[m]} T_j \le \gamma \right\}$$

$$= \sum_{u\in[m]} \Pr\left\{ \prod_{j\in[m]} T_j \le \gamma \,\middle|\, \mathbf{\Lambda} \in \mathrm{Int}(\mathcal{W}_u) \right\} \Pr\{\mathbf{\Lambda} \in \mathrm{Int}(\mathcal{W}_u)\} \tag{7.148}$$

$$= \sum_{u\in[m]} \Pr\left\{ \left(\prod_{j\in[u]} T_j^{(u)} \right) \cdot \underbrace{\left(\prod_{j\in[u+1,m]} \frac{\sum_{i\in[2,n]} |W_{i,j}|^2}{\sum_{i\in[n]} |W_{i,j}|^2} \right)}_{\le 1} \le \gamma \right\} \times \Pr\{\mathbf{\Lambda} \in \mathrm{Int}(\mathcal{W}_u)\} \tag{7.149}$$

$$\ge \sum_{u\in[m]} \Pr\left\{ \prod_{j\in[u]} T_j^{(u)} \le \gamma \right\} \Pr\{\mathbf{\Lambda} \in \mathrm{Int}(\mathcal{W}_u)\}, \tag{7.150}$$

where (7.149) follows since, by (7.147), $T_j = (\sum_{i\in[2,n]} |W_{i,j}|^2)/(\sum_{i\in[n]} |W_{i,j}|^2)$ for $j \in [u+1, m]$.

The following lemma enables the establishment of (7.144).

Lemma 7.4 *Let* $\mathbf{G} = [G_1, \cdots, G_u]^{\mathrm{T}} \in \mathbb{R}^u_{\ge}$ *be a random vector with continuously differentiable joint pdf. Let*

$$D_j := \frac{\sum_{i\in[2,n]} |W_{i,j}|^2}{|\sqrt{nG_j} + W_{1,j}|^2 + \sum_{i\in[2,n]} |W_{i,j}|^2}, \quad j \in [u], \tag{7.151}$$

where $\{W_{i,j}\}_{i\in[n], j\in[u]}$, *are i.i.d.* $\mathcal{CN}(0, 1)$*-distributed. Fix an arbitrary* $\xi_0 \in (0, 1)$. *Then, there exist a* $\delta > 0$ *and a finite constant* k *such that*

$$\inf_{\xi\in(\xi_0-\delta,\xi_0+\delta)} \left(\Pr\left\{ \prod_{j\in[u]} D_j \le \xi \right\} - \Pr\left\{ \prod_{j\in[u]} \frac{1}{1+G_j} \le \xi \right\} \right) > \frac{k}{n}. \tag{7.152}$$

Using Lemma 7.4 with $G_j = \tilde{\Lambda}_j^{(u)}$, it follows that there exist $\delta_u > 0$ and $0 \le k_u < \infty$, such that for every $\gamma \in (e^{-C_\varepsilon^{\mathrm{t}}-\delta_u}, e^{-C_\varepsilon^{\mathrm{t}}+\delta_u})$

$$\Pr\left\{ \prod_{j\in[u]} T_j^{(u)} \le \gamma \right\} \ge \Pr\left\{ \prod_{j\in[u]} \frac{1}{1+\tilde{\Lambda}_j^{(u)}} \le \gamma \right\} + O\left(\frac{1}{n}\right). \tag{7.153}$$

To show that $\{\tilde{\Lambda}_j^{(u)}\}_{j\in[u]}$, indeed satisfy the conditions in (7.152), using (7.147), (7.22), and (7.17) to obtain

$$\tilde{\Lambda}_j^{(u)} = \frac{\Lambda_j^{(u)}}{u}\left(\rho + \sum_{l\in[u]} \frac{1}{\Lambda_l^{(u)}}\right) - 1, \quad j \in [u]. \tag{7.154}$$

Since the joint pdf of $\mathbf{\Lambda}$ is continuously differentiable by assumption, the joint pdf of $\mathbf{\Lambda}^{(u)}$ is also continuously differentiable. Moreover, it can be shown that the transformation $\mathbf{\Lambda}^{(u)} \mapsto \tilde{\mathbf{\Lambda}}^{(u)}$ defined by (7.154) is a diffeomorphism of class C^2 [17, Page 147]. Therefore, the joint pdf of $\tilde{\mathbf{\Lambda}}^{(u)}$ is continuously differentiable.

Let $\delta_a := \min\{\delta_1 \cdots, \delta_m\}$, and use (7.153) in (7.150) to conclude that for every $\gamma \in (e^{-C_\varepsilon^{\mathrm{t}}-\delta_a}, e^{-C_\varepsilon^{\mathrm{t}}+\delta_a})$

$$\begin{aligned}
&\Pr\left\{\prod_{u\in[m]} T_j \le \gamma\right\} \\
&\ge \sum_{u\in[m]} \left\{\Pr\left\{\prod_{j\in[u]} \frac{1}{1+\tilde{\Lambda}_j^{(u)}} \le \gamma\right\} \Pr\{\mathbf{\Lambda} \in \mathrm{Int}(\mathcal{W}_u)\}\right\} + O\left(\frac{1}{n}\right) &(7.155)\\
&= \Pr\left\{\prod_{j\in[m]} \frac{1}{1+\Lambda_j v_j^*(\mathbf{\Lambda})} \le \gamma\right\} + O\left(\frac{1}{n}\right) &(7.156)\\
&=1 - \Pr\left\{\sum_{j\in[m]} \log(1+\Lambda_j v_j^*(\mathbf{\Lambda})) \le -\log\gamma\right\} + O\left(\frac{1}{n}\right) &(7.157)\\
&=1 - F_{\mathrm{t}}(-\log\gamma) + O\left(\frac{1}{n}\right), &(7.158)
\end{aligned}$$

where $F_{\mathrm{t}}(\cdot)$ is given in (7.15).

Choose γ_n as the solution of

$$1 - F_{\mathrm{t}}(-\log\gamma_n) + O\left(\frac{1}{n}\right) = 1 - \varepsilon + \frac{1}{n}. \tag{7.159}$$

Since $F_{\mathrm{t}}(C_\varepsilon^{\mathrm{t}}) = \varepsilon$ and $F_{\mathrm{t}}'(C_\varepsilon^{\mathrm{t}}) > 0$, it follows from Taylor's theorem that

$$-\log\gamma_n = C_\varepsilon^{\mathrm{t}} + O\left(\frac{1}{n}\right). \tag{7.160}$$

So, for sufficiently large n, γ_n in (7.160) belongs to the interval $(e^{-C_\varepsilon^{\mathrm{t}}-\delta_a}, e^{-C_\varepsilon^{\mathrm{t}}+\delta_a})$. Hence, by (7.142), (7.158), and (7.159), this γ_n satisfies (7.21).

The achievability proof of Theorem 7.3 is now completed.

7.3.3.2 Converse

As a first step towards establishing (7.33), the upper bound (7.29) is relaxed by lower-bounding its denominator. Recall that by the definition of $L_n^{\mathrm{tr}}(\mathbf{v}, \mathbf{\Lambda})$ in Sect. 7.3.2.2

$$\Pr\{L_n^{\mathrm{tr}}(\mathbf{v}, \mathbf{\Lambda}) \geq n\gamma_n(\mathbf{v})\} = \beta_{1-\varepsilon}(P_{\mathbb{Y},\mathbf{\Lambda}|J=j^*}, Q_{\mathbb{Y},\mathbf{\Lambda}|J=j^*}). \tag{7.161}$$

Using the following inequality: for every $\eta > 0$ [4, Eq. (102)]

$$\beta_{1-\varepsilon}(P, Q) \geq \frac{1}{\eta}\left(1 - P\left[\frac{dP}{dQ} \geq \eta\right] - \varepsilon\right). \tag{7.162}$$

Using (7.162) with $P = P_{\mathbb{Y},\mathbf{\Lambda}|J=j^*}$, $Q = Q_{\mathbb{Y},\mathbf{\Lambda}|J=j^*}$, $\eta = e^{n\gamma}$, and recalling that by the definition of $S_n^{\mathrm{tr}}(\mathbf{v}, \mathbf{\Lambda})$ in Sect. 7.3.2.2

$$1 - P_{\mathbb{Y},\mathbf{\Lambda}|J=j^*}\left[\frac{dP_{\mathbb{Y},\mathbf{\Lambda}|J=j^*}}{dP_{\mathbb{Y},\mathbf{\Lambda}|J=j^*}} \geq e^{n\gamma}\right] = \Pr\{S_n^{\mathrm{tr}}(\mathbf{v}, \mathbf{\Lambda}) \leq n\gamma\}. \tag{7.163}$$

For every $\gamma > 0$

$$\beta_{1-\varepsilon}(P_{\mathbb{Y},\mathbf{\Lambda}|J=j^*}, Q_{\mathbb{Y},\mathbf{\Lambda}|J=j^*}) \geq e^{-n\gamma}(\Pr\{S_n^{\mathrm{tr}}(\mathbf{v}, \mathbf{\Lambda}) \leq n\gamma\} - \varepsilon). \tag{7.164}$$

Using (7.164) and the estimate

$$\log c_{\mathrm{tr}}(n) = \frac{m}{2}\log n + O(1), \tag{7.165}$$

(which follows from (7.27), Assumption 1 in Theorem 7.3, and from algebraic manipulations), the right-hand side of 7.29 is upper-bounded as

$$R_{\mathrm{tr}}^*(n, \varepsilon) \leq \gamma - \frac{1}{n}\log\left(\inf_{\mathbf{v}(\cdot)}\Pr\{S_n^{\mathrm{tr}}(\mathbf{v}, \mathbf{\Lambda}) \leq n\gamma\} - \varepsilon\right) + \frac{m\log n}{2n} + O\left(\frac{1}{n}\right). \tag{7.166}$$

To conclude the proof, it remains to show that for every γ in a certain neighborhood of $C_\varepsilon^{\mathrm{t}}$ (recall that γ is a free optimization parameter),

$$\inf_{\mathbf{v}(\cdot)}\Pr\{S_n^{\mathrm{tr}}(\mathbf{v}, \mathbf{\Lambda}) \leq n\gamma\} \geq F_{\mathrm{t}}(\gamma) + O\left(\frac{1}{n}\right), \tag{7.167}$$

where $F_{\mathrm{t}}(\cdot)$ is the outage probability defined in (7.15) and the $O(1/n)$ term is uniform in γ. The desired result (7.33) follows then by substituting (7.167) into (7.166), setting γ as the solution of

$$F_{\mathrm{t}}(\gamma) - \varepsilon + O\left(\frac{1}{n}\right) = \frac{1}{n}, \tag{7.168}$$

and by noting that this γ satisfies

$$\gamma = C_{\varepsilon}^{\mathrm{t}} + O\left(\frac{1}{n}\right), \tag{7.169}$$

i.e., it belongs to the desired neighborhood of $C_{\varepsilon}^{\mathrm{t}}$ for sufficiently large n. Here, (7.169) follows by a Taylor series expansion [18, Theorem 5.15] of $F_{\mathrm{t}}(\gamma)$ around $C_{\varepsilon}^{\mathrm{t}}$, and because $F_{\mathrm{t}}(C_{\varepsilon}^{\mathrm{t}}) = \varepsilon$ and $F_{\mathrm{t}}'(C_{\varepsilon}^{\mathrm{t}}) > 0$ by assumption.

In the remainder of this section, the goal is to prove (7.167). The proof consists of the three steps sketched below.

Step 1: Given $\mathbf{v}$ and $\mathbf{\Lambda}$, the random variable $S_n^{\mathrm{tr}}(\mathbf{v}, \mathbf{\Lambda})$ (see (7.26) for its definition) is the sum of n i.i.d. random variables with mean

$$\mu(\mathbf{v}, \mathbf{\Lambda}) := \sum_{j \in [m]} \log\left(1 + \Lambda_j v_j(\mathbf{\Lambda})\right) \tag{7.170}$$

and variance

$$\sigma^2(\mathbf{v}, \mathbf{\Lambda}) := m - \sum_{j \in [m]} \frac{1}{\left(1 + \Lambda_j v_j(\mathbf{\Lambda})\right)^2}. \tag{7.171}$$

Fix an arbitrary power allocation function $\mathbf{v}(\cdot)$, and assume that $\mathbf{\Lambda} = \boldsymbol{\lambda}$. Let

$$u(\mathbf{v}, \boldsymbol{\lambda}) := \frac{\gamma - \mu(\mathbf{v}, \boldsymbol{\lambda})}{\sigma(\mathbf{v}, \boldsymbol{\lambda})}. \tag{7.172}$$

It follows from the Cramér-Esseen theorem [3, Theorem 15] that

$$\Pr\{S_n^{\mathrm{tr}}(\mathbf{v}, \mathbf{\Lambda}) \le n\gamma | \mathbf{\Lambda} = \boldsymbol{\lambda}\} \ge q_n(u(\mathbf{v}, \boldsymbol{\lambda})) + \frac{k_3}{n}, \tag{7.173}$$

where

$$q_n(x) := Q(-\sqrt{n}x) - \frac{[1 - nx^2]^+ e^{-\frac{nx^2}{2}}}{6\sqrt{n}} \tag{7.174}$$

and k_3 is a finite constant independent of $\boldsymbol{\lambda}$, $\mathbf{v}$ and γ.

Step 2: The right-hand side of (7.173) is made independent of $\mathbf{v}$ by minimizing $q_n(u(\mathbf{v}, \boldsymbol{\lambda}))$ over $\mathbf{v}$. Recall that $\mathbf{v}^* = \{v_j^*\}_{j\in[r]}$ in (7.22). Let

$$\hat{u}(\boldsymbol{\lambda}) := \begin{cases} \frac{\delta}{\sqrt{m}}, & \text{if } \mu(\mathbf{v}^*, \boldsymbol{\lambda}) \le \gamma - \delta \\ \frac{\gamma - \mu(\mathbf{v}^*,\boldsymbol{\lambda})}{\sigma(\mathbf{v}^*,\boldsymbol{\lambda}) + k(\gamma - \mu(\mathbf{v}^*,\boldsymbol{\lambda}))}, & \text{if } |\gamma - \mu(\mathbf{v}^*, \boldsymbol{\lambda})| < \delta \\ -\infty, & \text{if } \mu(\mathbf{v}^*, \boldsymbol{\lambda}) \ge \gamma + \delta, \end{cases} \tag{7.175}$$

where $\delta > 0$, $k < \infty$ and $\gamma \in (C_\varepsilon^{\mathrm{t}} - \delta, C_\varepsilon^{\mathrm{t}} + \delta)$. Specifically, the following result should be established: for every γ in a certain neighborhood of $C_\varepsilon^{\mathrm{t}}$,

$$\Pr\{S_n^{\mathrm{tr}}(\mathbf{v}, \boldsymbol{\Lambda}) \le n\gamma | \boldsymbol{\Lambda} = \boldsymbol{\lambda}\} \ge q_n(\hat{u}(\boldsymbol{\lambda})) + \frac{k_3}{n}. \tag{7.176}$$

Step 3: Average (7.176) over $\boldsymbol{\Lambda}$ and establish the bound (7.167).

The converse proof of Theorem 7.3 is now completed.

7.4 Chapter Summary

This chapter studies data transmission over MIMO AWGN channel with quasi-static fading. Since low-latency communication enforces the use of short blocklength codes, the corresponding fading model is quasi-static fading, where all symbols of the transmitted signal suffer the same fading. Since the fading coefficient can be arbitrarily small, the channel can be in deep fade, where reliable communication is impossible. Thus, the classical Shannon capacity with vanishing error probability equals zero. Instead, outage capacity is used, which quantifies the probability that reliable communication with respect to a certain transmission rate is possible. However, the outage capacity is an asymptotic notation that characterizes the theoretical benchmark when the blocklength is infinite. To account for the low-latency communication constraint, this chapter presents finite blocklength and second-order asymptotic bounds for all four possible combinations of CSI availability: noCSI, CSIR, CSIT, and CSITR. The theoretical results show that the second-order terms equal zero and thus the finite blocklength performance approximates the outage capacity very quickly. A main take-home message is that although the outage capacity is an asymptotic notion, it is valid for low-latency communication.

7.5 Exercises

7.1 What is MIMO? What are the advantages of MIMO in wireless communication? Why is it important for low-latency communication?

7.2 What is quasi-static fading? Why is it important for low-latency communication?

7.3 Why is having zero second-order term significant for quasi-static fading channels? What is the implication, and how is it related to outage capacity?

7.4 Prove Lemma 7.3 concerning the finite blocklength converse bound.

7.5 Prove Lemma 7.4 concerning the converse proof of second-order asymptotics.

7.6 Prove the inequality in (7.139) concerning a lower bound on the probability involving angle threshold decoding.

References

1. E. Biglieri, J. Proakis, S. Shamai (Shitz), Fading channels: information-theoretic and communications aspects. IEEE Trans. Inf. Theory **44**(6), 2619–2692 (1998)
2. D. Tse, P. Viswanath, *Fundamentals of Wireless Communication* (Cambridge University Press, Cambridge, 2005)
3. W. Yang, G. Durisi, T. Koch, Y. Polyanskiy, Quasi-static multiple-antenna fading channels at finite block length. IEEE Trans. Inf. Theory **60**(7), 4232–4265 (2014)
4. Y. Polyanskiy, H.V. Poor, S. Verdú, Channel coding rate in the finite block length regime. IEEE Trans. Inf. Theory **56**(5), 2307–2359 (2010)
5. V. Strassen, Asymptotische Abschätzungen in Shannon's Informationstheorie, in *Transactions of the Third Prague Conference on Information Theory etc.* (Czechoslovak Academy of Sciences, Prague, 1962), pp. 689–723
6. M. Hayashi, Information spectrum approach to second-order coding rate in channel coding. IEEE Trans. Inf. Theory **55**(11), 4947–4966 (2009)
7. G. Caire, G. Taricco, E. Biglieri, Optimum power control over fading channels. IEEE Trans. Inf. Theory **45**(5), 1468–1489 (1999)
8. E. Telatar, Capacity of multi-antenna Gaussian channels. Eur. Trans. Telecommun. **10**(6), 585–595, (1999)
9. N. Johnson, S. Kotz, N. Balakrishnan, *Continuous Univariate Distributions*, vol. 2, 2nd edn. (Wiley, London, 1995)
10. Y. Polyanskiy, *Channel Coding: Non-Asymptotic Fundamental Limits* (Princeton University, Princeton, 2010)
11. M. Effros, A. Goldsmith, Y. Liang, Generalizing capacity: new definitions and capacity theorems for composite channels. IEEE Trans. Inf. Theory **56**(7), 3069–3087 (2010)
12. A.T. James, Distribution of matrix variates and latent roots derived from normal samples. Ann. Math. Statist. **35**(2), 475–501 (1964)
13. C.E. Shannon, Probability of error for optimal codes in a Gaussian channel. Bell Syst. Tech. J. **38**(3), 611–656 (1959)
14. P. Absil, P. Koev, A. Edelman, On the largest principal angle between random subspaces. Linear Algebra Appl. **414**(1), 288–294 (2006)
15. J.C. Roh, B.D. Rao, Design and analysis of MIMO spatial multiplexing systems with quantized feedback. IEEE Trans. Signal Process. **54**(8), 2874–2886 (2006)
16. J. Neyman, E.S. Pearson, On the problem of the most efficient tests of statistical hypotheses. Philos. Trans. Roy. Soc. A **231**, 289–337 (1933)
17. J.R. Munkres, *Analysis on Manifolds* (Addison-Wesley, Reading, 1991)
18. W. Rudin, *Principles of Mathematical Analysis*, 3rd edn. (McGraw-Hill, New York, 1976)

Chapter 8
Degenerate MIMO: Lossy Multi-Connectivity

This chapter studies the problem of data transmission over parallel additive white Gaussian noise (AWGN) channels with independent quasi-static fading, which was named *multi-connectivity (MC)* [1, 2]. This problem finds applications in enhancing transmission reliability of wireless channels so that ultra-reliable and low-latency communication can be achieved in 5G, B5G, and beyond. For wireless channels with fading, a single channel can be in deep fade with very small fading coefficient and leads to communication outage. With the help of parallel and independent channels, as long as one of the available channel is in good condition, data can be transmitted reliably. Thus, multi-connectivity is a framework that guarantees transmission reliability via diversity, which can be achieved by transmitting one data packet using multiple transmitters at different locations or a single multi-antenna transmitter with multiple carrier frequencies. This chapter generalizes the pure multi-antenna channel coding problem in Chap. 7 (cf. [3]) to the joint source-channel coding (JSCC) problem [4] for a slightly degenerated case of independent channels, which is a special case of multi-antenna communication when channel state information (CSI) is available at both the transmitter and the receiver.

As shown in Chap. 7, for low-latency communication systems, quasi-static fading [5, Chapter 5.4] is the correct model since the blocklength of data packet is short and its transmission time is smaller than the channel coherence time of wireless channels. Under quasi-static fading, a popular research direction is the outage analysis of diversity and multiplexing tradeoff (DMT) under the assumption of an *infinite* signal-to-noise ratio (SNR) and a finite blocklength [6, 7]. However, any practical communication system can only have finite SNR, and outage analysis is not necessarily the valid low-latency performance benchmark.

The above problem can be solved using finite blocklength analyses introduced in earlier chapters of this book. In particular, the multi-connectivity problem with lossy reconstruction is a generalization of the joint source-channel coding problem in Theorem 2.5, where a single AWGN channel is generalized to multiple independent channels with quasi-static fading. The analysis of the fading is done similarly

L. Zhou, L. Bai, *Low-latency Communication*, Textbooks in Telecommunication Engineering, https://doi.org/10.1007/978-3-032-06627-5_8

to Chap. 7 for multi-antenna wireless communication. This chapter presents the analysis of multi-connectivity where one aims to transmit a memoryless source over parallel AWGN channels with independent quasi-static fading and noise. The results can be generalized to mismatched multi-connectivity using the techniques in Chap. 4 so that an arbitrary memoryless source can be transmitted over parallel channels with additive arbitrary noise and quasi-static fading.

8.1 Problem Formulation

Fix four alphabets $\mathcal{S}, \hat{\mathcal{S}}, \mathcal{X}, \mathcal{H}$, three integers $(k, n, \Psi) \in \mathbb{N}^2$, and a positive real number $P \in \mathbb{R}_+$. Let $P_S \in \mathcal{P}(\mathcal{S})$ denote the source distribution and assume that the source sequence S^k is generated i.i.d. from P_S. Assume that additive noise of each channel is independent, and for each $t \in [\Psi]$, the additive noise Z_t^n is i.i.d. from Gaussian distribution with mean zero and variance one, i.e., $Z_t^n \stackrel{\text{iid}}{\sim} \mathcal{N}(\cdot; 0, 1)$. Let $P_H \in \mathcal{P}(\mathcal{H})$ be the distribution of the fading coefficient, a.k.a., CSI.

As shown in Fig. 8.1, in lossy multi-connectivity, one aims to transmit a memoryless source $S^k \in \mathcal{S}^k$ over Ψ parallel AWGN channels with independent fading $(H_1, \ldots, H_\Psi) \in \mathcal{H}^\Psi$ generated from P_H. The memoryless source can be discrete with finite alphabets $(\mathcal{S}, \hat{\mathcal{S}})$ or Gaussian with $\mathcal{S} = \hat{\mathcal{S}} = \mathbb{R}$. Recall the definition of the distortion function $d(\cdot)$ in Sect. 2.2.1. When a discrete memoryless source (DMS) is considered, the distortion function $d(\cdot)$ is assumed bounded so that any distortion is finite, i.e., $\max_{(s,\hat{s})\in\mathcal{S}\times\hat{\mathcal{S}}} < \infty$. A typical example is the Hamming distortion measure with $d(s, \hat{s}) = \mathbb{1}(\hat{s} \neq s)$ for any $(s, \hat{s}) \in \mathcal{S} \times \hat{\mathcal{S}}$. When a Gaussian memoryless source (GMS) is considered, quadratic distortion is used, where $d(s, \hat{s}) = (\hat{s} - s)^2$ for any $(s, \hat{s}) \in \mathcal{S} \times \hat{\mathcal{S}}$.

For each $t \in [\Psi]$, an encoder f_t encodes S^k into a channel input $X_t^n \in \mathcal{X}^n$, which is corrupted by AWGN Z_t^n and random fading H_t, yielding noisy channel outputs Y_t^n, i.e.,

$$Y_t^n = H_t X_t^n + Z_t^n. \tag{8.1}$$

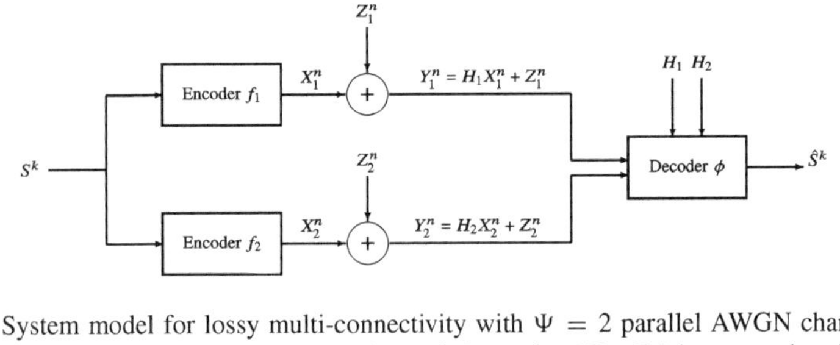

Fig. 8.1 System model for lossy multi-connectivity with $\Psi = 2$ parallel AWGN channels with independent quasi-static fading. The channel state information (H_1, H_2) is assumed available only at the receiver

The transmitter is subject to a maximal power constraint P, i.e.,

$$\frac{1}{n}\sum_{t\in[\Psi]}\|X_t^n\|^2 \le P. \tag{8.2}$$

Given noisy channel outputs $\mathbf{Y} := (Y_1^n, \ldots, Y_\Psi^n) \in (\mathbb{R}^n)^\Psi$ and CSI $\mathbf{H} = (H_1, \ldots, H_\Psi)$ for all channels, the receiver uses the decoder ϕ to generate a source estimate $\hat{S}^k$. Similarly to lossy source coding and JSCC in Chap. 2, the estimate $\hat{S}^k$ is allowed to be imperfect and the distortion is measured via $d(S^k, \hat{S}^k) := \frac{1}{k}\sum_{i\in[k]} d(S_i, \hat{S}_i)$.

To formally define a code, for each $t \in [\Psi]$, given any positive real number $P_t \in \mathcal{R}_+$, define the following set of codewords with power P_t:

$$\mathbb{R}(n, P_t) := \{x^n \in \mathcal{X}^n : \|x^n\|^2 = nP_t\}. \tag{8.3}$$

Definition 8.1 An (k, n, P)-code for lossy multi-connectivity consists of

- Ψ encoders such that for each $t \in [\Psi]$,

$$f_t : \mathcal{S}^k \to \mathbb{R}(n, P_t), \text{ and} \tag{8.4}$$

- A decoder

$$\phi : (\mathcal{R}^n)^\Psi \times \mathcal{H}^\Psi \to \hat{\mathcal{S}}^k, \tag{8.5}$$

such that the power allocation vector $\mathbf{P} := (P_1, \ldots, P_\Psi)$ satisfies the maximal power constraint, i.e.,

$$\sum_{t\in[\Psi]} P_t \le P. \tag{8.6}$$

Fix a target distortion level $D \in \mathbb{R}_+$. Consistent with lossy source coding and JSCC in Chap. 2, the reliability performance of an (k, n, P)-code is evaluated by the excess-distortion probability with respect to D:

$$\mathrm{P_e}(k, n, P, D) := \Pr\left\{d(S^k, \hat{S}^k) > D\right\}. \tag{8.7}$$

Note that the probability in (8.7) is calculated with respect to the distributions of the source sequence S^k, the AWGN $\mathbf{Z} := (Z_1^n, \ldots, Z_\Psi^n)$, and the CSI $\mathbf{H}$. The performance metric in (8.7) directly measures the communication quality, in contrast to artificially defined outage probability implicitly used in many studies.

Under low-latency constraint, consistent with JSCC analysis in Theorem 2.5, given blocklength n and tolerable excess-distortion probability $\varepsilon \in (0, 1)$, the theoretical benchmark is defined as

$$R^*(n, \varepsilon, P, D) := \sup \left\{ \log \frac{k}{n} : \exists \text{ an } (k, n, P)\text{-code s.t. } \mathrm{P_e}(k, n, P, D) \leq \varepsilon \right\}. \tag{8.8}$$

Analogously, given source sequence length k and blocklength n, one can study the following minimal excess-distortion probability

$$\mathrm{P_e^*}(k, n, P, D) := \inf\{\varepsilon : \exists \text{ an } (k, n, P)\text{-code s.t. } \mathrm{P_e}(k, n, P, D) \leq \varepsilon\}. \tag{8.9}$$

This chapter bounds $R^*(n, \varepsilon, P, D)$ and $\mathrm{P_e^*}(k, n, P, D)$. In particular, one has that $R^*(n, \varepsilon, P, D)$ is accurately approximated by the outage capacity and $\mathrm{P_e^*}(k, n, P, D)$ is approximated by the outage probability, which advocate the use of both asymptotic outage notion in low-latency communication over quasi-static fading channels.

8.2 Theoretical Results and Discussions

8.2.1 Preliminaries

Recall the definitions of the rate-distortion function $R(P_S, D)$ in (2.59) and the distortion-tilted information density in (2.62), i.e.,

$$R(P_S, D) = \inf_{P_{\hat{S}|S}: \mathbb{E}[d(S,\hat{S})] \leq D} I(S; \hat{S}), \tag{8.10}$$

$$\jmath(s|D, P_S) = -\log \mathbb{E}[\exp(\lambda^* D - d(s, \hat{S}))], \tag{8.11}$$

$$\lambda^* = -\left.\frac{\partial R(P_S, D')}{\partial D}\right|_{D'=D}, \tag{8.12}$$

where in (8.11), the expectation is calculated with respect to $P_{\hat{S}}^*$, which is the induced marginal distribution on $\hat{S}$ by the source distribution P_S and the optimal test channel $P_{\hat{S}|S}^*$ for the rate-distortion function $R(P_S, D)$. Given any source sequence s^k, define the *non-excess-distortion probability* as

$$\Phi(s^k, D) := \Pr_{P_{\hat{S}}^{*k}} \{d(s^k, \hat{S}^k) \leq D\}. \tag{8.13}$$

Recall the definition of the distortion-dispersion function $V(P_S, D)$ in (4.21):

$$V(P_S, D) = \mathrm{Var}[\jmath(S|D, P_S)]. \tag{8.14}$$

For any length-k sequence $s^k \in \mathcal{S}^k$, define the accumulated distortion-tilted information density as

$$\jmath(s^k|D, P_S) = \sum_{i\in[k]} \jmath(s_i|D, P_S). \tag{8.15}$$

For each $t \in [\Psi]$, given $P_t \in \mathbb{R}_+$, the fading coefficient $H_t \in \mathcal{H}$ and AWGN Z_t^n, define the following two functions

$$L_t^n(P_t, H_t, Z_t^n) := \frac{n}{2}\log(1+P_tH_t^2) + \frac{\sum_{j\in[n]} P_tH_t^2(1-Z_{t,j}^2) + 2H_t\sqrt{P_t}Z_{t,j}}{2(1+P_tH_t^2)}, \tag{8.16}$$

$$\tilde{L}_t^n(P_t, H_t, Z_t^n) := L_t^n(P_t, H_t, Z_t^n) - \log\frac{1+P_tH_t^2}{\sqrt{1+2P_tH_t^2}}. \tag{8.17}$$

As shall be shown, the above two functions appear in finite blocklength bounds, which mimic the accumulated mutual information density in (2.186) for channel coding in Chap. 2.

Recall that $\mathbf{P} = (P_1, \ldots, P_\Psi)$ denotes the power allocation vector and $\mathbf{H} = (H_1, \ldots, H_\Psi)$ denotes the CSI vector. Let $\mathcal{P}_{\max}(P) := \{\mathbf{P} : \sum_{t\in[\Psi]} P_t \le P\}$ and $\mathcal{P}_{\mathrm{eq}}(P) := \{\mathbf{P} : \sum_{t\in[\Psi]} P_t = P\}$ be the set of power allocation vectors satisfying the maximum and equal power constraints, respectively. Given fading parameters $\mathbf{H} = (H_1, \ldots, H_\Psi)$, define the following functions:

$$U_1(\mathbf{P}, \mathbf{H}) := \sum_{t\in[\Psi]} \frac{1}{2}\log(1+P_tH_t^2), \tag{8.18}$$

$$U_2(\mathbf{P}, \mathbf{H}) := \sum_{t\in[\Psi]} \frac{P_tH_t^2(P_tH_t^2+2)}{2(1+P_tH_t^2)}, \tag{8.19}$$

$$U_3(\mathbf{P}, \mathbf{H}) := \sum_{t\in[\Psi]} \frac{\sqrt{2}P_t^2H_t^4(7P_tH_t^2+12)}{(\sqrt{2(1+P_tH_t^2)})^3}. \tag{8.20}$$

As shall be illustrated in the proofs, the functions in (8.18) to (8.20) are critical for second-order asymptotics. In particular, $U_1(\mathbf{P}, \mathbf{H})$ is the total transmission rate of Ψ independent AWGN channels with quasi-static fading, $U_2(\mathbf{P}, \mathbf{H})$ is the channel dispersion function, and $U_3(\mathbf{P}, \mathbf{H})$ results from Taylor expansions in the proofs.

8.2.2 Finite Blocklength Bounds

This section presents finite blocklength bounds on the optimal excess-distortion probability. Recall from Sect. 2.2.3 that $\underline{D} = \min_{P_{\hat{X}|X}} \mathbb{E}[d(X, \hat{X})]$ and $\overline{D} = \min_{\hat{x} \in \hat{\mathcal{X}}} \mathbb{E}[d(X, \hat{x})]$. Fix any distortion level $D \in (\overline{D}, \underline{D})$. Recall that $|a|^+ = \max\{a, 0\}$ for any real number $a \in \mathbb{R}$. The first result is an achievability result, establishing the existence of a code with performance guarantee.

Theorem 8.1 *There exists a* (k, n, P)*-code such that*

$$\begin{aligned} \mathrm{P_e}(k, n, P, D) \leq \inf_{\mathbf{P} \in \mathcal{P}_{\max}(P)} \inf_{\gamma \in \mathbb{R}_+} \Bigg\{ & \exp(-n\gamma + 1) \\ & + \mathbb{E}\Bigg[\exp\Bigg(-\Bigg|\sum_{t \in [\Psi]} \tilde{L}_t^n(P_t, H_t, Z_t^n) - \log \frac{n\gamma}{\Phi(S^k, D)}\Bigg|^+\Bigg)\Bigg]\Bigg\}. \end{aligned} \tag{8.21}$$

Theorem 8.1 holds for any memoryless sources including DMS and GMS. To calculate the bound, one needs to first take a minimization over all possible power allocation vectors $\mathbf{P} \in \mathcal{P}_{\max}(P)$ and subsequently take another minimization over all positive real numbers $\gamma \in \mathbb{R}_+$. The key component of the bound is the expectation term that involves the accumulated mutual information density and the non-excess-distortion probability, which has the flavor of JSCC. The proof of Theorem 8.1 generalizes [4, Theorem 8].

Correspondingly, a converse bound is stated as follows, which lower bounds the minimal excess-distortion probability of any code.

Theorem 8.2 *Any* (k, n, P)*-code satisfies that*

$$\begin{aligned} \mathrm{P_e}(k, n, P, D) \geq \inf_{\mathbf{P} \in \mathcal{P}_{\max}(P)} \sup_{\gamma \in \mathbb{R}_+} \Bigg(& -\exp(-n\gamma) \\ & + \Pr\Bigg\{\sum_{t \in [\Psi]} L_t^n(P_t, H_t, Z_t^n) \leq \jmath(S^k | P_S, D) - n\gamma\Bigg\}\Bigg). \end{aligned} \tag{8.22}$$

The converse bound in Theorem 8.2 takes a similar form as the achievability bound in Theorem 8.1. The differences are twofold. Firstly, the minimization over $\gamma \in \mathbb{R}_+$ is replaced by a maximization. Secondly, the second term in the objective function is replaced by a probability term involving another accumulated mutual information density and the distortion-tilted information density. Theorem 8.2 generalizes [4, Theorem 1].

The finite blocklength bounds in Theorems 8.1 and 8.2 enable one to derive second-order asymptotics in the next section and allow one to understand the tightness of the second-order asymptotic bound.

8.2.3 Second-Order Asymptotics

The following conditions are required. Assume that (1) the distortion level $D \in (\underline{D}, \overline{D})$ such that the rate-distortion function $R(P_S, D)$ is positive and finite; (2) the distortion-dispersion function $V(P_X, D)$ is positive and finite; (3) the expected value $\mathbb{E}_{P_S \times P^*_{\hat{S}}}[d(X, \hat{X})^9]$ is finite; and (4) given any $\mathbf{P} \in \mathcal{P}_{\mathrm{eq}}(P)$, the fading parameters $\mathbf{H}$ satisfy the following moment constraints:

$$\mathbb{E}\left[\frac{U_3(\mathbf{P}, \mathbf{H})}{U_2(\mathbf{P}, \mathbf{H})}\right] < \infty, \ \mathbb{E}\left[\left(\frac{U_3(\mathbf{P}, \mathbf{H})}{U_2(\mathbf{P}, \mathbf{H})}\right)^2\right] < \infty. \tag{8.23}$$

Note that conditions (1) to (3) were imposed for lossy source coding in Sect. 2.2.3. Condition (4) is the additional requirement for fading parameters $\mathbf{H}$ so that the theoretical analyses can be done. Condition (4) is mild since most usually used fading models such as Rayleigh fading satisfies the condition [9].

The first main result concerns the minimal excess-distortion probability.

Theorem 8.3 *Assume that $k = \Theta(n)$. It follows that*

$$\mathrm{P}_{\mathrm{e}}^*(k, n, P, D) = \inf_{\mathbf{P} \in \mathcal{P}_{\mathrm{eq}}(P)} \mathbb{E}\left[\mathrm{Q}\left(\frac{nU_1(\mathbf{P}, \mathbf{H}) - kR(P_S, D)}{\sqrt{kV(P_S, D) + nU_2(\mathbf{P}, \mathbf{H})}}\right)\right] + O\left(\frac{\log n}{\sqrt{n}}\right). \tag{8.24}$$

Theorem 8.3 generalizes the second-order asymptotics in Theorem 2.5 for JSCC over a single non-fading channel to the case of multiple independent channels with quasi-static fading. Setting $\Psi = 1$ and allowing $\mathbf{H}$ to be constants, Theorem 8.3 reduces to JSCC with the characterization of the optimal excess-distortion probability, which can be transmitted into the results of the coding rate in Theorem 2.5.

As observed from Theorem 8.3, optimal second-order asymptotics is achieved with equal power constraint, i.e., all power should be used. This is intuitively true since using more power can achieve better performance for data transmission when the noise power and fading parameters are fixed.

An important implication of Theorem 8.3 is that the asymptotic notion of outage probability is a valid performance metric for low-latency communication. Specifically, it follows from [3, Lemma 17] that

$$\inf_{\mathbf{P} \in \mathcal{P}_{\mathrm{eq}}(P)} \mathbb{E}\left[\mathrm{Q}\left(\frac{nU_1(\mathbf{P}, \mathbf{H}) - kR(P_S, D)}{\sqrt{kV(P_S, D) + nU_2(\mathbf{P}, \mathbf{H})}}\right)\right] = \mathrm{P}_{\mathrm{out}}(k, n, D) + O\left(\frac{1}{n}\right), \tag{8.25}$$

where the outage probability is defined as

$$\mathrm{P}_{\mathrm{out}}(k, n, D) := \inf_{\mathbf{P} \in \mathcal{P}_{\mathrm{eq}}(P)} \Pr\left\{nU_1(\mathbf{P}, \mathbf{H}) \le kR(P_S, D)\right\}. \tag{8.26}$$

The optimal power allocation to achieve (8.26) is generally unknown. However, for Rayleigh fading channels, it follows from [5, Exercise 5.17] that equal power allocation is optimal, i.e., the infimum in (8.25) is achieved when $P_t = \frac{P}{\Psi}$ for each $t \in [\Psi]$.

The assumption that $k = \Theta(n)$ is valid. This is because in a JSCC problem, the number of source symbols that can be transmitted with n channel uses scales linearly with n, as shown in Theorem 2.5. The following corollary applies Theorem 8.3 and characterizes the maximal transmission rate $R^*(n, \varepsilon, P, D)$. Given any $P \in \mathcal{R}_+$ and any $\varepsilon \in (0, 1)$, define the outage capacity for lossy multi-connectivity problem as

$$\mathrm{C}_{\mathrm{out}}(P, D, \varepsilon) := \sup\left\{r \in \mathcal{R}_+ : \mathrm{P}_{\mathrm{out}}(r, 1, D) \leq \varepsilon\right\}. \tag{8.27}$$

Corollary 8.1 *Under conditions of Theorem 8.3, for any continuously differentiable fading distribution P_H,*

$$R^*(n, \varepsilon, P, D) = \mathrm{C}_{\mathrm{out}}(P, D, \varepsilon) + O\left(\frac{\log n}{n}\right). \tag{8.28}$$

Corollary 8.1 implies that the dispersion for lossy multi-connectivity is zero, which is consistent with the result in Chap. 7. Comparing the results in Sect. 2.2.3, Chap. 7, and Corollary 8.1, it follows that the dispersion equals zero when quasi-static fading channel is considered, regardless of whether one considers the pure data transmission problem or the JSCC problem. As a result, Corollary 8.1 justifies the validity of the asymptotic outage capacity for low-latency communications over quasi-static fading channels. This notion was previously used extensively in wireless communication systems without theoretical justification.

The results in Theorem 8.3 and Corollary 8.1 hold for any DMS under bounded distortion measure and GMS under quadratic distortion. For simplicity, only the proof of DMS will be presented. For GMS, the converse proof is exactly the same while the achievability proof requires to combine the proof here with those for GMS in Theorem 2.2. Taking one step forward and considering the mismatched JSCC result in Chap. 4, the results here can also be generalized to mismatched multi-connectivity by generalizing the mismatched code to parallel channels with independent quasi-static fading.

8.3 Numerical Illustrations

This section illustrates the theoretical results in Theorems 8.1, 8.2, and 8.3 via numerical simulations. Assume that Rayleigh fading with scale parameter one is considered so that for each $t \in [\Psi]$, $H_t \sim P_H$, where $P_H(h) = h \exp\left(-h^2/2\right)$ for any $h \in \mathcal{R}_+$. The following two kinds of memoryless sources are considered.

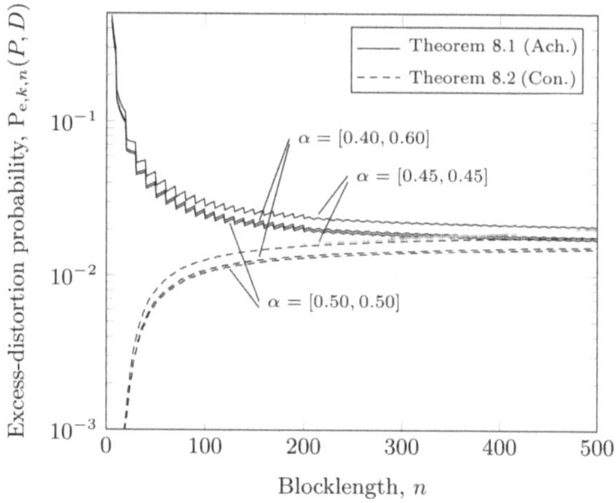

Fig. 8.2 Plot of finite blocklength bounds in Theorems 8.1 and 8.2 for the minimal excess-distortion probability of lossy multi-connectivity with quasi-static Rayleigh fading when $P = 5$, $D = 0.1$, and $\Psi = 2$ for BMS under Hamming distortion. The parameter $\alpha = (\alpha_1, \alpha_2)$ denotes the power allocation percentage for two channels

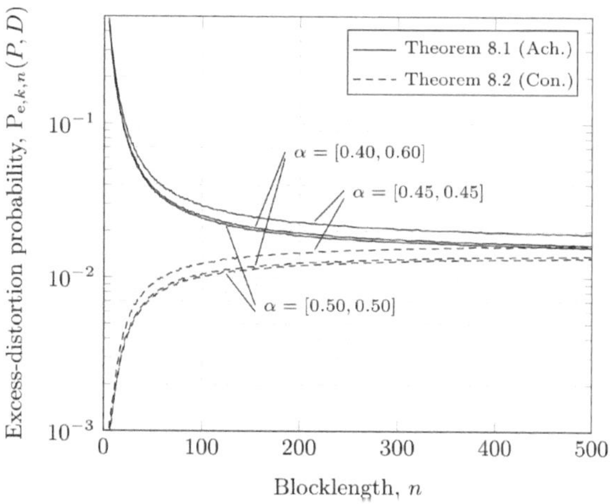

Fig. 8.3 Plot of finite blocklength bounds in Theorems 8.1 and 8.2 for the minimal excess-distortion probability of lossy multi-connectivity with quasi-static Rayleigh fading when $P = 5$, $D = 0.5$, and $\Psi = 2$ for GMS under quadratic distortion. The parameter $\alpha = (\alpha_1, \alpha_2)$ denotes the power allocation percentage for two channels

- Binary memoryless source (BMS) with bias p under Hamming distortion measure. The source alphabet is $\mathcal{S} = \{0, 1\}$ and the source distribution is $P_S(0) = p$ and $P_S(1) = 1 - p$. For any two sequences s^k and $\hat{s}^k$, the Hamming distortion between s^k and $\hat{s}^k$ is given by $d(s^k, \hat{s}^k) = \frac{1}{k} \sum_{i \in [k]} \mathbb{1}(s_i \neq \hat{s}_i)$.
- GMS under the quadratic distortion measure. The source distribution is $P_S = \mathcal{N}(\cdot; 0, \sigma_S^2)$ and the quadratic distortion measure for any two sequences s^k and $\hat{s}^k$ is$d(s^k, \hat{s}^k) = \frac{1}{k} \sum_{i \in [k]} (s_i - \hat{s}_i)^2$.

Consider a maximal power constraint of $P = 5$, a fixed rate of source symbols per channel use of $\frac{k}{n} = 1$ and set $\Psi \leq 3$.

In Figs. 8.2 and 8.3, the finite blocklength bounds in Theorems 8.1 and 8.2 are plotted with solid and dashed lines, respectively, via Monte-Carlo simulations when $\Psi = 2$ for different power allocations $\mathbf{P} = [P_1, P_2] = [\alpha_1, \alpha_2]P$, where the power allocation percentage $(\alpha_1, \alpha_2) \in \mathcal{R}_+$ satisfies $\alpha_1 + \alpha_2 \leq 1$. It is observed that the

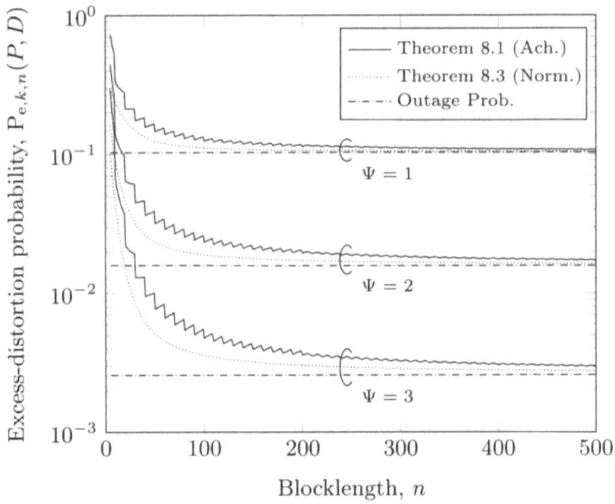

Fig. 8.4 Plot of the finite blocklength achievability bound in Theorem 8.1, the second-order asymptotic bound in Theorem 8.3 and the outage probability in (8.26) for lossy multi-connectivity with quasi-static Rayleigh fading when $P = 5$, $D = 0.1$, and $\Psi \in [3]$ for BMS under Hamming distortion

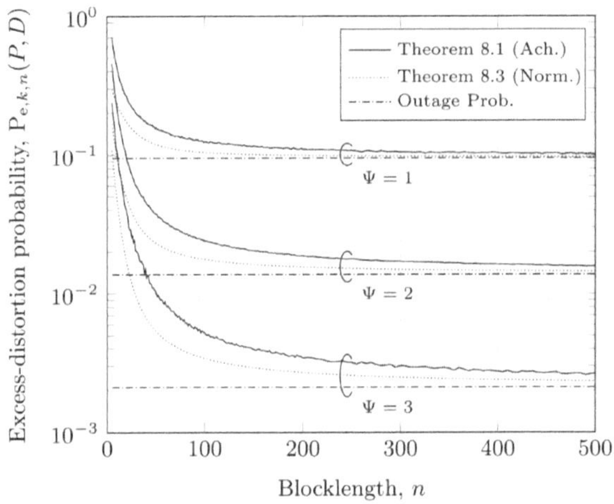

Fig. 8.5 Plot of the finite blocklength achievability bound in Theorem 8.1, the second-order asymptotic bound in Theorem 8.3 and the outage probability in (8.26) for lossy multi-connectivity with quasi-static Rayleigh fading when $P = 5$, $D = 0.5$, and $\Psi \in [3]$ for GMS under quadratic distortion

equal power allocation with $\alpha_1 = \alpha_2 = 0.5$ achieves the smallest excess-distortion probability for both the achievability and converse results. The conclusion can be verified for any possible values of $(\alpha_1, \alpha_2) \in \mathcal{R}_+$ such that $\alpha_1 + \alpha_2 \leq 1$. It is thus reasonable to conclude that the equal power allocation is optimal for low-latency communication. In further numerical simulations, equal power allocation is considered.

In Figs. 8.4 and 8.5, when there are at most three channels, i.e., $\Psi \in [3]$, the finite blocklength achievability bound in Theorem 8.1, the second-order asymptotic bound in Theorem 8.3 and the outage probability in (8.26) are plotted with solid lines, dotted lines and dash-dotted lines, respectively. It can be observed that (1) the achievable excess-distortion probability decreases significantly as the number of parallel channels, and (2) for each case, both the second-order asymptotic bound and the outage probability approximate the finite blocklength achievability well for blocklengths of hundreds.

8.4 Proof of Second-Order Asymptotics

8.4.1 Preliminaries

Fix two integers $(n,k)\in\mathbb{N}^2$. Recall the channel in (8.1). For each $t\in[\Psi]$, the conditional distribution of the channel output Y_t^n given the input X_t^n and the fading parameter h_t is a product of normal distributions, i.e., for any $(x_t^n,h_t,y_t^n)\in\mathcal{X}^n\times\mathcal{H}\times\mathbb{R}^n$,

$$P_{Y_t^n|X_t^nH_t}(y_t^n|x_t^n,h_t)=\prod_{i\in[n]}\frac{1}{\sqrt{2\pi}}\exp\left(-\frac{(y_{t,i}-H_tx_{t,i})^2}{2}\right). \tag{8.29}$$

Let $\{Q_{Y_t^n|H_t}\}_{t\in[\Psi]}\in(\mathcal{P}(\mathcal{Y}^n|\mathcal{H}))^{\Psi}$ be arbitrary. Furthermore, define the following fading information densities [8, Eq. (9)]

$$\tilde{\imath}(x_t^n;y_t^n|h_t):=\log\frac{P_{Y_t^n|X_t^nH_t}(y_t^n|x_t^n,h_t)}{Q_{Y_t^n|H_t}(y_t^n|h_t)}. \tag{8.30}$$

It follows that under the distribution $P_{Y_t^n|X_t^nH_t}$, the distributions of the fading information density $\tilde{\imath}(X_t^n;Y_t^n|H_t)$ and the accumulated information density $L_t^n(P_t,H_t,Z_t^n)$ in (8.16) are the same.

Fix any $t\in[\Psi]$, $P_t\in\mathbb{R}_+$, $x_t^n\in\mathcal{R}(n,P_t)$ and $h_t\in\mathcal{H}$.

Lemma 8.1 *Set $Q_{Y_t^n|H_t}(\cdot|H_t)\sim\mathcal{N}(\cdot;\mathbf{0}_n,(1+P_tH_t^2)\mathbf{I}_n)$. Under the distribution of $P_{Y_t^n|X_t^nH_t}(\cdot|x_t^n,H_t)$, the distribution of the fading information density $\tilde{\imath}(x_t^n;Y_t^n|H_t)$ depends on x_t^n only through its power $P_t=\frac{\|x_t^n\|^2}{n}$. Thus, $\tilde{\imath}(x_t^n;Y_t^n|H_t)$ and $L_t^n(P_t,H_t,Z_t^n)$ have the same distribution.*

Lemma 8.1 results from spherical symmetry of Gaussian i.i.d. random variables.

Recall the definitions of $V(P_S,D)$ in (8.14), $\mathrm{U}_1(\mathbf{P},\mathbf{H})$ in (8.18), $\mathrm{U}_2(\mathbf{P},\mathbf{H})$ in (8.19), and $\mathrm{U}_3(\mathbf{P},\mathbf{H})$ in (8.20). Define the following two functions:

$$V_{k+n\Psi}^{(\mathbf{P},\mathbf{H})}:=\frac{kV(P_S,D)+nU_2(\mathbf{P},\mathbf{H})}{k+n\Psi}, \tag{8.31}$$

$$T_{k+n\Psi}^{(\mathbf{P},\mathbf{H})}:=\frac{kT(P_S,D)+nU_3(\mathbf{P},\mathbf{H})}{k+n\Psi}. \tag{8.32}$$

Furthermore, define the following three functions of $\mathbf{H}$:

$$\Omega_1(k,n,P):=\sup_{\mathbf{P}\in\mathcal{P}_{\mathrm{eq}}(P)}\mathbb{E}\left[6T_{k+n\Psi}^{(\mathbf{P},\mathbf{H})}\left(\sqrt{V_{k+n\Psi}^{(\mathbf{P},\mathbf{H})}}\right)^{-3}\right], \tag{8.33}$$

$$\Omega_2(k,n,P) := \sup_{\mathbf{P}\in\mathcal{P}_{\mathrm{eq}}(P)} \mathbb{E}\left[\left(\sqrt{V_{k+n\Psi}^{(\mathbf{P},\mathbf{H})}}\right)^{-1}\right], \tag{8.34}$$

$$\Omega_3(k,n,P) := \sup_{\mathbf{P}\in\mathcal{P}_{\mathrm{eq}}(P)} \mathbb{E}\left[U_1(\mathbf{P},\mathbf{H})\left(\sqrt{V_{k+n\Psi}^{(\mathbf{P},\mathbf{H})}}\right)^{-1}\right]. \tag{8.35}$$

All above functions of fading parameters $\mathbf{H}$ are used in the proof of second-order asymptotics, mostly concerning the application of the Berry-Esseen theorem.

8.4.2 Achievability

Recall the definitions of the distortion-tilted information density $\jmath(s|P_S,D)$ in (2.62) and the non-excess-distortion probability $\Phi(s^k,D)$ in (8.13). For simplicity, let $\mathbf{Z}$ denote $(Z_1^n,\ldots,Z_\Psi^n)$ and let $\mathbf{z} := (z_1^n,\ldots,z_\Psi^n)$ denote a particular realization. In the achievability proof, the following lemma from [10, Lemma 2] is important.

Lemma 8.2 *For any DMS under any bounded distortion measure, there exist constants c_1, c_2, c_3 such that*

$$\Pr\left\{-\log\Phi(S^k,D) \le \sum_{i\in[k]} \jmath(S_i|P_S,D) + (c_1-0.5)\log k + c_2\right\} \ge 1 - \frac{c_3}{\sqrt{k}}. \tag{8.36}$$

Lemma 8.2 relates the non-excess-distortion probability with the excess-distortion information density.

Recall the definition of $\tilde{L}_t^n(\cdot)$ in (8.17). For simplicity, let

$$\Upsilon_2(S^k,D) := \sum_{i\in[k]} \jmath(S_i|P_S,D) + (c_1-0.5)\log k + c_2, \tag{8.37}$$

$$\Upsilon_1(\mathbf{P},\mathbf{H},\mathbf{Z},S^k,D) := \sum_{t\in[\Psi]} \tilde{L}_t^n(P_t,H_t,Z_t^n) - \log n - \log\gamma + \log\Phi(S^k,D), \tag{8.38}$$

$$\Upsilon(\mathbf{P},\mathbf{H},\mathbf{Z},S^k,D) := \Upsilon_1(\mathbf{P},\mathbf{H},\mathbf{Z},S^k,D) - \log\Phi(S^k,D) - \Upsilon_2(S^k,D). \tag{8.39}$$

Furthermore, given any power vector $\mathbf{P} = (P_1,\ldots,P_\Psi)$, define the set:

$$\mathcal{T}_{k,n}(\mathbf{P}) := \left\{(\mathbf{h},\mathbf{z},s^k) \in \mathcal{H}^\Psi \times \mathbb{R}^{n\Psi} \times \mathcal{S}^k :\ \Upsilon(\mathbf{P},\mathbf{h},\mathbf{z},s^k,D) > \log n\right\}. \tag{8.40}$$

Set $\gamma = \frac{\log n+1}{n}$. It follows from Theorem 8.1 and Lemma 8.2 that there exists an (k, n, P)-code such that for any power allocation vector $\mathbf{P} \in \mathcal{P}_{\max}(P)$, the excess-distortion probability satisfies

$$
\begin{aligned}
&\mathrm{P_e}(k, n, P, D)\\
&\le \mathbb{E}\left[\exp\left(-\left|\Upsilon_1(\mathbf{P}, \mathbf{H}, \mathbf{Z}, S^k, D)\right|^+\right)\right] + \frac{1}{n} &(8.41)\\
&= \mathbb{E}\left[\exp\left(-\left|\Upsilon_1(\mathbf{P}, \mathbf{H}, \mathbf{Z}, S^k, D)\right|^+\right) \times \mathbb{1}\left(-\log\Phi(S^k, D) \le \Upsilon_2(S^k, D)\right)\right]\\
&\quad + \mathbb{E}\left[\exp\left(-\left|\Upsilon_1(\mathbf{P}, \mathbf{H}, \mathbf{Z}, S^k, D)\right|^+\right)\right.\\
&\quad \left.\times\, \mathbb{1}\left(-\log\Phi(S^k, D) > \Upsilon_2(S^k, D)\right)\right] + \frac{1}{n} &(8.42)\\
&\le \mathbb{E}\left[\exp\left(-\left|\Upsilon(\mathbf{P}, \mathbf{H}, \mathbf{Z}, S^k, D)\right|^+\right) \times 1\left\{-\log\Phi(S^k, D) \le \Upsilon_2(S^k, D)\right\}\right]\\
&\quad + \Pr\left\{-\log\Phi(S^k, D) > \Upsilon_2(S^k, D)\right\} + \frac{1}{n} &(8.43)\\
&\le \mathbb{E}\left[\exp\left(-\left|\Upsilon(\mathbf{P}, \mathbf{H}, \mathbf{Z}, S^k, D)\right|^+\right)\right] + \frac{1}{n} + \frac{c_3}{\sqrt{k}} &(8.44)\\
&\le \frac{1}{n}\Pr\{(\mathbf{H}, \mathbf{Z}, S^k) \in \mathcal{T}_{k,n}(\mathbf{P})\} + \Pr\{(\mathbf{H}, \mathbf{Z}, S^k) \notin \mathcal{T}_{k,n}(\mathbf{P})\} + \frac{1}{n} + \frac{c_3}{\sqrt{k}} &(8.45)\\
&\le \frac{2}{n} + \frac{c_3}{\sqrt{k}} + \Pr\{(\mathbf{H}, \mathbf{Z}, S^k) \notin \mathcal{T}_{k,n}(\mathbf{P})\}, &(8.46)
\end{aligned}
$$

where (8.43) follows since (i) $\Upsilon(\mathbf{P}, \mathbf{H}, \mathbf{Z}, S^k, D) \le \Upsilon_1(\mathbf{P}, \mathbf{H}, \mathbf{Z}, S^k, D)$ if $-\log\Phi(S^k, D) \le \Upsilon_2(S^k, D)$ and $\exp(-|x|^+)$ is non-increasing in x, and (ii) $\exp(-|x|^+) \le 1$ for any x since $|x|^+ = \max\{0, x\}$, (8.44) follows from Lemma 8.2, and (8.45) follows since $\mathbb{E}[\exp(-|X|^+)] \le \frac{1}{a}\Pr\{X \ge \log a\} + \Pr\{X \le \log a\}$ for any $a \ge 1$.

It is thus left to upper bound the third term in (8.46). Define the following function of power allocation vector $\mathbf{P}$ and the fading parameter vector $\mathbf{H}$,

$$
B_1(\mathbf{P}, \mathbf{H}) := \sum_{t\in[\Psi]} \log\frac{1 + P_t H_t^2}{\sqrt{1 + 2P_t H_t^2}} + (c_1 - 0.5)\log k + c_2 + 2\log n + \log\gamma. \tag{8.47}
$$

It follows from the definitions of $L_t^n(\cdot)$ in (8.16), $\tilde{L}_t^n(\cdot)$ in (8.17), and $\Upsilon(\cdot)$ in (8.39) that

$$-\Upsilon(\mathbf{P},\mathbf{H},\mathbf{Z},S^k,D)+\log n=\sum_{i\in[k]}\jmath(S_i|P_S,D)-\sum_{t\in[\Psi]}L_t^n(P_t,H_t,Z_t^n)$$
$$-B_1(\mathbf{P},\mathbf{H}). \tag{8.48}$$

Thus,

$$\Pr\{(\mathbf{H},\mathbf{Z},S^k)\notin\mathcal{T}_{k,n}(\mathbf{P})\}$$
$$=\inf_{\mathbf{P}\in\mathcal{P}_{\max}(P)}\Pr\left\{\sum_{i\in[k]}\jmath(S_i|P_S,D)-\sum_{t\in[\Psi]}L_t^n(P_t,H_t,Z_t^n)-B_1(\mathbf{P},\mathbf{H})\geq 0\right\} \tag{8.49}$$
$$\leq\inf_{\mathbf{P}\in\mathcal{P}_{\mathrm{eq}}(P)}\Pr\left\{\sum_{i\in[k]}\jmath(S_i|P_S,D)-\sum_{t\in[\Psi]}L_t^n(P_t,H_t,Z_t^n)-B_1(\mathbf{P},\mathbf{H})\geq 0\right\}. \tag{8.50}$$

where (8.50) follows since $\mathcal{P}_{\mathrm{eq}}(P)\subset\mathcal{P}_{\max}(P)$.

For subsequent analyses, fix any $\mathbf{P}\in\mathcal{P}_{\mathrm{eq}}(P)$ and let

$$C(n,k):=B_1(\mathbf{P},\mathbf{H})-\sum_{t\in[\Psi]}\log\frac{1+P_tH_t^2}{\sqrt{1+2P_tH_t^2}}. \tag{8.51}$$

Recall that $\mathrm{Q}(\cdot)$ is the complementary Gaussian cdf. It follows from [4, Eq. (255)] that for any $(a,b)\in\mathbb{R}_+^2$,

$$\mathrm{Q}(a+b)\geq\mathrm{Q}(a)-\frac{b}{\sqrt{2\pi}}. \tag{8.52}$$

It follows from the definitions of $U_i(\mathbf{P},\mathbf{H})$ in (8.18) to (8.20) and the Berry-Esseen theorem (cf. Theorem 1.6) that

$$\Pr\left\{\sum_{i\in[k]}\jmath(S_i|P_S,D)-\sum_{t\in[\Psi]}L_t^n(P_t,H_t,Z_t^n)-B_1(\mathbf{P},\mathbf{H})\geq 0\right\}$$
$$\leq\Pr\Bigg\{\Bigg(\sum_{i\in[k]}\big(\jmath(S_i|P_S,D)-R(P_S,D)\big)$$
$$-\sum_{t\in[\Psi]}\left(L_t^n(P_t,H_t,Z_t^n)-\frac{n}{2}\log(1+P_tH_t^2)\right)\Bigg)$$
$$\geq nU_1(\mathbf{P},\mathbf{H})-kR(P_S,D)-B_1(\mathbf{P},\mathbf{H})\Bigg\} \tag{8.53}$$

$$\leq \mathbb{E}\left[\mathrm{Q}\left(\frac{nU_1(\mathbf{P},\mathbf{H}) - kR(P_S,D) - B_1(\mathbf{P},\mathbf{H})}{\sqrt{(k+n\Psi)V_{k+n\Psi}^{(\mathbf{P},\mathbf{H})}}}\right)\right] + \mathbb{E}\left[\frac{6T_{k+n\Psi}^{(\mathbf{P},\mathbf{H})}\left(V_{k+n\Psi}^{(\mathbf{P},\mathbf{H})}\right)^{-3/2}}{\sqrt{k+n\Psi}}\right] \tag{8.54}$$

$$\leq \mathbb{E}\left[\mathrm{Q}\left(\frac{nU_1(\mathbf{P},\mathbf{H}) - kR(P_S,D) - B_1(\mathbf{P},\mathbf{H})}{\sqrt{(k+n\Psi)V_{k+n\Psi}^{(\mathbf{P},\mathbf{H})}}}\right)\right] + \mathbb{E}\left[\frac{6T_{k+n\Psi}^{(\mathbf{P},\mathbf{H})}\left(V_{k+n\Psi}^{(\mathbf{P},\mathbf{H})}\right)^{-3/2}}{\sqrt{k+n\Psi}}\right] \tag{8.55}$$

$$\leq \mathbb{E}\left[\mathrm{Q}\left(\frac{nU_1(\mathbf{P},\mathbf{H}) - kR(P_S,D) - C(n,k)}{\sqrt{(k+n\Psi)V_{k+n\Psi}^{(\mathbf{P},\mathbf{H})}}}\right) + \frac{\sum_{t\in[\Psi]}\log\frac{1+P_tH_t^2}{\sqrt{1+2P_tH_t^2}}}{\sqrt{(k+n\Psi)V_{k+n\Psi}^{(\mathbf{P},\mathbf{H})}}}\right] + \frac{\Omega_1(k,n,P)}{\sqrt{k+n\Psi}} \tag{8.56}$$

$$\leq \mathbb{E}\left[\mathrm{Q}\left(\frac{nU_1(\mathbf{P},\mathbf{H}) - kR(P_S,D) - C(n,k)}{\sqrt{(k+n\Psi)V_{k+n\Psi}^{(\mathbf{P},\mathbf{H})}}}\right)\right] + \frac{\Omega_1(k,n,P)+\Omega_3(k,n,P)}{\sqrt{k+n\Psi}} \tag{8.57}$$

$$= \mathbb{E}\left[\mathrm{Q}\left(\frac{nU_1(\mathbf{P},\mathbf{H}) - kR(P_S,D)}{\sqrt{(k+n\Psi)V_{k+n\Psi}^{(\mathbf{P},\mathbf{H})}}}\right)\right] + \frac{\Omega_2(k,n,P)C(n,k)}{\sqrt{k+n\Psi}} + O\left(\frac{1}{\sqrt{n}}\right) \tag{8.58}$$

$$\leq \mathbb{E}\left[\mathrm{Q}\left(\frac{nU_1(\mathbf{P},\mathbf{H}) - kR(P_S,D)}{\sqrt{(k+n\Psi)V_{k+n\Psi}^{(\mathbf{P},\mathbf{H})}}}\right)\right] + O\left(\frac{\log n}{\sqrt{n}}\right), \tag{8.59}$$

where (8.56) follows from the definitions of $\Omega_1(k,n,P)$ in (8.33), $C(n,k)$ in (8.51) and by applying the inequality in (8.52), (8.57) follows from the fact that $\log\frac{1+P_tH_t^2}{\sqrt{1+2P_tH_t^2}} \leq \frac{1}{2}\log(1+P_tH_t^2)$ for any $t\in[\Psi]$ and the definition of $\Omega_3(k,n,P)$ in (8.35), (8.58) follows from the inequality in (8.52), the definitions of $\Omega_2(\cdot)$ in (8.34), the fact that $\Omega_1(k,n,P) = O(1)$ and $\Omega_3(k,n,P) = O(1)$ due to the assumptions in Theorem 8.3, and (8.59) follows from the inequality in (8.52) and the fact that $C(n,k) = O(\log n)$ due to its definition in (8.51).

The achievability proof of Theorem 8.3 is completed by combining (8.46), (8.50), and (8.59).

8.4.3 Converse

Similar to the definition of $\mathrm{P}_{\mathrm{e},k,n}^{*}(P,D)$ in (8.9), let $\mathrm{P}_{\mathrm{e},k,n}^{*,\mathrm{eq}}(P,D)$ denote the corresponding minimal excess-distortion probability for any (k,n,P)-code with equal power constraint. It follows that

$$\mathrm{P}^*_{\mathrm{e},k,n}(P,D) \geq \mathrm{P}^{*,\mathrm{eq}}_{\mathrm{e},k,n+1}(P,D). \tag{8.60}$$

This is because given any (k, n, P)-code with strict inequality in (8.6), one can construct a $(k, n+1, P)$-code with equality in (8.6). This section proves a lower bound on $\mathrm{P}^{*,\mathrm{eq}}_{\mathrm{e},k,n}(P,D)$ and uses (8.60) to establish the desired lower bound on $\mathrm{P}^*_{\mathrm{e},k,n}(P,D)$ in Theorem 8.2.

Recall the definition of $L_t^n(P_t, H_t, Z_t^n)$ in (8.16). For subsequent analyses, let

$$\underline{L}(P_t, H_t, Z_t^n) := L_t^n(P_t, H_t, Z_t^n) - \frac{n}{2}\log(1+P_tH_t^2) \tag{8.61}$$

$$= \frac{\sum_{j\in[n]} P_tH_t^2(1-Z_{t,j}^2) + 2H_t\sqrt{P_t}Z_{t,j}}{2(1+P_tH_t^2)}. \tag{8.62}$$

Note that $\mathbb{E}[\underline{L}(P_t, H_t, Z_t^n)] = 0$.

Given any power allocation vector $\mathbf{P} \in \mathcal{P}_{\mathrm{eq}}(P)$ and $\gamma > 0$, it follows from the definitions of $\jmath(S^k|P_S, D)$ in (8.15) that the probability term in Theorem 8.2 can be lower bounded as follows:

$$\Pr\left\{\sum_{t\in[\Psi]} L_t^n(P_t, H_t, Z_t^n) \leq \jmath(S^k|P_S, D) - n\gamma\right\}$$

$$= \Pr\left\{\sum_{t\in[\Psi]}\left(\frac{n}{2}\log(1+P_tH_t^2) + \underline{L}(P_t, H_t, Z_t^n)\right) \leq \sum_{i\in[k]} \jmath(S_i|P_S, D) - n\gamma\right\} \tag{8.63}$$

$$= \Pr\Bigg\{\sum_{i\in[k]}\left(\jmath(S_i|P_S, D) - R(P_S, D)\right) - \sum_{t\in[\Psi]}\underline{L}(P_t, H_t, Z_t^n)$$

$$\geq \sum_{t\in[\Psi]}\frac{n}{2}\log(1+P_tH_t^2) - kR(P_S, D) + n\gamma\Bigg\}. \tag{8.64}$$

It follows from Lemma 2.1 that $\mathbb{E}[\jmath(S_i, D)] = R(P_S, D)$. Thus, $\{\jmath(S_i, D) - R(P_S, D)\}_{i\in[k]}$ and $\left\{\frac{P_tH_t^2(1-Z_{t,j}^2+2H_t\sqrt{P_t}Z_{t,j})}{2(1+P_tH_t^2)}\right\}_{t\in[\Psi], j\in[n]}$ form a sequence of $(k + n\Psi)$ independent random variables with zero mean. It follows from the Berry-

Esseen theorem (cf. Theorem 1.6) that

$$\mathbb{E}\left[\left|\Pr\left\{\sum_{t\in[\Psi]} L_t^n(P_t,H_t,Z_t^n)\le \jmath(S^k,D)-n\gamma\right\}\right.\right.$$
$$\left.\left.-\,\mathrm{Q}\left(\frac{\sum_{t\in[\Psi]}\frac{n}{2}\log(1+P_tH_t^2)-kR(P_S,D)+n\gamma}{\sqrt{(k+n\Psi)V_{k+n\Psi}^{(\mathbf{P},\mathbf{H})}}}\right)\right|\right]$$
$$\le \mathbb{E}\left[\frac{6T_{k+n\Psi}^{(\mathbf{P},\mathbf{H})}\left(V_{k+n\Psi}^{(\mathbf{P},\mathbf{H})}\right)^{-3/2}}{\sqrt{(k+n\Psi)}}\right]. \tag{8.65}$$

Using Theorem 8.2 and the result in (8.65) and setting $\gamma=\frac{\log n}{n}$, it follows that

$$\mathrm{P}_{\mathrm{e},k,n}^{*,\mathrm{eq}}(P,D)+\frac{1}{n}$$
$$\ge \inf_{\mathbf{P}\in\mathcal{P}_{\mathrm{eq}}(P)}\left\{\mathbb{E}\left[\mathrm{Q}\left(\frac{nU_1(\mathbf{P},\mathbf{H})-kR(P_S,D)+\log n}{\sqrt{(k+n\Psi)V_{k+n\Psi}^{(\mathbf{P},\mathbf{H})}}}\right)-\frac{6T_{k+n\Psi}^{(\mathbf{P},\mathbf{H})}\left(V_{k+n\Psi}^{(\mathbf{P},\mathbf{H})}\right)^{-3/2}}{\sqrt{(k+n\Psi)}}\right]\right\} \tag{8.66}$$
$$\ge \inf_{\mathbf{P}\in\mathcal{P}_{\mathrm{eq}}(P)}\mathbb{E}\left[\mathrm{Q}\left(\frac{nU_1(\mathbf{P},\mathbf{H})-kR(P_S,D)+\log n}{\sqrt{(k+n\Psi)V_{k+n\Psi}^{(\mathbf{P},\mathbf{H})}}}\right)\right]$$
$$-\sup_{\mathbf{P}\in\mathcal{P}_{\mathrm{eq}}(P)}\mathbb{E}\left[\frac{6T_{k+n\Psi}^{(\mathbf{P},\mathbf{H})}\left(V_{k+n\Psi}^{(\mathbf{P},\mathbf{H})}\right)^{-3/2}}{\sqrt{(k+n\Psi)}}\right] \tag{8.67}$$
$$=\inf_{\mathbf{P}\in\mathcal{P}_{\mathrm{eq}}(P)}\mathbb{E}\left[\mathrm{Q}\left(\frac{nU_1(\mathbf{P},\mathbf{H})-kR(P_S,D)+\log n}{\sqrt{(k+n\Psi)V_{k+n\Psi}^{(\mathbf{P},\mathbf{H})}}}\right)\right]-\frac{\Omega_1(k,n,P)}{\sqrt{(k+n\Psi)}} \tag{8.68}$$
$$=\inf_{\mathbf{P}\in\mathcal{P}_{\mathrm{eq}}(P)}\mathbb{E}\left[\mathrm{Q}\left(\frac{nU_1(\mathbf{P},\mathbf{H})-kR(P_S,D)+\log n}{\sqrt{(k+n\Psi)V_{k+n\Psi}^{(\mathbf{P},\mathbf{H})}}}\right)\right]+O\left(\frac{1}{\sqrt{n}}\right), \tag{8.69}$$

where (8.68) follows from the definition of $\Omega_1(k,n,P)$ in (8.33) and (8.69) follows since $\Omega_1(k,n,P)=O(1)$ from the assumptions of Theorem 8.3.

It follows from (8.60) and (8.68) that

$$\mathrm{P}_{\mathrm{e},k,n}^{*}(P,D)$$
$$\ge \mathrm{P}_{\mathrm{e},k,n+1}^{*,\mathrm{eq}}(P,D) \tag{8.70}$$
$$\ge O\left(\frac{1}{\sqrt{n}}\right)+\inf_{\mathbf{P}\in\mathcal{P}_{\mathrm{eq}}(P)}\mathbb{E}\left[\mathrm{Q}\left(\frac{(n+1)U_1(\mathbf{P},\mathbf{H})-kR(P_S,D)+\log(n+1)}{\sqrt{(k+(n+1)\Psi)V_{k+(n+1)\Psi}^{(\mathbf{P},\mathbf{H})}}}\right)\right] \tag{8.71}$$

$$\geq \inf_{\mathbf{P}\in\mathcal{P}_{\text{eq}}(P)} \mathbb{E}\left[\mathrm{Q}\left(\frac{nU_1(\mathbf{P},\mathbf{H}) - kR(P_S, D) + \log(n+1)}{\sqrt{(k+n\Psi)V_{k+n\Psi}^{(\mathbf{P},\mathbf{H})}}}\right) - \frac{U_1(\mathbf{P},\mathbf{H})}{\sqrt{(k+n\Psi)V_{k+n\Psi}^{(\mathbf{P},\mathbf{H})}}}\right]$$

$$+ O\left(\frac{1}{\sqrt{n}}\right) \tag{8.72}$$

$$= \inf_{\mathbf{P}\in\mathcal{P}_{\text{eq}}(P)} \mathbb{E}\left[\mathrm{Q}\left(\frac{nU_1(\mathbf{P},\mathbf{H}) - kR(P_S, D) + \log(n+1)}{\sqrt{(k+n\Psi)V_{k+n\Psi}^{(\mathbf{P},\mathbf{H})}}}\right)\right]$$

$$- \frac{\Omega_3(k,n,P)}{\sqrt{k+n\Psi}} + O\left(\frac{1}{\sqrt{n}}\right) \tag{8.73}$$

$$= \inf_{\mathbf{P}\in\mathcal{P}_{\text{eq}}(P)} \mathbb{E}\left[\mathrm{Q}\left(\frac{nU_1(\mathbf{P},\mathbf{H}) - kR(P_S, D) + \log(n+1)}{\sqrt{(k+n\Psi)V_{k+n\Psi}^{(\mathbf{P},\mathbf{H})}}}\right)\right] + O\left(\frac{1}{\sqrt{n}}\right), \tag{8.74}$$

$$\geq \inf_{\mathbf{P}\in\mathcal{P}_{\text{eq}}(P)} \mathbb{E}\left[\mathrm{Q}\left(\frac{nU_1(\mathbf{P},\mathbf{H}) - kR(P_S, D)}{\sqrt{(k+n\Psi)V_{k+n\Psi}^{(\mathbf{P},\mathbf{H})}}}\right)\right] - \frac{\Omega_2(k,n,P)\log(n+1)}{\sqrt{k+n\Psi}}$$

$$+ O\left(\frac{1}{\sqrt{n}}\right) \tag{8.75}$$

$$= \inf_{\mathbf{P}\in\mathcal{P}_{\text{eq}}(P)} \mathbb{E}\left[\mathrm{Q}\left(\frac{nU_1(\mathbf{P},\mathbf{H}) - kR(P_S, D)}{\sqrt{(k+n\Psi)V_{k+n\Psi}^{(\mathbf{P},\mathbf{H})}}}\right)\right] + O\left(\frac{\log n}{\sqrt{n}}\right), \tag{8.76}$$

where (8.72) follows from (8.52), (8.73) follows from the definition of $\Omega_3(k, n, P)$ in (8.35) and the fact that $\frac{\log(1+1/n)}{\sqrt{(k+n\Psi)V_{k+n\Psi}^{(\mathbf{P},\mathbf{H})}}} \leq \frac{\log 2}{\sqrt{kV(P_S)}} = O(\frac{1}{\sqrt{n}})$ for any $n \geq 1$ since $k = \Theta(n)$, and (8.74) follows since $\Omega_3(k, n, P) = O(1)$ similarly to (8.69), (8.75) follows similarly to (8.72) and using the definition of $\Omega_2(\cdot)$ in (8.34), and (8.76) follows since $\Omega_2(k, n, P) = O(1)$ according to the assumption of Theorem 8.3.

The converse proof of Theorem 8.3 is now completed.

8.5 Alternative Achievability

The proof of Theorem 8.1 requires the knowledge of the source distribution P_S and the exact channel law in (8.1) to design the code. In the case of unknown P_S or channel law, the code cannot be used. To resolve this problem, this section presents a universal JSCC scheme, which does not require P_S and achieves the same second-order asymptotics in Theorem 8.3. The JSCC code dates back to Csiszár [11], who proposed a code using unequal error protection [12] and the method of types [13] to derive the error exponent for the JSCC problem. The idea of Csiszár is adapted to lossy multi-connectivity using a modified minimum distance decoder.

8.5.1 Coding Scheme

The coding scheme is analogous to the mismatched code in Chap. 4.1.2. Fix any integer $k \in \mathbb{N}$. Recall that $\mathcal{P}_k(\mathcal{S})$ denotes the set of all types on $\mathcal{S}$ with sequence length k and $P^*_{\hat{S}}$ is the induced marginal distribution on $\hat{\mathcal{S}}$ by the source distribution P_S and the optimizer of the rate-distortion function $R(P_S, D)$ in (8.10). For simplicity, assume that the elements (types) in the set $\mathcal{P}_k(\mathcal{S})$ are ordered as $Q_1, \ldots, Q_{|\mathcal{P}_k(\mathcal{S})|}$. Let $\{M_i\}_{i \in [|\mathcal{P}_k(\mathcal{S})|]}$ be a sequence of integers to be determined and define the following set of integers

$$\mathcal{D} := \{(r_1, r_2) \in \mathcal{R}^2_+ : r_1 \in [|\mathcal{P}_k(\mathcal{S})|], r_2 \in [M_{r_1}]\}. \tag{8.77}$$

The source and channel codebook is generated as follows. For each $i \in [|\mathcal{P}_k(\mathcal{S})|]$, generate M_i independent source codewords $\hat{\mathbf{S}}_i := \{\hat{S}^k(i, 1), \ldots, \hat{S}^k(i, M_i)\}$, where for each $j \in [M_i]$, $\hat{S}^k(i, j)$ is generated i.i.d. from $P^*_{\hat{S}}$. For each $i \in [|\mathcal{P}_k(\mathcal{S})|]$ and $t \in [\Psi]$, generate M_i independent channel codewords $\{X^n_t(i, 1), \ldots, X^n_t(i, M_i)\}$, where for each $j \in [M_i]$, $X^n_t(i, j)$ is generated from the uniform distribution over a sphere with radius $\sqrt{nP_t}$, i.e.,

$$X^n_t(i, j) \sim f^{\mathrm{sp}}_{X^n}(x^n) := \frac{\delta(\|x^n\|^2 - nP_t)}{A_n(\sqrt{nP_t})}, \tag{8.78}$$

where $\delta(\cdot)$ is the Dirac delta function, $A_k(r) := k\pi^{k/2}r^{k-1}/\Gamma(\frac{k+2}{2})$ is the surface area of an k-dimensional sphere with radius r, and $\Gamma(\cdot)$ is the Gamma function.

Fix threshold $\gamma_{k,n}$ to be specified. The encoding and decoding operation is as follows. Given a source sequence S^k, for each $t \in [\Psi]$, the encoder f_t transmits $X^n_t(I, J)$ if the type $\hat{S}^k = Q_I \in \mathcal{P}_k(\mathcal{S})$ and J is the index of the source codeword that minimizes distortion between the source sequence and the source codewords in the i-th sub-codebook $\mathbf{S}_i$, i.e.,

$$J = \operatorname*{arg\,min}_{\tilde{j} \in [M_I]} d(S^k, \hat{S}^k(I, \tilde{j})). \tag{8.79}$$

Given channel outputs $\mathbf{Y} = (Y^n_1, \ldots, Y^n_\Psi)$, the decoder ϕ generates source estimate $\hat{S}^k(\hat{I}, \hat{J})$ if $(\hat{I}, \hat{J})$ are the unique pair such that

$$\sum_{t \in [\Psi]} \tilde{\imath}(X^n_t(\hat{I}, \hat{J}); Y^n_t | H_t) - \log M_{\hat{I}} > \log \gamma_{k,n}. \tag{8.80}$$

Otherwise, the decoder declares an error.

Recall the definition of $\tilde{\imath}(X_t^n; Y_t^n|H_t)$ in (8.30):

$$\sum_{t\in[\Psi]} \tilde{\imath}(X_t^n(\hat{I}, \hat{J}); Y_t^n|H_t)$$
$$= -\frac{\sum_{t\in[\Psi]} \|Y_t^n - H_t X_t^n(\hat{I}, \hat{J})\|_2^2}{2} + \sum_{t\in[\Psi]} \left(\frac{n}{2}\log(1 + P_t H_t^2) + \frac{\|Y_t^n\|_2^2}{2(1 + P_t H_t^2)}\right). \tag{8.81}$$

Since the decoding metric $\tilde{\imath}(X_t^n; Y_t^n|H_t)$ does not depend on source distribution or the channel statistics, the above code is universal.

8.5.2 Performance Analysis

The excess-distortion probability of the above JSCC code satisfies

$$\begin{aligned} \mathrm{P_e}(k, n, P, D) \le & \sum_{i\in[|\mathcal{P}_k(\mathcal{S})|]} \Pr\{S^k \in \mathcal{T}_{Q_i}, \min_{\tilde{j}\in[M_i]} d(S^k, \hat{S}^k(i, \tilde{j})) > D\} \\ & + \sum_{i\in[|\mathcal{P}_k(\mathcal{S})|]} \Pr\{S^k \in \mathcal{T}_{Q_i}, (\hat{I}, \hat{J}) \neq (i, J)\}. \end{aligned} \tag{8.82}$$

Define the constant $\zeta := 4|\mathcal{S}||\hat{\mathcal{S}}| + 9$. For each $i \in [|\mathcal{P}_k(\mathcal{S})|]$, let

$$\log M_i := kR(Q_i, D) + \zeta \log k, \tag{8.83}$$

where Q_i is the i-element (type) in $\mathcal{P}_k(\mathcal{S})$. Using the type covering lemma for lossy source coding [14], it follows that for each $i \in [|\mathcal{P}_k(\mathcal{S})|$, when n is sufficiently large,

$$\Pr\left\{S^k \in \mathcal{T}_{Q_i}, \min_{\tilde{j}\in[M_i]} d(S^k, \hat{S}^k(i, \tilde{j})) > D\right\} = 0. \tag{8.84}$$

Given $s^k \in \mathcal{T}_{Q_i}$ and the i-th sub-codebook $\hat{\mathbf{s}}_i$, let $j(s^k, \hat{\mathbf{s}}_i)$ denote the index of the source codeword that minimizes the distortion with respect to s^k. The following lemma upper bounds the error probability of decoding $(i, j(s^k, \hat{\mathbf{s}}_i))$ given the source sequence s^k and the i-th sub-codebook $\mathbf{s}_i$. To present the lemma, define

$$\begin{aligned} B_2(\mathbf{P}, \mathbf{H}) := \sum_{t\in[\Psi]} \frac{1 + P_t H_t^2}{\sqrt{1 + 2P_t H_t^2}} \Bigg(& \frac{2(1 + P_t H_t^2)}{\sqrt{\pi}\sqrt{P_t H_t^2(P_t H_t^2 + 2)}} \\ & + \frac{12\sqrt{2}P_t H_t^2(7P_t H_t^2 + 12)}{(P_t H_t^2 + 2)\sqrt{P_t H_t^2(P_t H_t^2 + 2)}}\Bigg). \end{aligned} \tag{8.85}$$

Under the conditions of Theorem 8.3,there exists a constant c_6 such that

$$\sup_{\mathbf{P}\in\mathcal{P}_{\mathrm{eq}}(P)} \mathbb{E}[B_2(\mathbf{P},\mathbf{H})] \le c_6. \tag{8.86}$$

Lemma 8.3 *For any* $i \in [|\mathcal{P}_k(\mathcal{S})|]$*, given* $s^k \in \mathcal{T}_{Q_i}$ *and* $\hat{\mathbf{s}}_i$*,*

$$\begin{aligned}&\Pr\{(\hat{I},\hat{J}) \neq (i, j(s^k,\hat{\mathbf{s}}_i))|S^k = s^k,\ \hat{\mathbf{S}}_i = \hat{\mathbf{s}}_i\}\\&\le \Pr\left\{\sum_{t\in[\Psi]} \tilde{\imath}(X_t^n;Y_t^n|H_t) \le \log\gamma_{k,n} + \log M_i\right\} + \frac{|\mathcal{P}_k(\mathcal{S})|\mathbb{E}[B_2(\mathbf{P},\mathbf{H})]}{\gamma_{k,n}\sqrt{n}}.\end{aligned} \tag{8.87}$$

The proof of Lemma 8.3 is inspired by the dependence testing bound [15, Theorem 18] (see also [8, Theorem 1]) and the analysis of its variant to quasi-static fading channel in [9, Chapter 4].

Set $\gamma_{k,n} = |\mathcal{P}_k(\mathcal{S})| \le (k+1)^{|\mathcal{S}|}$. It follows from Lemma 8.3 and the Berry-Esseen theorem that for any power allocation vector $\mathbf{P} \in \mathcal{P}_{\mathrm{eq}}(P)$, the decoding error probability satisfies

$$\begin{aligned}&\sum_{i\in[|\mathcal{P}_k(\mathcal{S})|]} \Pr\{S^k \in \mathcal{T}_{Q_i},\ (\hat{I},\hat{J}) \neq (i, J)\}\\&= \sum_{i\in[|\mathcal{P}_k(\mathcal{S})|]}\sum_{s^k\in\mathcal{T}_{Q_i}} P_S^k(s^k)\sum_{\hat{\mathbf{s}}_i} P_{\hat{S}|S}^k(\hat{s}^k|s^k)\Pr\{(\hat{I},\hat{J}) \neq (i, j(s^k,\hat{\mathbf{s}}_i))|S^k = s^k, \hat{\mathbf{S}}_i = \hat{\mathbf{s}}_i\}\end{aligned} \tag{8.88}$$

$$\begin{aligned}&\le \sum_{i\in[|\mathcal{P}_k(\mathcal{S})|]}\sum_{s^k\in\mathcal{T}_{Q_i}} P_S^k(s^k)\Pr\left\{\sum_{t\in[\Psi]} \tilde{\imath}(X_t^n;Y_t^n|H_t) \le \log\gamma_{k,n} + \log M_i\right\}\\&\quad + \frac{|\mathcal{P}_k(\mathcal{S})|\mathbb{E}[B_2(\mathbf{P},\mathbf{H})]}{\gamma_{k,n}\sqrt{n}}\end{aligned} \tag{8.89}$$

$$\begin{aligned}&= \Pr\left\{\sum_{t\in[\Psi]} \tilde{\imath}(X_t^n;Y_t^n|H_t) \le \log\gamma_{k,n} + nR(\hat{P}_{S^k},D) + \zeta\log k\right\}\\&\quad + \frac{|\mathcal{P}_k(\mathcal{S})|\mathbb{E}[B_2(\mathbf{P},\mathbf{H})]}{\gamma_{k,n}\sqrt{n}}\end{aligned} \tag{8.90}$$

$$\begin{aligned}&\le \Pr\left\{\sum_{t\in[\Psi]} \tilde{\imath}(X_t^n;Y_t^n|H_t) \le |\mathcal{S}|\log(k+1) + nR(\hat{P}_{S^k},D) + \zeta\log k\right\}\\&\quad + \frac{\mathbb{E}[B_2(\mathbf{P},\mathbf{H})]}{\sqrt{n}}\end{aligned} \tag{8.91}$$

$$\leq \Pr\left\{S^k \in \mathcal{T}^k, \sum_{t\in[\Psi]} \tilde{\imath}(X_t^n; Y_t^n|H_t) \leq |\mathcal{S}| \log(k+1) + kR(\hat{P}_{S^k}, D) + \zeta \log k\right\}$$

$$+ \Pr\{S^k \notin \mathcal{T}^k\} + \frac{\mathbb{E}[B_2(\mathbf{P}, \mathbf{H})]}{\sqrt{n}} \tag{8.92}$$

$$\leq \Pr\left\{\sum_{t\in[\Psi]} L_t^n(P_t, H_t, Z_t^n) \leq \sum_{i\in[k]} \jmath(S_i, D) + O(\log k)\right\} + \frac{2|\mathcal{S}|}{k^2} + \frac{\mathbb{E}[B_2(\mathbf{P}, \mathbf{H})]}{\sqrt{n}} \tag{8.93}$$

$$= \mathbb{E}\Bigg[\Pr\Bigg\{\sum_{i\in[k]} (\jmath(S_i, D) - R(P_X, D)) - \sum_{t\in[\Psi]} \tilde{L}_t^n(P_t, H_t, Z^n)$$

$$\geq \Bigg(\sum_{t\in[\Psi]} \frac{n}{2} \log(1 + P_t H_t^2) - kR(P_S, D) + O(\log k)\Bigg)\Bigg\}\Bigg|\mathbf{H}\Bigg] + \frac{2|\mathcal{S}|}{k^2}$$

$$+ \frac{\mathbb{E}[B_2(\mathbf{P}, \mathbf{H})]}{\sqrt{n}} \tag{8.94}$$

$$\leq \mathbb{E}\left[\mathrm{Q}\left(\frac{nU_1(\mathbf{P}, \mathbf{H}) - kR(P_S, D) + O(\log n)}{\sqrt{kV(P_S, D) + nU_2(\mathbf{P}, \mathbf{H})}}\right)\right] + \frac{\Omega_1(k, n, P)}{\sqrt{(k + n\Psi)}} + \frac{2|\mathcal{S}|}{k^2}$$

$$+ \frac{\mathbb{E}[B_2(\mathbf{P}, \mathbf{H})]}{\sqrt{n}} \tag{8.95}$$

$$= \mathbb{E}\left[\mathrm{Q}\left(\frac{nU_1(\mathbf{P}, \mathbf{H}) - kR(P_S, D) + O(\log n)}{\sqrt{kV(P_S, D) + nU_2(\mathbf{P}, \mathbf{H})}}\right)\right] + o\left(\frac{1}{\sqrt{n}}\right) \tag{8.96}$$

$$\leq \mathbb{E}\left[\mathrm{Q}\left(\frac{nU_1(\mathbf{P}, \mathbf{H}) - kR(P_S, D)}{\sqrt{kV(P_S, D) + nU_2(\mathbf{P}, \mathbf{H})}}\right)\right] + O\left(\frac{\log n}{\sqrt{n}}\right), \tag{8.97}$$

where (8.90) follows from the choice of M_i in (8.90), (8.93) follows from Taylor expansion of the type-based rate-distortion function $R(Q_i, D)$ around the source distribution P_S (cf. [16, Eq. (91)]), the result in Lemma 8.1 and [17, Lemma 22], (8.94) follows similarly to (8.64), (8.95) follows from the Berry-Esseen theorem similarly to (8.65), the definition of $\Omega_1(k, n, P)$ in (8.33) and assumption that $k = O(n)$, (8.96) follows from the result in (8.86) and the result that $\Omega_1(k, n, P) = O(1)$ according to the assumptions of Theorem 8.3, and (8.97) follows similarly to (8.75) and (8.76) by using the inequality in (8.52) and the assumptions of Theorem 8.3.

The alternative achievability proof of Theorem 8.3 for DMS is completed by combining (8.82), (8.84), and (8.97).

8.6 Chapter Summary

This chapter studies the problem of joint source-channel coding over parallel channels with independent quasi-static fading and AWGN noise. The second-order asymptotic results imply asymptotic notions of outage probability and outage capacity are valid for low-latency communications over quasi-static fading channels. As the number of channels increases, transmission reliability improves. The contents in this chapter generalize the result in Theorem 2.5 for a single non-fading channel. Via the techniques in Chap. 4, the achievability analyses in this chapter can be generalized to transmit an arbitrary memoryless source over parallel channels with independent quasi-static fading and non-Gaussian noise. The take-home message is that quasi-static fading is the correct model for low-latency wireless communications, and asymptotic notions of outage capacity and outage probability are valid.

8.7 Exercises

8.1 What is multi-connectivity? Why is it important for wireless communication in achieving ultra-reliable and low-latency communication?

8.2 Prove the finite blocklength achievability bound in Theorem 8.1.

8.3 Prove the non-asymptotic converse bound in Theorem 8.2.

8.4 Prove the result in Lemma 8.1 concerning spherical symmetry of i.i.d. Gaussian random vectors.

8.5 Verify the result in (8.86) and prove the finite blocklength upper bound on the decoding error probability in Lemma 8.3.

8.6 Justify that $\Omega_1(k, n, P) = O(1)$ in (8.69) and $\Omega_2(k, n, P) = O(1)$ in (8.76) under the assumption of Theorem 8.3.

References

1. A. Wolf, P. Schulz, M. Dörpinghaus, J.C.S. Santos Filho, G. Fettweis, How reliable and capable is multi-connectivity? IEEE Trans. Commun. **67**(2), 1506–1520 (2019)
2. L. Zhou, A. Wolf, M. Motani, On lossy multi-connectivity: finite blocklength performance and second-order asymptotics. IEEE J. Sel. Areas Commun. **37**(4), 735–748 (2019)
3. W. Yang, G. Durisi, T. Koch, Y. Polyanskiy, Quasi-static multiple-antenna fading channels at finite blocklength. IEEE Trans. Inf. Theory **60**(7), 4232–4265 (2014)
4. V. Kostina, S. Verdú, Lossy joint source-channel coding in the finite blocklength regime. IEEE Trans. Inf. Theory **59**(5), 2545–2575 (2013)
5. D. Tse, P. Viswanath. *Fundamentals of Wireless Communication* (Cambridge University Press, Cambridge, 2005)
6. L. Zheng, D.N.C. Tse, Diversity and multiplexing: a fundamental tradeoff in multiple-antenna channels. IEEE Trans. Inf. Theory **49**(5), 1073–1096 (2003)
7. A. Khisti, S.C. Draper, The streaming-DMT of fading channels. IEEE Trans. Inf. Theory **60**(11), 7058–7072 (2014)

8. E. MolavianJazi, J.N. Laneman, A second-order achievable rate region for Gaussian multi-access channels via a central limit theorem for functions. IEEE Trans. Inf. Theory **61**(12), 6719–6733 (2015)
9. E. MolavianJazi, A unified approach to Gaussian channels with finite blocklength. Ph.D. Dissertation, Department Electrical Engineering, University of Notre Dame, 2014
10. V. Kostina, S. Verdú, Fixed-length lossy compression in the finite blocklength regime. IEEE Trans. Inf. Theory **58**(6), 3309–3338 (2012)
11. I. Csiszár, Joint source-channel error exponent. Prob. Contr. Inform. Theory **9**(5), 315–328 (1980)
12. Y.Y. Shkel, V.Y.F. Tan, S.C. Draper, Unequal message protection: asymptotic and non-asymptotic tradeoffs. IEEE Trans. Inf. Theory **61**(10), 5396–5416 (2015)
13. I. Csiszár, The method of types [information theory]. IEEE Trans. Inf. Theory **44**(6), 2505–2523 (1998)
14. K. Marton, Error exponent for source coding with a fidelity criterion. IEEE Trans. Inf. Theory **20**(2), 197–199 (1974)
15. Y. Polyanskiy, H.V. Poor, S. Verdú, Channel coding rate in the finite blocklength regime. IEEE Trans. Inf. Theory **56**(5), 2307–2359 (2010)
16. L. Zhou, V.Y.F. Tan, M. Motani, Second-order and moderate deviation asymptotics for successive refinement. IEEE Trans. Inf. Theory **63**(5), 2896–2921 (2017)
17. M. Tomamichel, V.Y.F. Tan, Second-order coding rates for channels with state. IEEE Trans. Inf. Theory **60**(8), 4427–4448 (2014)

Chapter 9
User Scheduling

This chapter studies an uplink multiuser access relay communication system. Such a system finds applications in multiuser relay systems such as hybrid satellite-terrestrial system, which aims to expand communication to a broader coverage area and meet the demand of higher throughput. For multiuser relay systems, the performance is strongly related to the design of multiuser access schemes [1], relay selection methods [2], and power allocation strategies [3]. For systems with a single relay, multiuser access schemes, especially user scheduling algorithms, are critical and should be optimized jointly with power allocation schemes. Furthermore, the development of integrated space-terrestrial information network (SIN) and the sixth generation (6G) communication architecture enforces the system to allow access to a massive number of devices. To meet the extremely high throughout and massive connectivity demands of future communication systems, it is vital to study efficient user scheduling algorithms with low-complexity multiuser relay communication systems.

To shed light on the design of such a system, this chapter studies user scheduling algorithms for an uplink multiuser access relay communication model consisting of a single receiver, a single relay, and multiple users. The relay uses decode-and-forward (DF) to facilitate the communication between users and the receiver. The channel between the receiver and the relay is assumed to be a quasi-static shadowed Rician fading channel, while the channels between the relay and users are assumed to experience independent quasi-static Rayleigh fading. The choice of quasi-static fading is validated by information theoretical analysis of low-latency communication in [4, 5]. In particular, under quasi-static fading channels, the asymptotic notion of outage capacity and outage probability provide tight approximations to the non-asymptotic performance at low latency.

This chapter presents the results in [6] and considers two cases of channel state information (CSI) availability. When instantaneous CSI is available, the relay is informed of the exact channel fading levels of each user, and the receiver is informed of the exact fading level from the terrestrial station. When channel distribution

L. Zhou, L. Bai, *Low-latency Communication*, Textbooks in Telecommunication Engineering, https://doi.org/10.1007/978-3-032-06627-5_9

information (CDI) is available, the relay and the receiver are only provided with the distribution of respective fading models. This chapter studies user scheduling algorithms for both cases and proposes user scheduling algorithms to maximize the sum transmission rate with CSI and to minimize the outage probability with CDI.

9.1 Problem Formulation

9.1.1 Communication Model

To discuss the user scheduling for the multiuser relay communication system, we consider an uplink model for the hybrid satellite-terrestrial communication system as shown in Fig. 9.1. There are N users with indices $\mathcal{N} = [N]$, one relay (the ground station), and one destination receiver (the satellite), all assumed to have a single antenna. Each user wishes to transmit messages to the receiver with the assist of the relay. Since the resource of the receiver is finite, assume that at

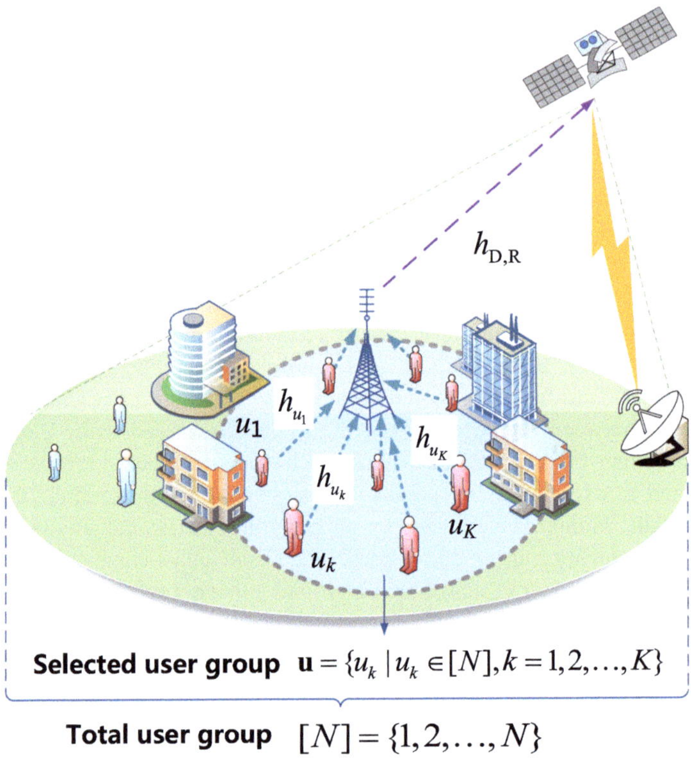

Fig. 9.1 System model of the hybrid satellite-terrestrial communication network

most K users are allowed to transmit messages in a time slot simultaneously. Let $\mathbf{u} = \{u_1, \ldots, u_K\} \in [N]_K$ denote the vector of the indices of the selected users by a scheduling algorithm.

At each time slot, after user scheduling, the communication between the selected K users and the destination proceeds in the following two phases. In the first phase, for each $k \in [K]$, the selected user u_k aims to send a message $M_{u_k} \in [1 : 2^{nR_{u_k}}]$ to the relay with power P_1, rate R_{u_k}, and blocklength n. Each message M_{u_k} is then encoded into a length-n codeword $X^n_{u_k}$. The received signal at the relay is

$$Y^n_{\mathrm{R}} = \sum_{k\in[K]} h_{u_k} X^n_{u_k} + Z^n_{u_k}, \tag{9.1}$$

where h_{u_k} is the channel fading coefficient between the user u_k and the relay, and $Z^n_{u_k}$ is the independent additive white Gaussian noise (AWGN) with zero mean and unit variance. Using Y^n_{R} and CSI for all K users, the terrestrial relay decodes the messages of each user as $\{\tilde{M}_{u_k}\}_{k\in[K]}$. In the second phase, the relay encodes its decoded messages $\{\tilde{M}_{u_k}\}_{k\in[K]}$ into a combined codeword X^n_{R} with power P_2 and sends it to the destination. The received signal at the destination is

$$Y^n_{\mathrm{D}} = h_{\mathrm{D,R}} X^n_{\mathrm{R}} + Z^n_{\mathrm{D,R}}, \tag{9.2}$$

where $h_{\mathrm{D,R}}$ is the channel fading coefficient between the relay and the destination, and $Z^n_{\mathrm{D,R}}$ is the independent AWGN with zero mean and unit variance. Finally, using the noisy outputs Y^n_{D} and CSI on the fading coefficient $h_{\mathrm{D,R}}$, the destination decodes the message $\{\hat{M}_{u_k}\}_{k\in[K]}$.

9.1.2 *Fading Channels and Assumptions*

Assume that the channel between each user $i \in [N]$ and the relay is an AWGN with Rayleigh fading, i.e., the fading coefficient satisfies $h_i \sim \mathcal{CN}(0, 2\sigma_i^2\mathbf{I})$. The fading random variables for different ground users are independent. Furthermore, assume that the channel gain $|h_{\mathrm{D,R}}|$ for the channel from the relay to the satellite follows a Shadowed-Rician (SR) distribution with parameters (Ω, b_0, m_s), i.e., $|h_{\mathrm{D,R}}| \sim \mathrm{SR}(\Omega, b_0, m_s)$, where Ω, b_0, and m_s are positive real numbers. Here, Ω denotes the average power of the LOS component, m_s is the fading order, and $2b_0$ represents the average power of the scattered component.

To assess the fundamental constraints of the uplink multiuser relay system, based on the availability of CSI, it is essential to analyze the instantaneous sum rate and outage probability, and to develop related optimization problems for scheduling user transmissions. In particular, the following two scenarios can be considered.

- **Case 1: Perfect CSI**
  Similar to most existing works, e.g., [7, 8], assume that perfect CSI of all channels are available. The CSI can be obtained by channel estimation during each transmission time slot via a backhaul channel [9]. In this setting, the achievable instantaneous sum rate is derived to propose efficient user scheduling algorithms that can maximize the sum rate.
- **Case 2: Statistical CSI**
  The assumption of perfect CSI is often unrealistic for practical communication systems. To address this issue, the scenario where only CDI is available is considered. Specifically, the relay has knowledge of the fading coefficient distributions for the channels from each user, while the destination is aware only of the distribution of the fading coefficient for the channel from the relay. In real-world systems, CDI can be obtained through channel parameter estimation techniques, such as the MUSIC algorithm presented in [10]. Under this framework, the outage probability is derived, and user scheduling algorithms are proposed to minimize it.

9.2 Scheduling Algorithms and Theoretical Results

When instantaneous CSI is available, combining capacity results for the multiple access channel and the relay channel [11], the optimal instantaneous achievable rate region of all ground users can be derived. Subsequently, under a homogeneous target rate constraint, an optimization problem is formulated to maximize the achievable sum rate via user scheduling algorithms. The optimal solution requires an exhaustive search and is too complicated for practical use. As a compromise, theoretical upper and lower bounds on the maximum achievable rate are derived, and then low-complexity scheduling algorithms are proposed. In particular, one can propose a greedy algorithm that achieves close-to-optimal performance and a further simplified sub-optimal algorithm that has almost linear complexity. The efficiency, computational complexity, and stability of user scheduling algorithms are analytically demonstrated and numerically verified.

When CDI is available, the instantaneous achievable rate region can not be obtained. Instead, under a homogeneous target rate constraint for all users, one can derive the outage probability of the uplink communication as a function of the rate constraint and CDI. Subsequently, an optimization problem is formulated to minimize the outage probability, and a theoretical lower bound is derived which serves as the benchmark. Furthermore, to obtain a feasible solution to the optimization problem, one can propose a user scheduling algorithm using the idea of alternative optimization (AO) [12, 13]. Via numerical simulations, one can find that the outage probability obtained from the user scheduling algorithm converges to the theoretical benchmark after a small number of iterations. The optimality and computational complexity of the scheduling algorithm are also numerically illustrated by comparing with the optimal exhaustive search method.

9.2.1 *The Case with Perfect CSI*

This section investigates the instantaneous achievable rate of the given system and formulates an optimization problem aimed at maximizing the sum rate. The solution to this problem corresponds to a user scheduling algorithm. However, finding the optimal solution requires an exhaustive search, which is highly complex and impractical for real-world communication systems. To overcome this, theoretical lower and upper bounds are first derived, and then low-complexity user scheduling algorithms are proposed to strike a balance between performance and computational complexity.

9.2.1.1 Analysis of the Instantaneous Achievable Sum Rate

For each $k \in [K]$, let $S_{u_k} := P_1|h_{u_k}|^2$ denote the received SNR for the channel between user $u_k \in [N]$ and the relay and let $S_{\mathrm{D,R}} := P_2|h_{\mathrm{D,R}}|^2$ denote the SNR for the channel between the relay and the destination. Using cut-set bounds for the relay channel and the multiple access channel in [11, Sec. 19.1 and Theorem 19.1], one can conclude that given selected users $\mathbf{u} = (u_1, \ldots, u_K)$, the instantaneous rate of each user R_{u_j} is upper bounded by

$$\sum_{j\in[K]} R_{u_j} \leq \min\left\{C\left(\sum_{j\in[K]} S_{u_j}\right), C(S_{\mathrm{D,R}})\right\}, \tag{9.3}$$

where $C(a) = \log(1+a)$ is the capacity of an AWGN channel with SNR $a \in \mathbb{R}_+$.

In a full-duplex DF relaying system with perfect CSI, the sum rate upper bound given by (9.3) can be achieved using block Markov coding and backward decoding ([11, Sec. 16.4.4 and Table 16.2]), when superposition encoding and successive cancellation decoding are applied at both the relay and the satellite. For further analysis, the scenario where the rate tuples $R_{u_k k\in[K]}$ are positioned at a corner point of the rate region is mainly focused on achieving the maximum sum rate. Specifically, let $\mathbf{u} := (u_1, \ldots, u_K) \in [N]_K$, and assume the decoding order of successive cancellation decoding at the relay follows $u_1 \to u_2 \to \ldots \to u_K$. Under this condition, the signal-to-interference plus noise ratio (SINR) for user u_k at the relay is given by

$$\gamma_{u_k} = \frac{S_{u_k}}{\sum_{i\in[k+1:K]} S_{u_i} + 1}, \; k \in [K-1] \text{ and } \gamma_{u_K} = S_{u_K}. \tag{9.4}$$

Since the decoding order of successive cancellation at the destination may differ from that at the relay, one can introduce a bijective mapping $\Psi \in \mathcal{F}(K, N)$ to relate the decoding order $\mathbf{u}$ at the relay to the one at the satellite. Specifically, for any $k \in [K]$, if u_k corresponds to the k-th decoded message at the relay, it becomes the

$\Psi(u_k)$-th decoded message at the destination. Consequently, the combination of the decoding order $\mathbf{u}$ and the bijective mapping $\Psi \in \mathcal{F}(K, N)$ fully characterizes the set of selected users and the decoding sequences at both the relay and the destination. Note that the transmission power P_2 is distributed across all K messages. Therefore, the SINR for each message transmitted from the relay to the satellite is given by

$$\gamma_{\Psi(u_k)} = \begin{cases} \frac{\alpha_{\Psi(u_k)}}{\sum_{i\in[\Psi(u_k)+1:K]} \alpha_i + \frac{1}{S_{\mathrm{D,R}}}} & \text{if } \Psi(u_k) \neq K, \\ \alpha_{\Psi(u_k)} S_{\mathrm{D,R}} & \text{otherwise}, \end{cases} \tag{9.5}$$

where $\alpha_{\Psi(u_k)}$ is the power splitting ratio of the message from user u_k and satisfies $\sum_{k\in[K]} \alpha_{\Psi(u_k)} = 1$. Therefore, given $\mathbf{u}$ and Ψ, the instantaneous achievable rate of user u_k satisfies

$$R_{u_k} = \min\left\{C(\gamma_{u_k}), C(\gamma_{\Psi(u_k)})\right\}, \ k \in [K]. \tag{9.6}$$

9.2.1.2 Maximize the Instantaneous Achievable Sum Rate

Under the assumption that at most K users are allowed to transmit in a time slot, given a per user rate constraint $R_k \in \mathbb{R}_+$, an optimization problem is formulated as in Problem 1 to maximize the achievable instantaneous sum rate as follows:

Problem 1:

$$R^*_{\mathrm{sum}} := \max_{\substack{\mathbf{u}\in([N]_K),\Psi\in\mathcal{F}(K,N),\\ (\alpha_{\Psi(u_1)},\ldots,\alpha_{\Psi(u_K)})\in[0,1]^K}} \min\left\{C\left(\sum_{j\in[K]} S_{u_j}\right), C(S_{\mathrm{D,R}})\right\} \tag{9.7}$$

$$\text{s.t. } R_{u_k} \geq R_{\mathrm{target}}, \tag{9.8}$$

$$\sum_{k\in[K]} \alpha_{\Psi(u_k)} = 1, \tag{9.9}$$

$$S_{u_i} \geq S_{u_j} \text{ for } (i, j) \in [K]_2 \text{ and } i > j. \tag{9.10}$$

The objective function of Problem 1 is the maximum achievable sum rate and should be optimized over all possible schedule user decoding orders $\mathbf{u}$ at the relay, all possible decoding orders at the destination determined by $\mathbf{u}$ and the bijective mapping Ψ, all power allocation vectors $(\alpha_{\Psi(u_1)}, \ldots, \alpha_{\Psi(u_K)})$ at the relay. The first constraint in (9.8) makes sure that every scheduled user satisfies the rate constraint R_{target}, the second constraint in (9.9) is the power constraint at the relay to make sure the total transmission power at the relay is P_2, and the third constraint in (9.10) guarantees that the decoding order at the relay: $u_1 \to u_2 \to \ldots \to u_K$ is in the descend order of received SINR values.

Remark that Problem 1 can be decomposed into two sub-problems. Specifically, one have sequential sub-problems Problems 2(a) and 2(b), where Problem 2(b) uses the solution of Problem 2(a) to proceed.

$$\textbf{Problem2(a)}: \max_{\mathbf{u}\in[N]_K} C\left(\sum_{u_k\in\mathcal{U}} S_{u_k}\right) \tag{9.11}$$

$$\text{s.t. } C(\gamma_{u_k}) \geq R_{\text{target}},\ k \in [K], \tag{9.12}$$

$$S_{u_i} \geq S_{u_j} \text{ for } (i,j) \in [N]_2 \text{ and } i > j. \tag{9.13}$$

$$\textbf{Problem2(b)}: \max_{\Psi\in\mathcal{F}(K,N),(\alpha_{\Psi(u_1)},\ldots,\alpha(\Psi(u_K)))\in[0,1]^K} C(S_{\text{D,R}}) \tag{9.14}$$

$$\text{s.t. } C(\gamma_{\Psi(u_k)}) \geq R_{\text{target}}, k \in [K], \tag{9.15}$$

$$\sum_{k\in[K]} \alpha_{\Psi(u_k)} = 1. \tag{9.16}$$

By solving Problem 2(a), the optimal set of scheduled users and its corresponding decoding order $\mathbf{u}$ are determined in the relay to maximize the sum rates of K users under the rate constraint R_{target}. Given the decoding order $\mathbf{u}$, Problem 2(a) can be solved to determine the optimal power allocation in the relay, ensuring that the achievable rates from the relay to the satellite satisfy R_{target} for each scheduled user u_k. Note that the objective function of Problem 2(a) depends solely on the fading coefficient $h_{\text{D,R}}$ between the relay and the satellite. Consequently, it is decoupled from the user scheduling and power allocation strategies at the relay.

With the constraint in (9.15), the power allocation vector $\{\alpha_{\Psi(u_k)}\}_{k\in[K]}$ satisfies $\frac{\alpha_{\Psi(u_k)}}{\sum_{i\in[\Psi(u_k)+1,K]}\alpha_i+\frac{1}{S_{\text{D,R}}}} \geq 2^{R_{\text{target}}}-1$ for $\Psi(u_k) \in [K-1]$ and $S_{\text{D,R}}\alpha_K \geq 2^{R_{\text{target}}}-1$. Thus, $\alpha_{\Psi(u_k)} \geq \frac{2^{R_{\text{target}}}-1}{S_{\text{D,R}}}2^{R_{\text{target}}(K-\Psi(u_k))}$. Combined with the power constraint that $\sum_{k\in[K]}\alpha_{\Psi(u_k)} = 1$, it can be concluded that Problem 2(b) is feasible and one can always find a feasible power allocation vector $\{\alpha_{\Psi(u_k)}\}_{k\in[K]}$ if the number of users supported with rate R_{target} satisfies

$$K \leq \frac{\log(S_{\text{D,R}}+1)}{R_{\text{target}}}. \tag{9.17}$$

Note that (9.17) also implies that $S_{\text{D,R}}$ should be larger than $2^{R_{\text{target}}}-1$ to make sure that at least one user can be served with desired rate, i.e., $K \geq 1$.

9.2.1.3 Upper and Lower Bounds on the Instantaneous Achievable Sum Rate

Assume that (9.17) holds so that Problem 2(b) is feasible and the achievable rate between the relay and the satellite is $C(S_{\mathrm{D,R}})$. Then, to bound R^*_{sum}, it suffices to derive upper and lower bounds for the optimal value of Problem 2(a). For any decoding order $\mathbf{u} \in [N]_K$, let $S_{\max} := \max_{k\in[N]} S_k$ and $S_{\min} := \min_{k\in[N]} S_k$. Furthermore, let $\mathcal{S} = \{S_i\}_{i\in[N]}$, given any $a \in \mathbb{R}_+$ and any set $\mathcal{A}$, let $\phi(a, \mathcal{A}) := \min_{S_i\in\mathcal{A}: S_i\geq a} S_i$. Finally, given any $R_{\mathrm{target}} \in \mathbb{R}_+$, define the following quantities

$$\gamma_{\mathrm{target}} := 2^{R_{\mathrm{target}}} - 1, \tag{9.18}$$

$$\hat{S}^{\mathrm{lb}}_{u_K} := \phi\left(\gamma_{\mathrm{target}}, \mathcal{S}\right), \tag{9.19}$$

$$\hat{S}^{\mathrm{lb}}_{u_k} := \phi\left(\gamma_{\mathrm{target}}\left(\sum_{i\in[k+1:K]} \hat{S}^{\mathrm{lb}}_{u_i} + 1\right) \mathcal{S} \setminus \{\hat{S}^{\mathrm{lb}}_{u_j}\}_{j\in[k+1,K]}\right), \tag{9.20}$$

$$S^{\mathrm{ub}}_{u_k} := \frac{S_{\max}}{(\gamma_{\mathrm{target}} + 1)^{k-1}}, \tag{9.21}$$

$$S^{\mathrm{ub}}_{u_K} := \frac{S_{\max}}{(\gamma_{\mathrm{target}} + 1)^{K-2}\gamma_{\mathrm{target}}} - 1, \tag{9.22}$$

where $k \in [K-1]$.

Theorem 9.1 *Given any target rate R_{target} and corresponding γ_{target}, the following results hold.*

1. ***Problem 2(a)** is feasible if and only if*

$$\frac{\gamma_{\mathrm{target}}}{S_{\max}} \leq \min\left\{1, \frac{1}{\left(\sum_{i\in[k+1:K]} \hat{S}^{\mathrm{lb}}_{u_i} + 1\right)}\right\}. \tag{9.23}$$

2. *If **Problem 2(a)** is feasible, then*

$$R^*_{\mathrm{sum}} \geq \min\left\{C\left(\sum_{k\in[K]} S^{\mathrm{lb}}_{u_k}\right), C(S_{\mathrm{D,R}})\right\}, \tag{9.24}$$

where $S^{\mathrm{lb}}_{u_k} = \hat{S}^{\mathrm{lb}}_{u_k}$ for $k \in [2, K]$ and $S^{\mathrm{lb}}_{u_1} = S_{\max}$.

3. *R^*_{sum} is upper bounded by*

$$R^*_{\mathrm{sum}} \leq \min\left\{C\left(\sum_{k\in[K]} S^{\mathrm{ub}}_{u_k}\right), C(S_{\mathrm{D,R}})\right\}. \tag{9.25}$$

The proof of Theorem 9.1 is left as an exercise.

9.2.1.4 A Greedy Iterative User Scheduling Algorithm

To efficiently solve Problem 2(a) with reduced complexity, a user scheduling algorithm is proposed as an alternative to achieve near-optimal performance compared to exhaustive search. Specifically, the algorithm operates in two phases: Given the available CSI, the first step is to determine the maximum number of users K that the system can support. In the second phase, a greedy iterative user scheduling algorithm is employed to select K users for transmission. The detailed procedure is as follows.

1. **Determine the maximum number of supported users** K
   Given an R_{target} and the terrestrial SNR values $\mathcal{S}$, the number of supported users should first be determined. This is because to ensure reliable communication of a multiple-access system, individual rate decreases with the increase in user numbers [11]. In previous analysis, an upper bound on K in (9.17) that depends on the SNR $S_{\text{D,R}}$ of the satellite-relay link is derived. However, K is also related to channel qualities of users. Combined with the effects of both, K is determined via the following steps:

   [Step 1:] Initialize $K = \min\left\{N, \left\lfloor \frac{\log(S_{D,R}+1)}{R_{\text{target}}} \right\rfloor\right\}$.
   [Step 2:] Given K, if (9.23) in Theorem 9.1 holds, end the algorithm; otherwise, reduce the value of K by 1 and repeat Step 2 until $K = 0$.

2. **Schedule users via a greedy iterative scheduling algorithm**
   Note that K has been chosen to ensure that Problem 2(a) is feasible. Solving Problem 2(a) is then equivalent to proposing a user scheduling algorithm based on SNR values of all terrestrial channels. However, Problem 2(a) is a combinatorial optimization problem (COP), which is NP hard and may not have an analytical expression for optimal solution. To proceed, a greedy algorithm that iteratively selects users is proposed to achieve a near-optimal performance.

   Firstly, several definitions are given to describe the scheduling algorithm. Given $\{S_i\}_{i\in[N]}$ of all user-relay channels, let $\{S_i^{\uparrow}\}_{i\in[N]}$ be the increasing order and for each $n \in [N]$, let $L_{\min}(n)$ be the sum of the weakest n SINRs of users, i.e., $L_{\min}(n) := \sum_{i\in[n]} S_i^{\uparrow}$. Furthermore, given target rate R_{target} and the corresponding $\gamma_{\text{target}} = 2^{R_{\text{target}}} - 1$, define a sequence $\mathbf{T} = (T_1, \ldots, T_K)$ where

$$T_1 = \frac{S_{\max}}{\gamma_{\text{target}}}, \tag{9.26}$$

$$T_2 = \min\left\{T_1 - S_{u_2^*}, \frac{S_{u_2^*}}{\gamma_{\text{target}}}\right\}, \tag{9.27}$$

Algorithm 1 Greedy iterative user scheduling algorithm (GIUS)

Input: N, K, terrestrial channel coefficients $\{h_k\}_{k\in[K]}$, transmission power P_1, target rate R_{target}
Output: Scheduled users $\mathbf{u}^* = \{u_1^*, u_2^*, \ldots, u_K^*\}$.
  $\mathcal{U}_1 = [N]$; Calculate S_i for $i \in [K]$, and $S_{\max}$, $S_{\min}$, γ_{target}.
  Choose the first user as $u_1^* = \arg\max_{i\in[N]} S_i$, $k \leftarrow 2$,
  $\mathcal{U}_2 = \mathcal{U}_1 \setminus \{u_1^*\}$,
  while $k \le K$ **do**
    Calculate the set $\mathcal{V}_k$ from (9.29)
    if The set $\mathcal{V}_k$ is empty **then**
      $k \leftarrow k - 1$
    else
      Select the k-th user as $u_k^* = \arg\max_{i\in\mathcal{V}_k} S_i$
      $\mathcal{U}_k \leftarrow \mathcal{U}_k \setminus \{u_k^*\}$, $\mathcal{U}_{k+1} \leftarrow \mathcal{U}_k$
      $k \leftarrow k + 1$
    end if
  end while

$$T_{k-1} = \min\left\{T_{k-2} - S_{u_{k-1}^*}, \frac{S_{u_{k-1}^*}}{\gamma_{\text{target}}}\right\}, \quad k \in [3:K]. \tag{9.28}$$

The greedy user selection algorithm then operates as follows. For each $k \in [K]$, let $\mathcal{U}_k$ denote the set of users that can be selected. To select the first user, let $\mathcal{U}_1 = [N]$ and the first user u_1^* is selected such that its received SNR equals $S_{\max}$, i.e., $u_1^* = \arg\max_{i\in[N]} S_i$. For the subsequent user selection, given each $k \in [K]$, let

$$\mathcal{V}_k := \left\{i \in \mathcal{U}_k : S_i \in [S_{\min}, \min\{T_{k-1} - L_{\min}(k) - 1, S_{u_{k-1}^*}\}]\right\}. \tag{9.29}$$

The k-th user u_k^* is then selected as $u_k^* = \arg\max_{i\in\mathcal{V}_k} S_i$, and $\mathcal{U}_k = \mathcal{U}_{k-1} \setminus \{u_k^*\}$. Note that it should be considered whether $\mathcal{V}_k$ is empty for any $k \in [K]$. If $\mathcal{V}_k$ is empty, one can update $\mathcal{U}_{k-1}$ as $\mathcal{U}_k$ and trace back to $k-1$ to select another user other than u_{k-1}^*. The user selection stops until all K users $\mathbf{u}^* = (u_1^*, u_2^*, \ldots, u_K^*)$ are selected. The greedy iterative user scheduling (GIUS) scheme is summarized in Algorithm 1.

9.2.1.5 A Low Complexity User Scheduling Algorithm

The complexity of the user scheduling scheme in Algorithm 1 is closely tied to the channel qualities of terrestrial users. It scales polynomially with the total number of users N, but in the worst case, it scales exponentially with the selected number of users K. This level of complexity may still be impractical for power-limited terrestrial stations.

Algorithm 2 Lower-bound-based user scheduling algorithm (LBUS)

Input: N, K, terrestrial channel coefficients $\{h_k\}_{k\in[K]}$, transmission power P_1, target rate R_{target}
Output: Scheduled users $\mathbf{u}^* = \{u_1^*, u_2^*, \ldots, u_K^*\}$.
Calculate S_i for $i \in [K]$, and $S_{\max}$, $S_{\min}$, γ_{target}.
Choose the first user as $u_1^* = \arg\max_{i\in[N]} S_i$; Initialize the candidate user set as $\mathcal{U}=[N] \setminus \{u_1^*\}$.
Calculate the candidate set of K-th user as $\mathcal{U}_K := \{i \in \mathcal{U} : S_K^{\min} \leq S_i \leq S_K^{\max}\}$, where $S_K^{\min} = \hat{S}_{u_K}^{\text{lb}}$, $S_K^{\max} = \frac{S_{\max}}{(\gamma_{\text{target}}+1)^{K-2}\gamma_{\text{target}}} - 1$.
while $\mathcal{U}_K \neq \emptyset$ **do**
u_K^* is selected by $u_K^* = \arg\max_{i\in\mathcal{U}_K} S_i$.
Update candidate user set as $\mathcal{U}$ as $\mathcal{U} \setminus \{u_K^*\}$
for $i = 2, \ldots, K-1$ **do**
User u_k^* is chosen as

$$u_k^* = \mathop{\arg\min}_{i\in\mathcal{U}:S_i\geq\max\left\{\gamma_{\text{target}}\left(\sum_{i\in[k+1,K]} S_{u_i^*}+1\right), S_{u_{k+1}^*}\right\}} S_i. \tag{9.30}$$

Update the candidate user set as $\mathcal{U} \setminus \{u_k^*\}$.
end for
if (9.30) does not have a solution for any $k \in [2:K]$ **then**
Update the candidate user set as $\mathcal{U} = [N] \setminus \{u_1^*\}$
Update $\mathcal{U}_K$ as $\mathcal{U}_K \setminus \{u_K^*\}$.
else
Break.
end if
end while

In claim (ii) of Theorem 9.1, a lower bound on the achievable sum rate is derived using $\{S_{u_k}^{\text{lb}}\}_{k\in[K]}$. Based on this concept, a lower-bound-based user scheduling algorithm (LBUS) is proposed. This algorithm selects all users according to the theoretical lower bounds $\{S_{u_k}^{\text{lb}}\}_{k\in[K]}$, except for u_K^*. The user scheduling scheme is outlined in Algorithm 2.

9.2.2 *The Case with CDI*

When the number of users N is large, obtaining reliable estimates of the fading coefficients for all channels becomes nearly impossible. Therefore, the scenario where only the CDI of each channel is available at both the relay and the satellite is considered. In this case, given a target rate, closed-form expressions for the outage probability are derived for the two phases of communication. An optimization problem is then formulated to minimize the outage probability. To address this, a low-complexity solution is proposed, which corresponds to an efficient user scheduling algorithm for terrestrial users.

9.2.2.1 Assumptions and Preliminaries

The set of all users $\mathcal{N} = [N]$ are divided into $M \leq N$ groups, where users with the same CDI are grouped together. Specifically, for any two users $(i, j) \in [N]_2$, they lie in the same group if the fading coefficients (h_i, h_j) for their terrestrial channels to the relay have the same distribution. This is valid, since this way, users close in locations are grouped together.

For clarity, consider any $k \in [M]$, and let $\mathcal{M}_k$ represent the set of users in group k, where the channel fading coefficients are Rayleigh distributed with a parameter σ_k^2. Specifically, $h_j \sim \mathcal{CN}(0, 2\sigma_k^2\mathbf{I})$ for all $j \in \mathcal{M}_k$, and σ_k^2 corresponds to the CDI for users in group $\mathcal{M}_k$. It is important to note that when only CDI is available, the task of user selection reduces to choosing user groups. During each time slot, K users $(K \leq M)$ are chosen to transmit messages. Define $\mathbf{w} := \{w_1, \ldots, w_K\} \in [M]_K$ as the set of indices corresponding to the selected user groups, and let $\mathbf{v} := (v_1, \ldots, v_K) \in [N]_K$ denote the index vector of the chosen users, where $v_k \in \mathcal{M}_{w_k}$. The users v_k from the group $\mathcal{M}_{w_k}$ are selected in a time-division manner.

The communication proceeds in the two phases as described in Sect. 9.1. After user scheduling, K users $\mathbf{v}$, each from a group are selected. For each $k \in [K]$, user v_k sends a message M_{v_k} to the satellite with power P_1 with the aid of the terrestrial relay, who works in full-duplex DF mode with power P_2. Successive cancellation decoding is done at the relay and the satellite. The decoding order at the relay is given by $v_1 \to v_2 \to \ldots \to v_K$, which satisfies $\sigma_{v_1}^2 > \sigma_{v_2}^2 > \ldots > \sigma_{v_K}^2$. This assumption simplifies the theoretical analysis in the following section. The decoding order at the satellite is related to that at the relay via the bijective mapping $\Psi' \in \mathcal{F}(K, M)$.

9.2.2.2 Expressions of Outage Probabilities

When only CDI is available, the outage probability—defined as the probability that the transmission fails to achieve a given rate—is used as the performance metric. For a given user selection vector $\mathbf{v} = (v_1, \ldots, v_K)$, an outage event for user v_j occurs if any of the preceding $j - 1$ users experience an outage. Specifically, considering a target rate R_{target} and the corresponding threshold γ_{target} (cf. (9.19)), the outage probability for user v_k during the first phase satisfies the following condition for each $k \in [K - 1]$:

$$\mathbb{P}^1_{v_k,\text{out}} = 1 - \mathbb{P}\left\{S_{v_j} \geq \Big(\sum_{i\in[j+1:k]} S_{v_i} + 1\Big)\gamma_{\text{target}}\,,\, j \in [k]\right\}, \tag{9.31}$$

and

$$\mathbb{P}^1_{v_K,\text{out}} = 1 - \mathbb{P}\Bigg\{ S_{v_j} \geq \Big(\sum_{i \in [j+1:k]} S_{v_i} + 1 \Big) \gamma_{\text{target}}, \, \forall\, j \in [K-1] \\ \text{and } S_{v_K} \geq \gamma_{\text{target}} \Bigg\}, \tag{9.32}$$

where $S_{v_k} = P_1 |h_{v_k}|^2$ is the received SNR of the signals from user v_k.

Recall that DF is used at the relay, and a power allocation vector is used to split the power P_2 at the relay among K messages of users. Recall that $\gamma_{\Psi(v_k)}$ is the received SNR of user v_k defined in (9.5). For the second phase, the outage probability of each user v_k is

$$\mathbb{P}^2_{v_k,\text{out}} = 1 - \mathbb{P}\left\{ \gamma_{\Psi(v_k)} \geq \gamma_{\text{target}} \, , \, j \in [k] \right\}, \; k \in [K]. \tag{9.33}$$

Note that (9.33) is subject to the power allocation $\{\alpha_{\psi(v_k)}\}_{k \in [K]}$ and the received SNR $S_{\text{D,R}}$ for the satellite-relay channel. From arguments around (9.17), if the number of selected users satisfies $K \leq \frac{\log(S_{\text{D,R}}+1)}{R_{\text{target}}}$, one can always find a power allocation scheme that guarantees $\gamma_{\Psi(v_k)}$ is larger than the threshold γ_{target}. Thus, the outage probability of each user at the second hop satisfies $\mathbb{P}^2_{v_k,\text{out}} = 1 - \mathbb{P}\left\{ K \leq \frac{\log(S_{\text{D,R}}+1)}{R_{\text{target}}} \right\}$ for $k \in [K]$, i.e., $\mathbb{P}^2_{v_k,\text{out}} = \mathbb{P}^2_{\text{out}} = \mathbb{P}\{S_{\text{D,R}} < 2^{K R_{\text{target}}} - 1\}$ for $k \in [K]$.

In Theorem 9.2, a closed-form expression for the outage probability for user v_K at the relay, $\mathbb{P}_{v_K,\text{out}}$, is derived. One can see that the user scheduling algorithm relies only on $\mathbb{P}_{v_K,\text{out}}$.

Theorem 9.2 *The following claims hold.*

1. *Given the selected users* $\mathbf{v} = \{v_1, v_2, \ldots, v_K\}$ *and the corresponding CDI* $\{\sigma^2_{v_k}\}_{k \in [K]}$ *for their terrestrial channels to the relay, let* $\lambda_{v_k} := \frac{1}{2P_1\sigma^2_{v_k}}$ *for* $k \in [K]$. *The outage probability* $\mathbb{P}^1_{v_K,\text{out}}$ *of user* v_K *during the first phase of uplink communication satisfies*

$$\mathbb{P}^1_{v_K,\text{out}} = 1 - \frac{\lambda_{v_K} A_{K-1}}{\gamma_{\text{target}} B_{K-1} + \lambda_{v_K}} e\Big(- \gamma_{\text{target}} \left((1+\gamma_{\text{target}}) B_{K-1} + \lambda_{v_K} \right) \Big), \tag{9.34}$$

where

$$A_1 = 1, \, B_1 = \lambda_{v_1}, \tag{9.35}$$

$$A_k = \frac{\lambda_{v_k} A_{k-1}}{\gamma_{\text{target}} B_{k-1} + \lambda_{v_k}}, \tag{9.36}$$

$$B_k = (1+\gamma_{\text{target}}) B_{k-1} + \lambda_{v_k}, \, k \in [2, K-1]. \tag{9.37}$$

2. *The outage probability* $\mathbb{P}_{\text{out}}^2$ *for the second phase communication satisfies*

$$\mathbb{P}_{\text{out}}^2 = \left(\frac{2b_0 m_s}{2b_0 m_s + \Omega}\right)^{m_s} \sum_{n \in \mathbb{N}} \frac{(m_s)_n}{(n!)^2} \left(\frac{\Omega}{2b_0 m_s + \Omega}\right)^n \Gamma\left(n+1, \frac{2^{KR_{\text{target}}} - 1}{2b_0 P_2}\right), \tag{9.38}$$

where $(x)_n$ *is the Pochhammer symbol and* $\Gamma(a, x) = \int_0^x t^{a-1} \exp(-t) \mathrm{d}t$ *is the unnormalized incomplete Gamma function.*[1]

3. *Combining the above two claims, the outage probability for user* u_K *satisfies* $\mathbb{P}_{v_K,\text{out}} = 1 - (1 - \mathbb{P}_{v_K,\text{out}}^1)(1 - \mathbb{P}_{\text{out}}^2)$.

The proof of Theorem 9.2 is left as an exercise.

9.2.2.3 Minimize the Outage Probability

Note that the outage probability $\mathbb{P}_{\text{out}}^2$ is solely determined by the SNR of the satellite-relay channel and is independent of user scheduling during the first phase. Therefore, to minimize the overall outage probability of uplink communication, it is sufficient to focus on minimizing the outage probability in the first phase. The resulting optimization problem can be formulated as follows:

$$\textbf{Problem 3}: \quad p_{\text{out}}^* := \min_{\substack{\mathbf{v} \in [N]_K, \mathbf{w} \in [M]_K \\ v_k \in \mathcal{M}_{w_k}, k \in [K]}} \max_{k \in [K]} \{\mathbb{P}_{v_k,\text{out}}\} \tag{9.39}$$

$$\text{s.t.} \quad \mathbb{P}_{v_k,\text{out}} = 1 - (1 - \mathbb{P}_{v_k,\text{out}}^1)(1 - \mathbb{P}_{\text{out}}^2), \tag{9.40}$$

$$\lambda_{v_i} < \lambda_{v_j}, \text{ for } (i, j) \in [K]_2 \text{ s.t. } i < j. \tag{9.41}$$

Recall that successive cancellation decoding is used at the relay. Given a decoding order $\mathbf{v} = (v_1, \ldots, v_K)$, the outage event of user v_k occurs if any user v_i with $i < k$ suffers an outage. Thus, $\mathbb{P}_{v_K,\text{out}}^1 \geq \mathbb{P}_{v_k,\text{out}}^1$ for $k \in [K-1]$ and thus $\max_{k \in [K]} \mathbb{P}_{v_k,\text{out}} = \mathbb{P}_{v_K,\text{out}}$.

Problem 3 is an NP-hard problem, and it is non-trivial to obtain a closed-form expression for the optimal solution. Note that the objective function $\mathbb{P}_{v_K,\text{out}}$ of Problem 3 depends only on $\{\lambda_{v_k}\}_{k \in [K]}$. As a compromise, a lower bound on the objective function is firstly derived by assuming that λ_{v_k} is consistent that satisfies $\lambda_{v_k} \geq \lambda_{\min}$, where $\lambda_{\min} := \min_{k \in [M]} \{\lambda_k\}$ and $\lambda_k := \frac{1}{2P_1\sigma_k^2}$ for each $k \in [M]$. The relaxed optimization problem is as follows.

[1] The infinite series expression of Eq. (9.38) is the series expansion of the confluent hypergeometric function ${}_1F_1(a, b, x)$ in [14] and [15]. Its convergence and other properties are available in [14] and [15].

$$\textbf{Problem 4}: \quad \min_{\lambda_{v_k} \geq \lambda_{\min}, k \in [K]} \mathbb{P}_{v_K,\text{out}} \tag{9.42}$$

$$\text{s.t. } \mathbb{P}_{v_K,\text{out}} = 1 - (1 - \mathbb{P}^1_{v_K,\text{out}})(1 - \mathbb{P}^2_{\text{out}}) \tag{9.43}$$

Theorem 9.3 *Problem 4 has unique optimal solution* $\{\lambda^{\text{op}}_{v_k}\}$, *which satisfies*

$$\begin{aligned} \lambda^{\text{op}}_{v_1} &= \frac{1}{2\sigma^2_{max}}, \lambda^{\text{op}}_{v_K} = -\frac{\gamma_{\text{target}} B_{K-1}}{2} + \frac{\sqrt{\gamma^2_{\text{target}} B^2_{K-1} + 4B_{K-1}}}{2} \\ \lambda^{\text{op}}_{v_k} &= \frac{1}{\left(\sum_{i \in [k:K]} \frac{1}{\lambda^{\text{op}}_{v_k} + D_{k,i}} + \gamma_{\text{target}}(1 + \gamma_{\text{target}})^{K-k}\right)}, \quad k \in [2 : K-1], \end{aligned} \tag{9.44}$$

where $D_{k,i} = \frac{1}{\gamma_{\text{target}}(1+\gamma_{\text{target}})^{i-1-k}}(\gamma_{\text{target}} B_{i-1} + \lambda^{\text{op}}_{v_i}) - \lambda^{\text{op}}_{v_k}$ *for* $i \in [k+1 : K]$ *and* $D_{k,k} = \gamma_{\text{target}} B_{k-1}$. *Note that* $\{D_{k,i}\}_{i \in [k:K]}$ *are independent of* $\lambda^{\text{op}}_{v_k}$, *and* B_k *was defined in* (9.37).

The proof of Theorem 9.3 is left as an exercise. Theorem 9.3 provides a lower bound to the minimal outage probability via solutions to Problem 4 and thus can be used as a benchmark to evaluate the performance of any user scheduling algorithm.

9.2.2.4 A Low Complexity User Scheduling Algorithm

Note that theoretical solutions in (9.44) are higher degree equations and don't have closed-form expressions. In this section, to minimize the outage probability, a solution to Problem 3 which corresponds to a user scheduling algorithm is proposed. Recall that for each $k \in [M]$, σ^2_k is the CDI for users in group $\mathcal{M}_k$ and $\lambda_k = \frac{1}{2P_1\sigma^2_k}$. For each $k \in [2, K-1]$, define a function

$$h(\lambda_{w_k}) = \frac{1}{\lambda_{w_k}} - \sum_{i \in [k:K]} \frac{1}{\lambda_{w_k} + D_{k,i}} - \gamma_{\text{target}}(1 + \gamma_{\text{target}})^{K-k}. \tag{9.45}$$

Note that Theorem 9.3 provides an optimal solution $\{\lambda^{\text{op}}_{v_k}\}_{k \in [K]}$ to Problem 4, a relaxed version of Problem 3. To obtain a feasible solution, the user scheduling algorithm proceeds by choosing the λ_{v_k} close to $\lambda^{\text{op}}_{v_k}$. As stated before, the user selection is essentially user group selection.

The user groups $\mathbf{w}^* = (w^*_1, \ldots, w^*_K)$ are selected as follows. The first user group w^*_1 is selected as the one with the maximal CDI σ^2_k, i.e., $w^*_1 = \arg\min_{k \in [M]}\{\lambda_k\}$.

Algorithm 3 AO iterative user group selection algorithm (AOIUS)

Input: User group numbers M, user group indices $\mathcal{M}_k$ ($k \in [M]$), CDI parameters $\{\sigma_k^2\}_{k\in[M]}$, the transmit power P_1, the target SINR γ_{target} , the threshold δ.
Output: Scheduled user groups $\mathbf{w}^* = \{w_1^*, w_2^*, \ldots, w_K^*\}$.
  Calculate $\lambda_k = \frac{1}{2P_1\sigma_k^2}$ for $k \in [M]$; Choose the first user group as $w_1^* = \arg\min_{k\in[M]}\{\lambda_k\}$.
  Initialize $\mathcal{W} = \{w_1^*, w_2^0, w_3^0, \ldots, w_K^0\}$ by randomly selecting from $[M]$;
  Set $\mathbb{P}_{\text{out}}^1(0) = 1$. Calculate the outage probability $\mathbb{P}_{\text{out}}^1(1)$ according to (9.34); $i = 1$.
  while $|\mathbb{P}_{out}^1(i) - \mathbb{P}_{\text{out}}^1(i-1)| > \delta$ **do**
    for $k = 2, \ldots, K-1$ **do**
      $\mathcal{W} \leftarrow \mathcal{W} \setminus w_k^{i-1}$.
      Calculate the zero point z_{w_k} of $h(\lambda_{w_k})$.
      Calculate w_k^i by $w_k^i = \arg\min_{\substack{j\in[M]\setminus\mathcal{W} \\ \lambda_{w_{k-1}^i}<\lambda_j<\lambda_{w_{k+1}^{i-1}}}} |\lambda_j - z_{w_k}|$.
      $\mathcal{W} \leftarrow \mathcal{W} \cup \{w_k^i\}$.
      Update the value of $\{B_j\}_{j\in[k:K]}$ according to the current $\mathcal{W}$ based on (9.34).
    end for
    $\mathcal{W} \leftarrow \mathcal{W} \setminus w_K^{i-1}$.
    Calculate w_K^i by $w_K^k = \arg\min_{\substack{j\in[M]\setminus\mathcal{W} \\ \lambda_{w_{K-1}^i}<\lambda_j<\lambda_{\max}}} |\lambda_j - z_{w_K}|$, where $z_{w_K} = -\frac{\gamma_{\text{target}}B_{K-1}}{2} + \frac{\sqrt{\gamma_{\text{target}}^2 B_{K-1}^2 + 4B_{K-1}}}{2}$.
    $\mathcal{W} \leftarrow \mathcal{W} \cup \{w_K^i\}$.
    $i \leftarrow i + 1$. Calculate $\mathbb{P}_{\text{out}}^1(i)$ and B_j ($j \in [K]$) based on (9.34).
  end while
  $\mathbf{w}^* \leftarrow \mathcal{W}$.

When only two users are allowed, the second one is selected as

$$w_2^* = \arg \min_{k\in[M]\setminus w_1^*} |\lambda_k - z_{w_2}|, \tag{9.46}$$

where $z_{w_2} := -\frac{\gamma_{\text{target}}\lambda_{w_1^*}}{2} + \frac{\sqrt{\gamma_{\text{target}}^2 \lambda_{w_1^*} 2 + 4\lambda_{w_1^*}}}{2}$.

When more than two users are allowed, i.e., $K > 2$, the selection of user groups is more involved and an iterative user group selection scheme is proposed based on alternate optimization, which is summarized in Algorithm 3 and named AOIUS. Remark that AOIUS converges because the outage probability monotonically decreases during iteration rounds.

9.3 Numerical Illustrations

9.3.1 The Case with Perfect CSI

In this section, simulation results of proposed algorithms, i.e., the greedy algorithm GIUS and the lower complexity algorithm LBUS, are provided under the assump-

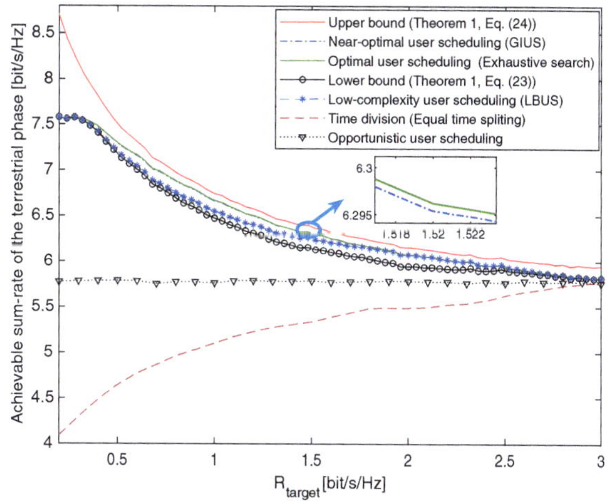

Fig. 9.2 Achievable sum rate of user scheduling schemes for different number of users as a function of the target rate R_{target}, $N = 10$

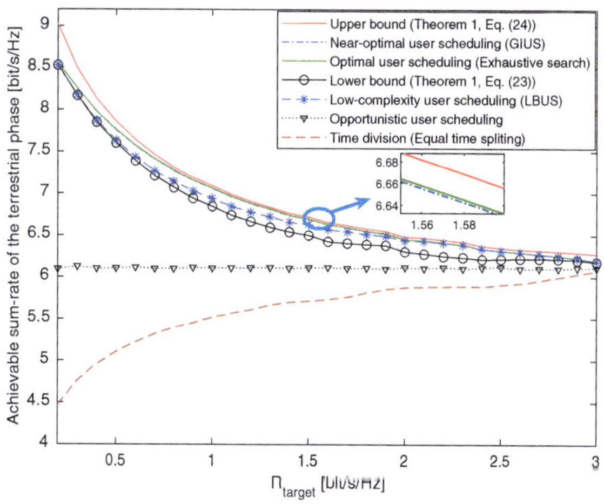

Fig. 9.3 Achievable sum rate of user scheduling schemes for different number of users as a function of the target rate R_{target}, $N = 20$

tion of knowledge of perfect CSI for channels. Assume that $\{S_i\}_{i\in[N]}$ are generated i.i.d. from a Rayleigh distribution with the variance $P_1\sigma^2 = 10$. To better evaluate the performance of different user scheduling schemes, assume that the channel between the satellite and the relay is strong enough so that the sum rate is only restricted by user scheduling schemes.

9.3.1.1 Achievable Sum Rate

In Figs. 9.2 and 9.3, the simulated performances of proposed user scheduling algorithms GIUS and LBUS, versus the optimal exhaustive search, a time division multiple access scheme (TDMA) where each ground user accesses the channel by an equal time duration, an opportunistic user scheduling scheme where the system always selects the user with the strongest channel gain [16], and the derived upper and lower bounds in Theorem 9.1 are plotted. For each user scheduling algorithm, one can obtain each data point by averaging the simulated results of 5000 independent simulations. Figure 9.2 shows that the GIUS user scheduling algorithm achieves sum rate with negligible loss compared with exhaustive search and thus demonstrates the efficiency of the GIUS algorithm. Comparing the results in Figs. 9.2 and 9.3, it can be concluded that the gap between the performance of GIUS and the upper bound becomes smaller as the total number of users N increases. This phenomenon implies that GIUS tends to be "optimal" for large-scale user groups. It is reasonable because the received SNR values of selected users will tend to the optimal solution $\{S_{u_k}^{\text{ub}}\}_{k\in[K]}$ when N is large enough.

Besides, Fig. 9.2 also shows that the performance of LBUS with nearly linear complexity has worse performance than GIUS, but still better than TDMA and opportunistic user scheduling. This implies that the LBUS algorithm, although simple, is still an effective non-orthogonal user scheduling algorithm. Finally, remark that as R_{target} increases, the performance of all user scheduling algorithms becomes identical. This is because when R_{target} is sufficiently large, only one user can be served with the desired target rate, and the user scheduling reduces to the orthogonal case.

9.3.1.2 Computational Complexity

In this section, the computational complexity of user scheduling algorithms with respect to the total number of users N and the scheduled number of users K is discussed. From the analysis in Chap. 9.2.2, one can conclude that the user scheduling algorithm LBUS has nearly linear complexity with respect to N as it only needs to select the K-th user. The computational complexity of GIUS is more complicated. Without loss of generality, assume that the transmit SNR is normalized as 1 and assume that the channel from each user to the terrestrial station is subject to Rayleigh fading with the same variance σ^2, i.e., $h_i \sim \mathcal{CN}(0, 2\sigma^2\mathbf{I})$ for all $i \in [N]$. For each $k \in [K]$, let the number of candidates of GIUS be N_k and its statistical average with respect to channel statistics is

$$\mathbb{E}[N_k] = N\int_{S_{\min}}^{\min\left\{T(k-1)-L_{\min}(k)-1,\,S_{u_{k-1}^*}\right\}} \frac{1}{2\sigma^2}\exp\left(-\frac{x}{2\sigma^2}\right)\mathrm{d}x. \tag{9.47}$$

Specifically, one has

$$\begin{aligned}\mathbb{E}[N_k] &= N\int_{S_{\min}}^{\min\left\{T(k-1)-L_{\min}(k)-1,\,S_{u_{k-1}^*}\right\}} \frac{1}{2\sigma^2}\exp\left(-\frac{x}{2\sigma^2}\right)dx & (9.48)\\ &\leq N\int_{S_{\min}}^{T(k-1)-L_{\min}(k)-1} \frac{1}{2\sigma^2}\exp\left(-\frac{x}{2\sigma^2}\right) & (9.49)\end{aligned}$$

$$= N\left(\exp\left(-\frac{S_{\min}}{2\sigma^2}\right) - \exp\left(-\frac{1}{2\sigma^2}\left(T(k-1) - L_{\min}(k) - 1\right)\right)\right) \tag{9.50}$$

$$\leq N\left(\exp\left(-\frac{S_{\min}}{2\sigma^2}\right) - \exp\left(-\frac{1}{2\sigma^2}\left(\frac{S_{\max}}{\gamma_{\text{target}}} - \sum_{i=2}^{k-1} S_{u_{k-1}^*} - L_{\min}(k) - 1\right)\right)\right). \tag{9.51}$$

For N large enough, $S_{u_{k-1}^*} \approx S_{u_{k-1}}^{\text{ub}}$, because the received SNRs of selected users will tend to the theoretical optimal solution given by (9.22) in this case. If $\frac{S_{\max}}{\gamma_{\text{target}}^{k-2}(\gamma_{\text{target}}-1)} - L_{\min}(k) - 1 > S_{\min}$, it follows from (9.48) to (9.51) that

$$\mathbb{E}[N_k] \leq N \exp\left(-\frac{S_{\min}}{2\sigma^2}\right)\left(1 - \exp\left(-\frac{1}{2\sigma^2}\left(\frac{S_{\max}}{\gamma_{\text{target}}^{k-2}(\gamma_{\text{target}} - 1)} - L_{\min}(k) - 1 - S_{\min}\right)\right)\right) \tag{9.52}$$

$$\leq N \exp\left(-\frac{S_{\min}}{2\sigma^2}\right)\left(1 - \exp\left(\frac{1}{2\sigma^2}\left(L_{\min}(k) + 1 + S_{\min}\right)\right)\right), \tag{9.53}$$

when k is large enough. Thus, the average number of candidate users $\mathbb{E}[N_k]$ decreases exponentially with k, which is the total number of selected users so far. This implies that the set of candidate users shrinks significantly after selecting each user, and this is the reason why GIUS has much lower complexity compared to exhaustive search.

Next, the complexity of user scheduling algorithms is numerically simulated. The simulation runs 2000 independent simulations of exhaustive search, GIUS, and LBUS algorithms with the target rates $R_{\text{target}} = 0.6/1.2\text{bit/s/Hz}$. Then the average computing times of each user scheduling algorithm are calculated. The simulated results are plotted in Figs. 9.4 and 9.5, where Fig. 9.4 demonstrates the running time and Fig. 9.5 plots the achievable sum rate.

Figure 9.4 shows that when $R_{\text{target}} = 1.2\text{bit/s/Hz}$, the complexity of exhaustive search is much higher than the algorithms GIUS and LBUS. Furthermore, the gaps between the computational complexity of exhaustive search and user scheduling algorithms in this chapter grow when the total number of users grows. On the other hand, Fig. 9.5 shows that for the same case, the achievable sum rates of GIUS algorithm and exhaustive search are almost the same. This implies that GIUS achieves close-to-optimal performance with much reduced complexity. When the target rate decreases to $R_{\text{target}} = 0.6\text{bit/s/Hz}$, the number of selected users K grow. In this case, it is out of computing power to simulate exhaustive search, and one can only compare the performance of proposed algorithms. As expected, LBUS has a much reduced complexity, especially for $K \geq 20$. Surprisingly, the loss of

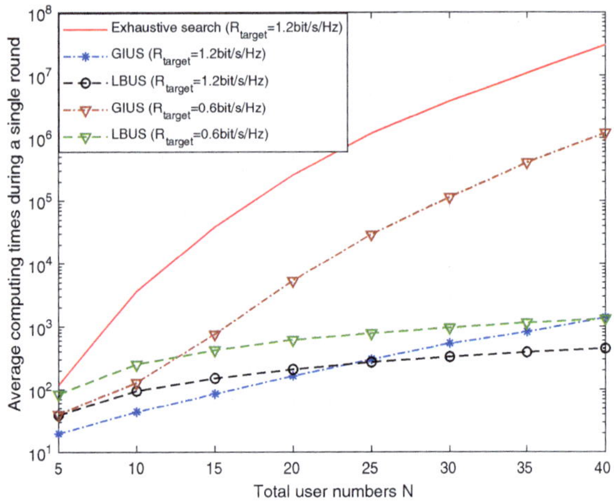

Fig. 9.4 The complexity comparison between three user scheduling algorithms

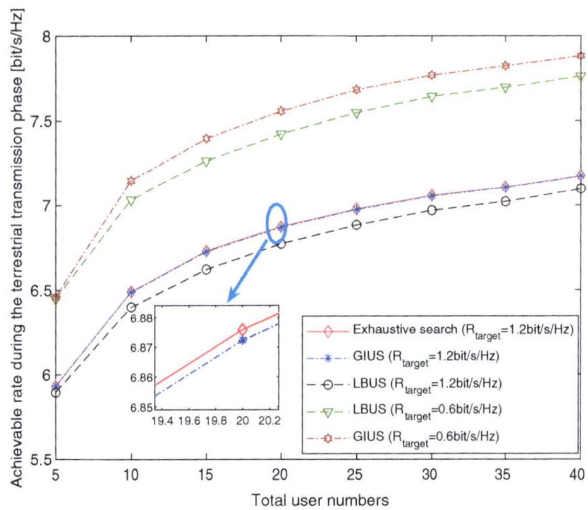

Fig. 9.5 The performance comparison between three user scheduling algorithms

achievable rate is tolerable and implies that the linear complexity algorithm LBUS is very useful when the number of selected users is small.

To further compare the computational complexity of the proposed algorithms GIUS and LBUS, both algorithms for different number of selected users K are simulated by changing the target rates R_{target}, when the total number of users $N = 40$. The average computing times are plotted in Fig. 9.6. From Fig. 9.6, one can find that the computational complexity of GIUS is much more sensitive to the number of selected users K (and thus the target rate R_{target}) while LBUS roughly changes linearly with the number of selected users. In particular, the average computing times of GIUS grow quickly and reach its maximum point roughly for $K = 20$, while LBUS has much smaller running time for almost all cases.

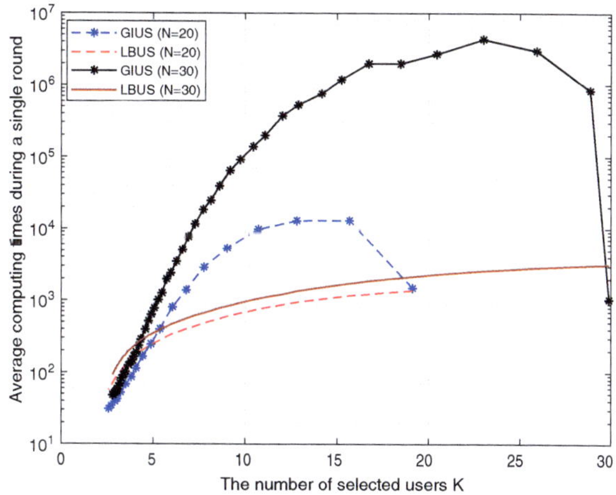

Fig. 9.6 The average computing times versus K

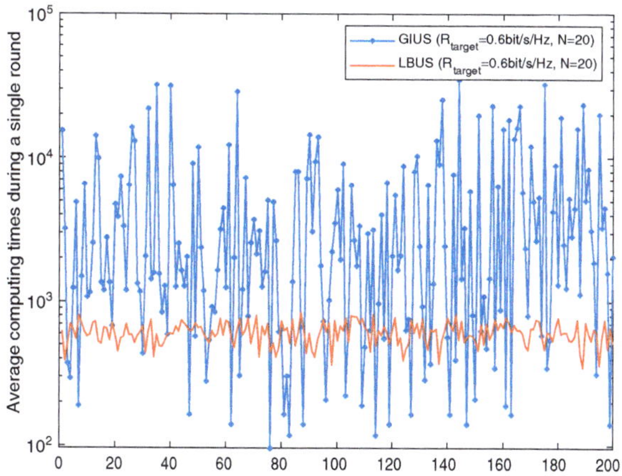

Fig. 9.7 The average computing times versus channel variation

The stability of computational complexity of user scheduling algorithms is also compared by setting $N = 40$ and $R_{\text{target}} = 0.4$ bit/s/Hz. Both algorithms are simulated 50 times independently and then the running time is recorded. The results are plotted in Fig. 9.7. One can find that GIUS strongly depends on the channel realizations, while LBUS is more stable. This is because the distributions of channels will determine how the searching field shrinks during each iteration of GIUS.

9.3.2 *The Case with CDI*

To evaluate the efficiency of the proposed algorithms, the performance of AOIUS is systematically simulated in terms of outage probability and computational complex-

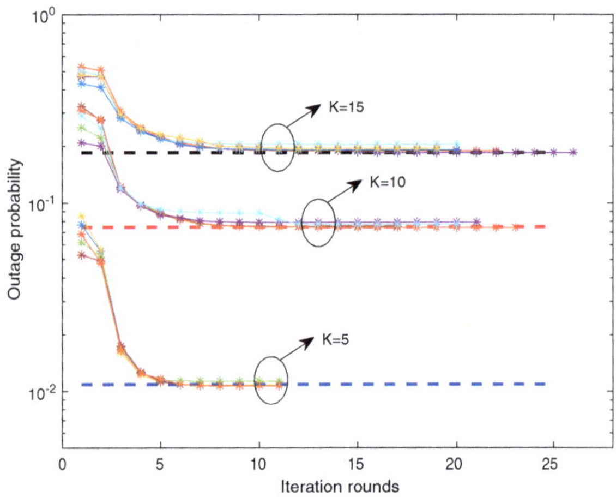

Fig. 9.8 Illustration of the quick convergence of AOIUS

ity. The channel between the relay and the satellite is assumed to experience frequent heavy shadowing, with parameters specified as $(b_0, m_s, \Omega) = (0.063, 0.739, 8.97 \times 10^{-4})$ [17].

To eliminate random variations, the value of $P_1\sigma_k^2$ for group k is assumed to follow a uniform distribution over $(-\infty, 20\,\text{dB}]$. Based on this, the average outage probability is computed over 200 realizations of the average channel parameters. The relay's transmit power is set as $P_2 = 30\,\text{dB}$.

It is worth noting that the effects of noise and large-scale fading in both the terrestrial and satellite channels are excluded from the simulation, as these factors can be incorporated into the transmit powers P_1 and P_2 through normalization.

9.3.2.1 Convergence

The results of convergence of AOIUS are illustrated in Fig. 9.8. Without loss of generality, assume that the channel gain between the satellite and the relay is strong enough so that $\mathbb{P}_{out}^2 \approx 1$. This assumption doesn't affect the convergence analysis. The terrestrial average received SNR $P_1\sigma_k^2$ of group k is assumed uniformly distributed over $(-\infty, 20\text{ dB}]$. The target rate is set as $R_{\text{target}} = 0.02$ bit/s/Hz, and the total number of user groups M is assumed to be 7000. The threshold δ for the stopping criterion is set as zero, which means the algorithm stops only when the final outage probability doesn't change. Let the number of allowed users $K = 5, 10, 15$. For each K, AOIUS is simulated five times for independent channel realizations, based on which the calculated outage probabilities are plotted versus iterations. From Fig. 9.8, one can find that using AOIUS, the outage probability of scheduled users monotonically decreases with iteration times and finally converges to a fixed value, which is the theoretical lower bound on the outage probability obtained from Theorem 9.3. This implies that the proposed user scheduling algorithm is close to optimal in this numerical example.

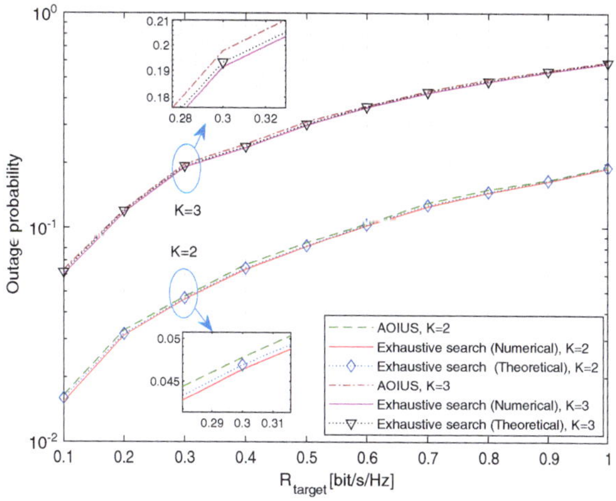

Fig. 9.9 Simulated outage probabilities

9.3.2.2 Outage Probability

To demonstrate the efficiency of AOIUS, one can simulate AOIUS for the cases of $K = 2, 3$ and compare the results with the optimal exhaustive search algorithm. Set $M = 10$, and $\delta = 0$ as the stopping criterion. The results are plotted in Fig. 9.9. The difference between theoretical and numerical simulation results is that the theoretical results correspond to simulations of user schedule algorithms where the outage probability is calculated from Theorem 9.2 and numerical correspond to Monte Carlo simulation of the outage probability, which is calculated as the ratio of outage event for 10^4 independent realizations of fading channels. For both user scheduling algorithms, the matching of theoretical and numerical simulation results verifies the correctness of the calculation of the outage probability in Theorem 9.2. Furthermore, since exhaustive search is optimal, the almost negligible gap between the outage probabilities of AOIUS and exhaustive search implies that the user scheduling algorithm is efficient in minimizing the outage probability. Finally, one can comment that the outage probability increases with the number of selected users K. This is because interference among users is more severe when more users transmit messages simultaneously.

9.3.2.3 Computational Complexity

The optimal exhaustive search is not practical since its complexity scales exponentially as the number of selected users grows. The complexity of AOIUS depends on channel distribution, the target rate, the total number of user groups M, and the number of selected user groups K. The complexities of AOIUS and exhaustive search are shown in Fig. 9.10, where each data point corresponds to the average computing time of a single run of user scheduling algorithms over 100 realizations

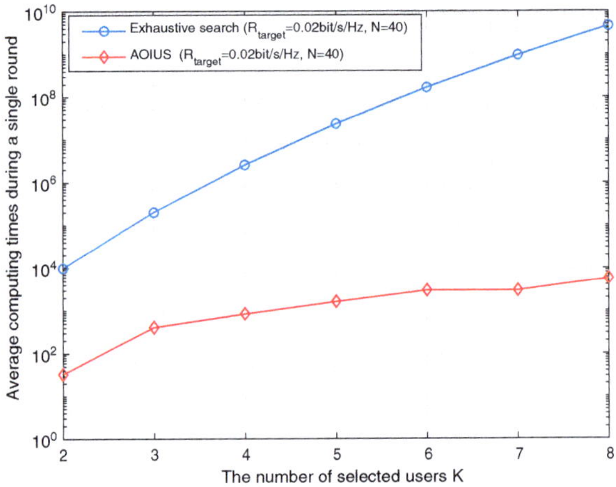

Fig. 9.10 Computing times of exhaustive search and AOIUS

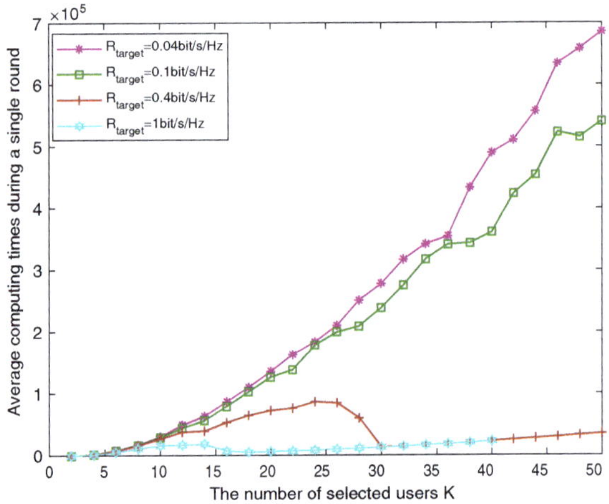

Fig. 9.11 Computing times of AOIUS

of $\{\sigma_i\}_{i\in[M]}$. In this case, set $M = 40$ and $R_{\text{target}} = 0.02$ bit/s/Hz. From Fig. 9.10, one can find that the complexity of AOIUS is much lower and stable, especially for a large number of selected users K.

Finally, to explore the effect of the target rate R_{target} and the selected number of user groups K on the computational complexity of AOIUS, the complexity of AOIUS is simulated for $M = 5000$ user groups and target rates $R_{\text{target}} = 0.04/0.1/0.4/1$ bit/s/Hz, respectively, and plot the results in Fig. 9.11. From Fig. 9.11, one can observe that the computation complexity of AOIUS roughly grows linearly with the number of selected users when $R_{\text{target}} = 0.04/0.1$ bit/s/Hz. However, when $R_{\text{target}} = 0.4/1$ bit/s/Hz, the computational complexity does not scale much with K. This is because when the target rate is relatively high, the outage event occurs frequently. In extreme case, AOIUS stops at the very beginning of the algorithm as the outage probability is always equal to 1. Thus, the computational complexities for the cases of $R_{\text{target}} = 0.4/1$ bit/s/Hz drop down when K is larger than a certain threshold (denoted by K_0) and then grow very slowly as K tends to infinity. This implies that the system cannot accommodate as more than K_0 users with the desired target rate. However, the exact characterization of K_0 is challenging and left as future work.

9.4 Chapter Summary

This chapter studies the uplink multiuser relay communication system, which consists of a single receiver, a relay, and a finite number of users. Under mild assumptions on the channel models, efficient user scheduling algorithms are proposed to optimize the communication performance: maximizing the sum rate when perfect CSI is available and minimizing the outage probability when only CDI is known. Specifically, when perfect CSI is available, theoretical upper and lower bounds on the instantaneous sum rate are first derived. A greedy iterative user scheduling algorithm is proposed, which achieves near-optimal performance with reduced computational complexity relative to the optimal algorithm. Furthermore, a lower-bound-based scheduling algorithm is introduced. While it exhibits slightly worse performance compared to the greedy algorithm, it offers significantly lower complexity, particularly when scheduling a large number of users within a single time slot. When only CDI is available, analytical expressions for the outage probabilities of the two-phase communication process are derived. Subsequently, an alternative optimization-based user scheduling algorithm is proposed, effectively balancing performance and complexity. A takeaway message is that low-complexity user scheduling algorithms can achieve remarkably excellent performance with judicious design and strike a good tradeoff between complexity and performance.

9.5 Exercises

9.1 What are multiuser relay systems? Why is user scheduling important for multiuser relay systems?

9.2 Why should one consider the case of channel distribution information for practical communication systems?

9.3 Why could the user scheduling algorithm AOIUS in Algorithm 3 converge?

9.4 What is the reason to decompose Problem 1 into two sub-problems with loss of performance?

9.5 Prove the claims in Theorem 9.1 concerning bounds on the sum rate for the case of perfect CSI.

9.6 Justify the expression for outage probabilities in Theorem 9.2.

9.7 Justify the solution in Theorem 9.3 is unique.

References

1. P.K. Upadhyay, P.K. Sharma, Max-max user-relay selection scheme in multiuser and multirelay hybrid satellite-terrestrial relay systems. IEEE Commun. Lett. **20**(2), 268–271 (2015)
2. C. Zhang, H. Lin, Y. Huang, L. Yang, Performance of integrated satellite-terrestrial relay network with relay selection and outdated CSI. IEEE Access **8**, 169652–169662 (2020)
3. P. Hajipour, A. Shahzadi, S. Ghazi-Maghrebi, Improved performance for a heterogeneous satellite-cooperative network with best relay node selection. China Commun. **16**(5), 93–105 (2019)
4. W. Yang, G. Durisi, T. Koch, Y. Polyanskiy, Quasi-static multiple-antenna fading channels at finite blocklength. IEEE Trans. Inf. Theory **60**(7), 4232–4265 (2014)
5. L. Zhou, A. Wolf, M. Motani, On lossy multi-connectivity: Finite blocklength performance and second-order asymptotics. IEEE J. Sel. Ares. Commun. **37**(4), 735–748 (2019)
6. L. Zhu, L. Bai, L. Zhou, J. Choi, Efficient user scheduling for uplink hybrid satellite-terrestrial communication. IEEE Trans. Wireless Commun. **22**(3), 1885–1899 (2023)
7. M. Lin, Z. Lin, W.-P. Zhu, J.-B. Wang, Joint beamforming for secure communication in cognitive satellite terrestrial networks. IEEE J. Sel. Ares. Commun. **36**(5), 1017–1029 (2018)
8. J. Zhang, X. Zhang, M.A. Imran, B. Evans, Y. Zhang, W. Wang, Energy efficient hybrid satellite terrestrial 5G networks with software defined features. J. Commun. Networks **19**(2), 147–161 (2017)
9. Z. Lin, M. Lin, J.-B. Wang, T. De Cola, J. Wang, Joint beamforming and power allocation for satellite-terrestrial integrated networks with non-orthogonal multiple access. IEEE J. Sel. Topics Signal Process. **13**(3), 657–670 (2019)
10. R. Schmidt, Multiple emitter location and signal parameter estimation. IEEE Trans. Antennas Propag. **34**(3), 276–280 (1986)
11. A. El Gamal, Y.-H. Kim, *Network Information Theory* (Cambridge University Press, Cambridge, 2011)
12. R. Hunger, W. Utschick, D.A. Schmidt, M. Joham, Alternating optimization for MMSE broadcast precoding, in *IEEE ICASSP* (2006)
13. S.S. Christensen, R. Agarwal, E. De Carvalho, J.M. Cioffi, Weighted sum-rate maximization using weighted MMSE for MIMO-BC beamforming design. IEEE Trans. Wireless Commun. **7**(12), 4792–4799 (2008)
14. A.T. James, Distributions of matrix variates and latent roots derived from normal samples. Ann. Math. Stat. **35**(2), 475–501 (1964)
15. M. Abramowitz, I. A. Stegun, R.H. Romer, *Handbook of Mathematical Functions with Formulas, Graphs, and Mathematical Tables* (US Government Printing Office, 1948)
16. K. An, M. Lin, T. Liang, On the performance of multiuser hybrid satellite-terrestrial relay networks with opportunistic scheduling. IEEE Commun. Lett. **19**(10), 1722–1725 (2015)
17. A. Abdi, W.C. Lau, M.-S. Alouini, M. Kaveh, A new simple model for land mobile satellite channels: first-and second-order statistics. IEEE Trans. Wireless Commun. **2**(3), 519–528 (2003)

Chapter 10
Covert Communication

Security has been a dominant concern for wireless communications in both military and industrial scenarios [1]. In wireless channels, a hostile adversary could easily eavesdrop on communication between two legitimate users. Traditional ways to secure communications include encrypting messages at the application layer [2]. The encryption algorithms, such as key encryption and key exchange protocols, are mostly based on computational complexity of challenging mathematical problems. However, with the development of quantum computing, the computation power increases, and the security of cryptographic methods based on computational complexity is challenged.

To combat the adversary with unlimited computation power, Shannon [3] pioneered the study of physical layer security by proposing the "one-time-pad" that achieves perfect secrecy. Though impractical, the idea of "one-time-pad" forms the basic concept of physical layer security. Subsequently, Wyner [4] and Csiszár-Körner [5] studied the wiretap channel model that uses coding techniques to achieve secure communication when the channel between legitimate users is stronger than the channel between the legitimate transmitter and the eavesdropper. When the channel condition is not satisfied, secret key generation was studied to guarantee communication security [6, 7]. The advances of the above attempts in wiretap channel and secret key generation were summarized in the textbook of Bloch and Barros [8].

However, the above physical layer security studies only safeguard the content of legitimate communication, which might be inadequate for sensitive scenarios, where the behavior of communication matters. To tackle the above problem, *covert communication* [9], also known as low probability of detection communication, has been proposed. As an emerging technique in physical layer security, covert communication aims to protect the behavior of legitimate transmission so that any adversary cannot detect whether any communication occurs. The square root law of covert communication [9] states that the number of bits, which can be covertly and reliably transmitted, scales with the square root of the number of channel uses.

L. Zhou, L. Bai, *Low-latency Communication*, Textbooks in Telecommunication Engineering, https://doi.org/10.1007/978-3-032-06627-5_10

Thus, the traditional channel capacity equals 0 in covert communication. The square root law holds for various channel models, including additive white Gaussian noise (AWGN) channels and discrete memoryless channels [9–12]. The exact coefficient of the square root law is named the capacity of covert communication, which can be understood as a second-order term in the traditional finite blocklength analyses. Based on the studies for point-to-point AWGN channels, covert communication has been explored for wireless communication systems, particularly in multiple-input multiple-output (MIMO) AWGN channels and massive MIMO scenarios [13]. Massive MIMO, with its large antenna arrays and enhanced spatial diversity, provides new opportunities for securely transmitting information while minimizing the risk of detection.

This chapter presents theoretical benchmarks and corresponding proofs for AWGN channels, MIMO AWGN channels, and massive MIMO channels, based on the results in [9, 10, 13, 14].

10.1 AWGN Channel

10.1.1 Problem Formulation

Let $(\mathcal{X}, \mathcal{Y})$ be two alphabets, $P_X \in \mathcal{P}(\mathcal{X})$ be a distribution, and $P_{Y|X} \in \mathcal{P}(\mathcal{Y}|\mathcal{X})$ be a conditional distribution. Fix two integers $(n, M) \in \mathbb{N}^2$. Assume that W is uniformly distributed over $\mathcal{M}$. As shown in Fig. 10.1, the task of covert communication is that Alice aims to transmit a message W reliably over a memoryless channel $P_{Y|X}^n$ without being detected by the malicious user Willie. Specifically, Alice transmits the channel input X^n, which is corrupted by two memoryless channels. The legitimate receiver Bob aims to decode W using the channel output Y^n through a noisy channel $P_{Y|X}^n$. The malicious user Willie aims to detect whether Alice is transmitting using the channel output Y^n through the same noisy channel $P_{Y|X}^n$. Specifically, Willie uses binary hypothesis testing (BHT) to determine whether Alice is communicating with Bob. The null hypothesis H_0 indicates that Alice is transmitting, while the alternative hypothesis H_1 indicates that Alice is not.

The channel noise is an AWGN with zero-mean and power $\sigma^2 \in \mathbb{R}_+$. The noisy channels from Alice to Bob and Willie are

$$Y^n = X^n + Z^n. \tag{10.1}$$

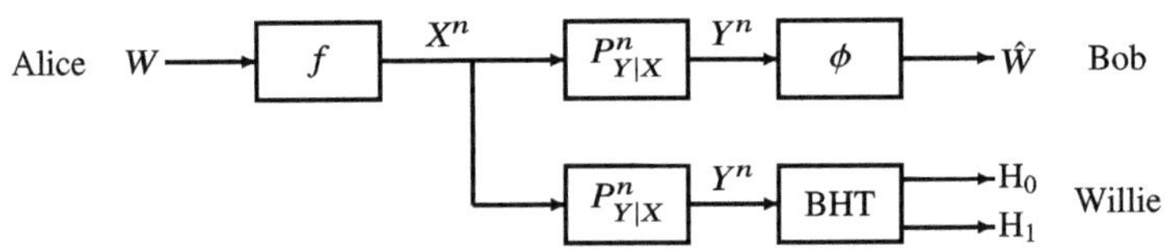

Fig. 10.1 System model for covert communication

Let the probability distribution of the channel input X^n be P_{X^n}, the probability distributions at Willie when Alice is transmitting or not as Q_1, Q_0, respectively. When Willie applies BHT, two kinds of errors occur: Type-I error, where H_0 is rejected while it is true, and Type-II error, where H_0 is accepted while it is false. Let the probabilities of Type-I and Type-II error be α and β, respectively. The goal of covert communication is to maximize $\alpha + \beta$ so that $\alpha + \beta = 1$, which corresponds to the scenario where Willie random guesses whether the communication occurs. It is desired that the error probability of the best test performed by Willie is the same as random guess. To achieve such a goal, it is preferable to design a coding scheme that makes the distributions of the channel output at Willie indistinguishable at both states H_0 and H_1.

Two metrics are adopted to evaluate the covertness level via similarity between distributions: the total variation distance (TVD) and the KL Divergence (KLD). Given any two distributions (P_X, Q_X) defined on the finite alphabet $\mathcal{X}$, it follows from (1.6) and (1.8) that the TVD and KLD are

$$\mathrm{TV}(P_X, Q_X) = \sum_{x \in \mathcal{X}} \frac{1}{2} |P_X(x) - Q_X(x)|, \tag{10.2}$$

$$D(P_X \| Q_X) = \sum_{x \in \mathrm{supp}(P_X)} P_X(x) \log \frac{P_X(x)}{Q_X(x)}, \tag{10.3}$$

respectively. Note that TVD is symmetric while KLD is not. Thus, $\mathrm{TV}(P_X, Q_X) = \mathrm{TV}(Q_X, P_X)$ while $D(P_X \| Q_X) \neq D(Q_X \| P_X)$.

It follows from [15, Theorem 13.1.1] that for optimal BHT tests that the sum of Type-I and Type-II error probabilities is related to TVD:

$$\alpha + \beta = 1 - \mathrm{TV}(Q_1, Q_0). \tag{10.4}$$

Furthermore, the Pinsker's inequality [16, Lemma 11.6.1] relates KLD and TVD:

$$\frac{1}{2} D(P_X \| Q_X) \geq \mathrm{TV}(P_X, Q_X)^2. \tag{10.5}$$

If KLD tends to 0, TVD tends to 0. Thus, both TVD and KLD can be used as covertness measures.

KLD is chosen as the covert metric due to its excellent decomposability, and the discrimination between Q_1 and Q_0 can be defined as

$$D(Q_1 \| Q_0) \leq \delta. \tag{10.6}$$

10.1.2 Theoretical Benchmark

Fix integers $(n, M) \in \mathbb{N}_+^2$ and positive real numbers $(\delta, \varepsilon) \in \mathbb{R}_+^2$.

Definition 10.1 An (n, M)-code consists of an encoder $f : \mathcal{M} \to \mathcal{X}^n$ and a decoder $\phi : \mathcal{Y}^n \to \mathcal{M}$.

The transmission rate is $R_n := \frac{\log M}{n}$. It is not assumed that the eavesdropper observes a noisier channel than the intended receiver; instead, they both observe the same channel outputs. The reason for dropping the wiretap structure is that, unlike in secret communication where the assumption that the eavesdropper observes a noisier channel allows one to obtain information theoretic secrecy without using a secret key, in LPD problems the wiretap assumption does not bring essential new insights. In particular, the square-root law does not rely on the wiretap structure. Thus, the wiretap structure is not assumed. In the above setting, LPD communication is impossible without a secret key. Assume that such a key is available, and are not concerned with its length.

To evaluate the performance of any (n, M)-code, the error probability and the covertness level are considered. Specifically, the error probability of the legitimate receiver Bob is

$$\mathrm{P}_{\mathrm{e},n} := \Pr\{\hat{W} \neq W\}. \tag{10.7}$$

The covert level $\mathrm{P}_{\mathrm{c},n}$ can be measured via either TVD $\mathrm{TV}(\mathrm{Q}_1 \| \mathrm{Q}_0)$ or KLD $D(Q_1 \| Q_0)$. The study of covert communication characterizes the tradeoff among the transmission rate, the error probability, and the covertness level. For low-latency covert communication, the theoretical benchmark is the maximal transmission rate subject to an error probability of at most ε and a covert constraint of at most δ, i.e.,

$$R_{\mathrm{c}}^*(n, \varepsilon, \delta) := \sup\left\{R_n : \exists \text{ an } (n, M)\text{-code s.t. } \mathrm{P}_{\mathrm{e},n} \leq \varepsilon \text{ and } \mathrm{P}_{\mathrm{c},n} \leq \delta\right\}. \tag{10.8}$$

It follows from [9–11] that for almost any discrete memoryless channel (DMC) and AWGN channels

$$R_{\mathrm{c}}^*(n, \varepsilon, \delta) = \frac{b}{\sqrt{n}} + \frac{c}{n^{\frac{3}{4}}} + O\left(\frac{\log n}{n}\right), \tag{10.9}$$

where $(b, c) \in \mathbb{R}_+^3$ are the coefficients of the second- and third-order terms. This section presents the characterization and proofs of b for AWGN channels.

10.1.3 Second-Order Asymptotics and Discussions

Assume that (i) Alice knows a lower bound for the power of the AWGN on Willie's channel σ^2, and (ii) Alice and Bob share a secret key of unlimited length, which is used to determine the transmission period.

The square root law [9] states as follows.

Theorem 10.1 (Square Root Law) *Under both TVD and KLD covertness metrics,*

$$\lim_{\varepsilon \to 0} R_{\mathrm{c}}^*(n, \varepsilon, \delta) = O\left(\frac{1}{\sqrt{n}}\right). \tag{10.10}$$

In contrast to channel coding where the capacity is a constant, the capacity of covert communication is zero. The exact term before $\frac{1}{\sqrt{n}}$ was characterized in [9] under TVD covertness metric and [10] under KLD covertness metric, respectively.

Theorem 10.2 (Second-Order Asymptotics) *Under KLD covertness metric,*

$$\lim_{\varepsilon \to 0} R_{\mathrm{c}}^*(n, \varepsilon, \delta) = \sqrt{\frac{\delta}{n}} + o\left(\frac{1}{\sqrt{n}}\right). \tag{10.11}$$

In covert communication, the square root law demonstrates that under TVD or KLD covertness metric, the covert transmission rate $R_{\mathrm{c}}^*(n, \varepsilon, \delta)$ has a first-order term of zero, with the second-order term in the order of $O\left(\frac{1}{\sqrt{n}}\right)$. The square root can be understood as a cost of covertness, where the transmitter must send signals at an extremely low power to avoid detection by the malicious user Willie.

The second-order asymptotics quantifies the coefficient of the square root law. Under KLD covertness measure, the covert transmission rate is roughly $\sqrt{\frac{\delta}{n}}$, where δ is the tolerable covertness level. As δ increases, the covertness guarantee decreases and the transmission rate increases. This is consistent with the intuition since looser security requirement allows larger transmission rate.

10.1.4 Proof

10.1.4.1 Achievability

The achievability part is proved by showing that for Gaussian input distributions, the covert constraint is met and the decoding error probability tends to 0 as the blocklength n tends to infinity.

Consider a random codebook where the codewords are i.i.d. generated with the distribution $\mathcal{N}(0, \rho_n)$ where

$$\rho_n := 2\sigma^2 \sqrt{\frac{\delta}{n}}. \tag{10.12}$$

Fix any $i \in [n]$. In this case, each channel input X_i is generated i.i.d. from $\mathcal{N}(\cdot; 0, \rho_n)$. It follows from the channel law that the channel output Y follows $\mathcal{N}(0, \rho_n + \sigma^2)$ under hypothesis H_1 and follows $\mathcal{N}(0, \sigma^2)$ under hypothesis H_0.

Due to the i.i.d. nature of the channel input and the additive noise, the covertness under the KLD metric satisfies

$$
\begin{aligned}
& D(Q_1^n \| Q_0^n) \\
& = nD(Q_1 \| Q_0) && (10.13) \\
& = n \int_{y\in\mathbb{R}} \frac{\exp\left(-\frac{y^2}{2(\rho_n+\sigma^2)}\right)}{\sqrt{2\pi(\rho_n+\sigma^2)}} \log \frac{\frac{1}{\sqrt{2\pi(\rho_n+\sigma^2)}}\exp\left(-\frac{y^2}{2(\rho_n+\sigma^2)}\right)}{\frac{1}{\sqrt{2\pi\sigma^2}}\exp\left(-\frac{y^2}{2\sigma^2}\right)} \mathrm{d}y && (10.14) \\
& = n \int_{y\in\mathbb{R}} \frac{\exp\left(-\frac{y^2}{2(\rho_n+\sigma^2)}\right)}{\sqrt{2\pi(\rho_n+\sigma^2)}} \log\left(\sqrt{\frac{\sigma^2}{\rho_n+\sigma^2}} \exp\left(\frac{\rho_n y^2}{2\sigma^2(\rho_n+\sigma^2)}\right)\right)\mathrm{d}y && (10.15) \\
& = -\frac{n}{2}\log\left(1+\frac{\rho_n}{\sigma^2}\right) + \frac{n\rho_n}{2\sigma^2(\rho_n+\sigma^2)} \int_{y\in\mathbb{R}} \frac{y^2}{\sqrt{2\pi(\rho_n+\sigma^2)}} \\
& \quad \times \exp\left(-\frac{y^2}{2(\rho_n+\sigma^2)}\right)\mathrm{d}y && (10.16) \\
& = -\frac{n}{2}\log\left(1+\frac{\rho_n}{\sigma^2}\right) + \frac{n\rho_n}{2\sigma^2} && (10.17) \\
& \le -\frac{n}{2}\left(\frac{\rho_n}{\sigma^2} - \frac{\rho_n^2}{2\sigma^4}\right) + \frac{n\rho_n}{2\sigma^2} && (10.18) \\
& = \frac{n\rho_n^2}{4\sigma^4} && (10.19) \\
& \le \delta, && (10.20)
\end{aligned}
$$

where (10.14) follows from the definition of AWGN channel, (10.17) follows since the integral of (10.16) is $\mathbb{E}[Y^2] = \sigma^1 + \rho_n$, and (10.18) follows since for any positive real number $x \in \mathbb{R}_+$,

$$
\log(1+x) \ge x - \frac{x^2}{2}, \tag{10.21}
$$

and (10.20) follows from the definition of ρ_n in (10.12).

In the aforementioned analysis of random codebook, the expected performance of the random codebook was primarily verified. Since the KL divergence is a nonlinear function, the average performance (expected performance) of random codebook can directly ensure that there exists at least one specific deterministic codebook, within the codebooks following the described Gaussian probability distribution, that satisfies the covert constraint defined by the KL divergence.

It follows from [10, Eq. (15)] that the rate $R_c^*(n, \varepsilon, \delta)$ in (10.11) is achievable if

$$\lim_{n\to\infty} R_c^*(n, \varepsilon, \delta) \le P - \liminf_{n\to\infty} \frac{1}{n} \log \frac{P_{Y^n|X^n}(Y^n|X^n)}{Q_1(Y^n)}, \tag{10.22}$$

where $P - \liminf$ is the limit inferior in probability, i.e., for a random variable J^n, $P - \liminf_{n\to\infty} J^n$ denotes the largest real number $j \subset \mathbb{R}_+$ such that $\lim_{n\to\infty} \Pr\{J^n \ge j\} \to 1$. The ratio between $P_{Y^n|X^n}$ and Q_1 is evaluated as

$$\frac{P_{Y^n|X^n}(Y^n|X^n)}{Q_1(Y^n)} = \frac{\prod_{i=1}^n \frac{1}{\sqrt{2\pi\sigma^2}} e^{-\frac{(Y_i-X_i)^2}{2\sigma^2}}}{\prod_{i=1}^n \frac{1}{\sqrt{2\pi(\rho_n+\sigma^2)}} e^{-\frac{Y_i^2}{2(\rho_n+\sigma^2)}}} \tag{10.23}$$

$$= \left(\frac{\rho_n + \sigma^2}{\sigma^2}\right)^{\frac{n}{2}} \exp\left(\frac{\sum_{i\in[n]} Y_i^2}{2(\rho_n + \sigma^2)} - \frac{\sum_{i\in[n]} Z_i^2}{2\sigma^2}\right). \tag{10.24}$$

Hence,

$$\frac{1}{\sqrt{n}} \log \frac{P_{Y^n|X^n}(Y^n|X^n)}{Q_1(Y^n)} = \frac{\sqrt{n}}{2} \log\left(\frac{\rho_n + \sigma^2}{\sigma^2}\right) + \frac{1}{\sqrt{n}}\left(\frac{\sum_{i\in[n]} Y_i^2}{2(\rho_n + \sigma^2)} - \frac{\sum_{i\in[n]} Z_i^2}{2\sigma^2}\right). \tag{10.25}$$

The expectation of (10.25) is

$$\mathbb{E}\left[\frac{1}{\sqrt{n}} \log \frac{P_{Y^n|X^n}(Y^n|X^n)}{Q_1(Y^n)}\right] = \frac{\sqrt{n}}{2} \log\left(\frac{\rho_n + \sigma^2}{\sigma^2}\right) + \frac{1}{\sqrt{n}}\left(\frac{\sum_{i\in[n]} \mathbb{E}[Y_i^2]}{2(\rho_n + \sigma^2)} - \frac{\sum_{i\in[n]} \mathbb{E}[Z_i^2]}{2\sigma^2}\right) \tag{10.26}$$

$$= \frac{\sqrt{n}}{2} \log\left(\frac{\rho_n + \sigma^2}{\sigma^2}\right) + 0 \tag{10.27}$$

$$\ge \frac{\sqrt{n}}{2}\left(\frac{\rho_n}{\sigma^2} - \frac{\rho_n^2}{2\sigma^4}\right) \tag{10.28}$$

$$= \sqrt{\delta} - \frac{\delta}{\sqrt{n}}, \tag{10.29}$$

where (10.27) follows since the second moments of Y_i^2 and Z_i^2 are $\rho_n + \sigma^2$ and σ^2, respectively, (10.28) follows from using (10.21) again, and (10.29) follows from the definition of ρ_n in (10.12).

Thus, the result in (10.29) implies

$$\lim_{n\to\infty} \mathbb{E}\left[\frac{1}{\sqrt{n}} \log \frac{P_{Y^n|X^n}(Y^n|X^n)}{Q_1(Y^n)}\right] \geq \sqrt{\delta}. \tag{10.30}$$

By Chebyshev's inequality, for any real number $a \in \mathbb{R}_+$,

$$\Pr\left\{\left|\frac{1}{\sqrt{n}} \log \frac{P_{Y^n|X^n}(Y^n|X^n)}{Q_1(Y^n)} - \sqrt{\delta}\right| \geq a\right\} \leq \frac{1}{a^2} \mathsf{Var}\left(\frac{1}{\sqrt{n}} \log \frac{P_{Y^n|X^n}(Y^n|X^n)}{Q_1(Y^n)}\right). \tag{10.31}$$

Thus, if the variance in the right-hand side of (10.31) tends to 0 as n tends to infinity,

$$P - \liminf_{n\to\infty} \frac{1}{\sqrt{n}} \log \frac{P_{Y^n|X^n}(Y^n|X^n)}{Q_1(Y^n)} \geq \sqrt{\delta}, \tag{10.32}$$

and by invoking (10.22),

$$\lim_{n\to\infty} \sqrt{n} R_{\mathrm{c}}^*(n, \varepsilon, \delta) \geq \sqrt{\delta}. \tag{10.33}$$

The variance can be calculated as

$$\begin{aligned}
&\mathsf{Var}\left(\frac{1}{\sqrt{n}} \log \frac{P_{Y^n|X^n}(Y^n|X^n)}{Q_1(Y^n)}\right) \\
&= \mathsf{Var}\left(\frac{1}{\sqrt{n}}\left(\sum_{i\in[n]} \frac{Y_i^2}{2(\rho_n + \sigma^2)} - \sum_{i\in[n]} \frac{Z_i^2}{2\sigma^2}\right)\right) && (10.34)\\
&= \frac{1}{n} \sum_{i\in[n]} \mathsf{Var}\left(\frac{Y_i^2}{2(\rho_n + \sigma^2)} - \frac{Z_i^2}{2\sigma^2}\right) && (10.35)\\
&= \frac{1}{n} \mathsf{Var}\left(\frac{Y^2}{2(\rho_n + \sigma^2)} - \frac{Z^2}{2\sigma^2}\right) && (10.36)\\
&= \mathbb{E}\left[\left(\frac{Y^2}{2(\rho_n + \sigma^2)} - \frac{Z^2}{2\sigma^2}\right)^2\right] && (10.37)\\
&= \frac{1}{4(\rho_n + \sigma^2)^2} \mathbb{E}\left[\left(X^2 + 2XZ - \frac{\rho_n Z^2}{\sigma^2}\right)^2\right] && (10.38)\\
&\leq \frac{1}{4\sigma^4} \mathbb{E}\left[\left(X^2 + 2XZ - \frac{\rho_n Z^2}{\sigma^2}\right)^2\right]. && (10.39)
\end{aligned}$$

It can be verified that expanding the square inside the expectation of (10.39), the summand of expectation terms tends to 0 as n tends to infinity. The achievability proof is thus completed by combining (10.31), (10.32), and (10.33).

10.1.4.2 Converse

Fix $(\varepsilon, \delta) \in \mathbb{R}^2$. Fix any (n, M)-code with channel input X^n and channel output Y^n for reliable communication, which satisfies the error probability constraint $\Pr\{\hat{W} \neq W\} \leq \varepsilon$ and the covertness constraint $\mathrm{P}_{\mathrm{c},n} \leq \delta$.

It follows from the Fano's inequality in Theorem 1.3 that

$$1 + \varepsilon \log M \geq H(X^n|Y^n) \tag{10.40}$$

$$\geq H(X^n) - I(X^n; Y^n), \tag{10.41}$$

where (10.41) follows from the definition of mutual information. Thus,

$$I(X^n; Y^n) \geq H(X^n) - \varepsilon \log M - 1 \tag{10.42}$$

$$= (1 - \varepsilon) \log M - 1, \tag{10.43}$$

where (10.43) follows since (i) X^n is a function of the uniformly drawn message $W \in [M]$, which implies that $\log M = H(X^n)$, and (ii) the error probability constraint states that $\Pr\{\hat{W} \neq W\} \leq \varepsilon$.

Let $\bar{X} := \frac{1}{n} \sum_{i\in[n]} X_i$ and $\bar{Y}$ be the output of passing $\bar{X}$ over an AWGN channel with noise power σ^2. Let the second moment of $\bar{X}$ be ρ_n. It follows that

$$I(X^n; Y^n) = \sum_{i\in[n]} I(X^n; Y_i|Y^{i-1}) \tag{10.44}$$

$$= \sum_{i\in[n]} H(Y_i|Y^{i-1}) - H(Y_i|X^n, Y^{i-1}) \tag{10.45}$$

$$= \sum_{i\in[n]} H(Y_i|Y^{i-1}) - H(Y_i|X_i) \tag{10.46}$$

$$\leq \sum_{i\in[n]} I(X_i; Y_i) \tag{10.47}$$

$$\leq nI(\bar{X}; \bar{Y}) \tag{10.48}$$

$$\leq \frac{n}{2} \log\left(1 + \frac{\rho_n}{\sigma^2}\right) \tag{10.49}$$

$$\leq \frac{n\rho_n}{2\sigma^2}, \tag{10.50}$$

where (10.44) follows since $\sum_{i\in[n]} I(X^n; Y_i|Y^{i-1}) = \sum_{i\in[n]} H(X^n|Y^{i-1}) - H(X^n|Y^i) = H(X^n) - H(X^n|Y^n) = I(X^n; Y^n)$, (10.45) follows from the definition of mutual information, (10.46) follows since Y_i does not depend on X^{i-1} and Y^{i-1}, (10.47) follows from the definition of mutual information and the fact that $H(Y_i|Y^{i-1}) \leq H(Y_i)$, (10.48) follows from concavity of mutual information in the input distribution P_{X_i}, (10.49) follows since (i) the second moment of $\bar{X}$ is ρ_n, and (ii) the zero-mean Gaussian input distributions maximize $I(X; Y)$ for a fixed second moment ρ_n, and (10.50) follows from the fact that for any positive real number $x \in \mathbb{R}_+$, $\log(1+x) \leq x$.

Let Q^n be the distribution of Y^n. For each $i \in [n]$, let $Q_{n,i}$ be the marginal distribution of Y_i. Furthermore, let $\bar{Q}_n$ denote the average output distribution $\mathcal{Y}$, which is induced by the input $\bar{X}$ and the AWGN channel. The KLD satisfies

$$D(Q^n \| Q_0^n) = -H(Y^n) + \mathbb{E}_{Q^n}\left[\log \frac{1}{Q_0^n(Y^n)}\right] \tag{10.51}$$

$$= -\sum_{i\in[n]} H(Y_i|Y^{i-1}) + \mathbb{E}_{Q^n}\left[\log \frac{1}{Q_0(Y_i)}\right] \tag{10.52}$$

$$= -\sum_{i\in[n]} H(Y_i|Y^{i-1}) + \mathbb{E}_{Q_{n,i}}\left[\log \frac{1}{Q_0(Y_i)}\right] \tag{10.53}$$

$$\geq -\sum_{i\in[n]} H(Y_i) + \mathbb{E}_{Q_{n,i}}\left[\log \frac{1}{Q_0(Y_i)}\right] \tag{10.54}$$

$$= \sum_{i\in[n]} D(Q_{n,i} \| Q_0) \tag{10.55}$$

$$\geq nD(\bar{Q}_n \| Q_0), \tag{10.56}$$

where (10.56) follows from the convexity of relative entropy.

Furthermore,

$$D(\bar{Q}_n \| Q_0) = \mathbb{E}_{\bar{Q}_n}\left[\log \frac{\bar{Q}_n(Y)}{Q_0(Y)}\right] \tag{10.57}$$

$$= \mathbb{E}_{\bar{Q}_n}\left[\log \bar{Q}_n(Y)\right] + \mathbb{E}_{\bar{Q}_n}\left[\log \frac{1}{Q_0(Y)}\right] \tag{10.58}$$

$$= -H(\bar{Q}_n) + \mathbb{E}_{\bar{Q}_n}\left[\log \frac{1}{Q_0(Y)}\right], \tag{10.59}$$

where (10.57) follows from the definition of KLD, (10.58) follows from the properties of logarithm and the additivity of expectation, and (10.59) follows from the definition of entropy. Since the channel input and the noise are independent of each other, the second moment of $\bar{Q}_n$ is $\rho_n + \sigma^2$. Recall that Q_0 is the channel output when no communication takes place, whose distribution is $\mathcal{N}(\cdot; 0, \sigma^2)$.

The second term in (10.59) satisfies

$$\mathbb{E}_{\bar{Q}_n}\left[\log\frac{1}{Q_0}\right] = \mathbb{E}_{\bar{Q}_n}\left[\log\left(\sqrt{2\pi\sigma^2}\exp\left\{\frac{Y^2}{2\sigma^2}\right\}\right)\right] \tag{10.60}$$

$$= \frac{1}{2}\log(2\pi\sigma^2) + \mathbb{E}_{\bar{Q}_n}\left[\frac{Y^2}{2\sigma^2}\right] \tag{10.61}$$

$$= \frac{1}{2}\log(2\pi\sigma^2) + \frac{\rho_n + \sigma^2}{2\sigma^2}, \tag{10.62}$$

where (10.62) follows from since second moment of the distribution $\bar{Q}_n$ is $\rho_n + \sigma^2$. Combining (10.59) and (10.62) leads to

$$D(\bar{Q}_n \| Q_0) = -H(\bar{Q}_n) + \mathbb{E}_{\bar{Q}_n}\left[\log\frac{1}{Q_0}\right] \tag{10.63}$$

$$\geq -\frac{1}{2}\log(2\pi e(\rho_n + \sigma^2)) + \frac{1}{2}\log(2\pi\sigma^2) + \frac{\rho_n + \sigma^2}{2\sigma^2} \tag{10.64}$$

$$= -\frac{1}{2}\log\left(1 + \frac{\rho_n}{\sigma^2}\right) + \frac{\rho_n}{2\sigma^2} \tag{10.65}$$

$$= \frac{\rho_n^2}{4\sigma^4} + o(\rho_n^2), \tag{10.66}$$

where (10.64) follows since the zero-mean Gaussian distribution maximizes the entropy when the second moment is fixed [16, Theorem 8.6.5], and (10.66) follows from the Taylor Series of expansion of $\log(1 + x)$ around $x = 0$, which implies $\log(1 + x) = x - \frac{x^2}{2} + o(x^2)$.

Using the covertness constraint under KLD in (10.66),

$$\frac{\delta}{n} \geq D(\bar{Q}_n \| Q_0) \geq \frac{\rho_n^2}{4\sigma^4} + o(\rho_n^2), \tag{10.67}$$

it follows that

$$\rho_n \leq 2\sigma^2\sqrt{\frac{\delta}{n}} + o\left(\frac{1}{\sqrt{n}}\right). \tag{10.68}$$

Combining (10.43), (10.50), and (10.68) yields

$$(1-\varepsilon)\log M - \log_2 e \leq I(X^n; Y^n) \leq \frac{n\rho_n}{2\sigma^2} \leq \sqrt{n\delta} + o\left(\frac{1}{\sqrt{n}}\right). \tag{10.69}$$

The converse proof is completed by using the definition of $R_{\mathrm{c}}^*(n, \varepsilon, \delta)$ in (10.8).

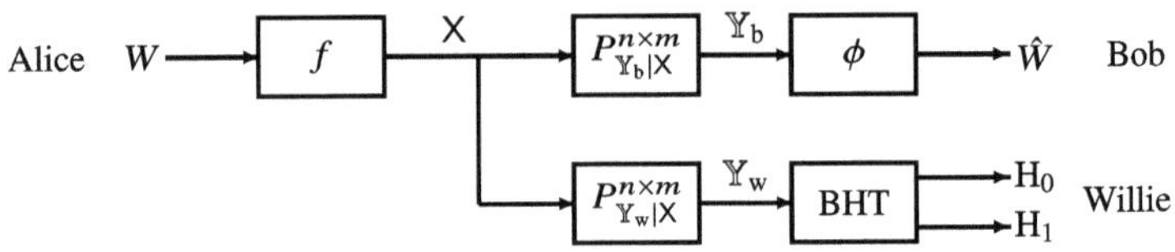

Fig. 10.2 System model for MIMO covert communication

10.2 MIMO AWGN Channel

10.2.1 Problem Formulation

This section considers a MIMO AWGN channel under the KLD covertness measure in (10.3) as Fig. 10.2. A legitimate transmitter Alice communicates with a legitimate receiver Bob in the presence of an adversary Willie. For simplicity, assume that all parties have $m \in \mathbb{N}$ antennas. The channel input-output relation is given by

$$\mathbb{Y}_b = \mathsf{X}\mathsf{H}_b + \mathbb{Z}_b, \tag{10.70}$$

$$\mathbb{Y}_w = \mathsf{X}\mathsf{H}_w + \mathbb{Z}_w. \tag{10.71}$$

Here, $\mathsf{X} \in \mathbb{R}_+^{n\times m}$ is the signal transmitted over n channel uses; $\mathbb{Y}_b \in \mathbb{R}_+^{n\times m}$ is the corresponding received signal at Bob and $\mathbb{Y}_w \in \mathbb{R}_+^{n\times m}$ is the corresponding received signal at Willie; the deterministic matrices H_b, H_w are Bob's and Willie's channel matrices, assumed known to everyone and both matrices have full rank; $\mathbb{Z}_b, \mathbb{Z}_w \in \mathbb{R}_+^{n\times m}$ denote the additive noise at receivers, which have i.i.d. $\mathcal{N}(0, \sigma_b^2)$ and $\mathcal{N}(0, \sigma_w^2)$ entries, respectively. Compared to the MIMO system model in Chap. 7, this section considers a non-fading scenario with an equal number of antennas. Therefore, eliminating the influence of the channel coefficient matrix is reasonable, further simplifying the complex Gaussian space to the Gaussian space.

Recall the definitions provided in Sect. 10.1.1. Denote the probability distributions of the channel input as $P_{\mathsf{X}}^{n\times m}$ with second moment $\rho_{mn} \in \mathbb{R}_+$, the probability distributions at Bob when Alice is transmitting or not as $P^{n\times m}_{\mathbb{Y}_b|1}$, $P^{n\times m}_{\mathbb{Y}_b|1}$, respectively, the probability distributions at Willie when Alice is transmitting or not as $P^{n\times m}_{\mathbb{Y}_w|1}$, $P^{n\times m}_{\mathbb{Y}_w|1}$, respectively. Let $\mathsf{Y} \in \mathbb{R}_+^{n\times m}$. For convenience in computation, let $\sigma_b = \sigma_w = 1$.

In covert communication, the legitimate receiver Bob wants to reliably decode X from the received vector Y. The adversary Willie tries to determine from the observations of Y whether Alice is communicating.

Definition 10.2 An (n, M) code for a MIMO AWGN channel consists of an encoder at the transmitter Alice $f : \mathcal{M} \to \mathcal{X}^{n\times m}$ that maps the message to a codeword satisfying the covert constraint δ and a decoder at the legitimate user Bob $\phi : \mathcal{Y}^{n\times m} \to \mathcal{M}$ satisfying the maximal error probability ε as Definition 10.1.

Analogously to (10.8), the low-latency theoretical benchmark for MIMO AWGN channels is

$$R_{\mathrm{c}}^{*}(n, \varepsilon) := \sup \left\{ R_n : \exists \text{ an } (n, M)\text{-code s.t. } \mathrm{P}_{\mathrm{e},n} \leq \varepsilon \text{ and } \mathrm{P}_{\mathrm{d},n} \leq \delta \right\}. \quad (10.72)$$

10.2.2 Second-Order Asymptotics and Discussions

Theorem 10.3 *Under KLD covertness metric,*

$$\lim_{n \to \infty} \lim_{\varepsilon \to 0} \sqrt{n} R_{\mathrm{c}}^{*}(n, \varepsilon, \delta) = \sqrt{m\delta}. \quad (10.73)$$

The similar result also holds under TVD covertness metric, which was first derived in [17] and also stated in [18, Theorem 3.2].

In a MIMO scenario, the square root law incorporates the spatial degrees of freedom due to the multiple antennas. For a system with m transmit and receive antennas, the covert transmission rate is given by (10.73). Compared to the single-antenna result of $\sqrt{\frac{\delta}{n}}$, the MIMO system achieves a gain of $\sqrt{m}$ in the covert transmission rate. As shown in Fig. 10.3, when the number of antenna is 1, the result in the MIMO scenario matches that of a single-antenna system. As the number of antennas increases, the covert communication rate increases significantly. As the covert constraint is strengthened, i.e., as δ decreases, the covert transmission rate decreases (Fig. 10.4).

Theorem 10.3 demonstrates that the spatial degrees of freedom in MIMO systems significantly improve the performance of covert communication under the same covertness constraint. The use of multiple antennas enhances covert communication by increasing the total channel capacity while simultaneously maintaining the

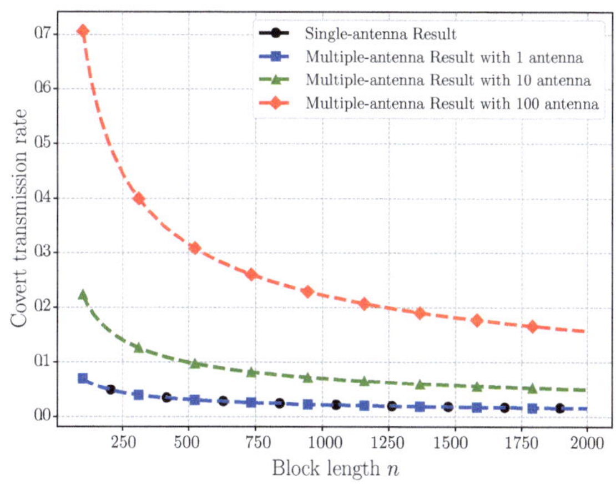

Fig. 10.3 Numerical plots of the covert transmission rate versus channel uses n and the number of antennas m, where $\delta = 0.5$

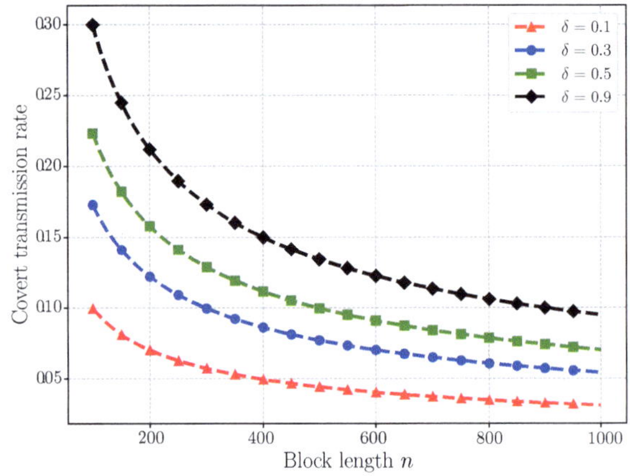

Fig. 10.4 Numerical plots of the covert transmission rate versus channel uses n and the detection probability δ, where $m = 10$

interference required to confuse Willie. However, the gain is sub-linear, as it scales with $\sqrt{m}$, rather than the number of transmit and receiver antennas m. This behavior is inherent to covert communication, where the transmission rate remains dominated by the second-order term, regardless of the number of antennas. Nevertheless, the result indicates that MIMO systems effectively mitigate the cost of covertness, providing a scalable way to improve covert transmission performance. This insight is particularly valuable for designing efficient covert communication systems in scenarios where maintaining undetectability is critical.

10.2.3 Proof

10.2.3.1 Achievability

The achievability part is proved by calculating the power bound using Gaussian random coding to satisfy the covert constraint. The channel is decomposed into m parallel sub-channels defined by the GSVD precoding. In a non-fading scenario, it is reasonable to allocate equal power to each transmitting antenna. Consider a random codebook where the codewords are i.i.d. generated with the distribution $\mathcal{N}(0, \rho_{mn})$. The covert constraint is met by

$$D(P^{n\times m}_{\mathbb{Y}_w|1} \| P^{n\times m}_{\mathbb{Y}_w|0}) = E_{P^{n\times m}_{\mathbb{Y}_w|1}}\left[\log \frac{\prod_{i\in[n]}\prod_{j\in[m]} \frac{1}{\sqrt{2\pi(1+\rho_{mn})}} \exp\left(-\frac{y_{ij}^2}{2(1+\rho_{mn})}\right)}{\prod_{i\in[n]}\prod_{j\in[m]} \frac{1}{\sqrt{2\pi}} \exp\left(-\frac{y_{ij}^2}{2}\right)}\right] \tag{10.74}$$

$$= E_{P^{n\times m}_{\mathbb{Y}_{\mathrm{w}}|1}}\left[mn \cdot \log \frac{\frac{1}{\sqrt{2\pi(1+\rho_{mn})}}\exp\left(-\frac{y^2}{2(1+\rho_{mn})}\right)}{\frac{1}{\sqrt{2\pi}}\exp\left(-\frac{y^2}{2}\right)}\right] \tag{10.75}$$

$$= -\frac{mn}{2}\log(1+\rho_{mn}) + E_{P^{n\times m}_{\mathbb{Y}_{\mathrm{w}}|1}}\left[mn \cdot \frac{y^2\rho_{mn}}{2(1+\rho_{mn})}\right] \tag{10.76}$$

$$= -\frac{mn}{2}\log(1+\rho_{mn}) + \frac{mn\rho_{mn}}{2} \tag{10.77}$$

$$= \frac{mn}{2}(\rho_{mn} - \log(1+\rho_{mn})), \tag{10.78}$$

where (10.75) follows from the assumption that the constructed codewords are i.i.d. generated and (10.77) follows from the output with the distribution $\mathcal{N}(0, 1+\rho_{mn})$. To obtain the power bound satisfying the covert constraint, it is necessary to ensure

$$D(P^{n\times m}_{\mathbb{Y}_{\mathrm{w}}|1} \| P^{n\times m}_{\mathbb{Y}_{\mathrm{w}}|0}) \leq \delta. \tag{10.79}$$

The following lemma will be used in the derivation of the bounds.

Lemma 10.1 *With any $\eta > 1$, for sufficiently small $a > 0$, the following inequality holds:*

$$\frac{a^2}{4\eta} < \frac{a - \log(1+a)}{2} < \frac{a^2}{4}. \tag{10.80}$$

From Lemma 10.1, an upper bound of (10.78) will lead to a lower bound of the solution of (10.79), which is sufficient for the covert constraint:

$$\rho_{\mathrm{suf}} = \sqrt{\frac{4\delta}{mn}}. \tag{10.81}$$

The achievability bound will be evaluated under the covert constraint:

$$\log M(n, \varepsilon, \delta) \geq \log M(n, \varepsilon, \rho_{\mathrm{suf}}) \tag{10.82}$$

$$\geq \frac{mn}{2}\log\left(1 + \sqrt{\frac{4\delta}{mn}}\right) \tag{10.83}$$

$$= \frac{mn}{2}\left(\sqrt{\frac{4\delta}{mn}} - \frac{2\delta}{mn} + o\left(\frac{1}{n}\right)\right) \tag{10.84}$$

$$= \sqrt{mn\delta} + O(1), \tag{10.85}$$

where (10.82) follows from the result in (10.81), (10.83) follows by substituting the power bound into $\log M = \frac{mn}{2}\log(1+\mathrm{SNR})$ [16, Eq. 9.17] while the channel can be viewed as m parallel sub-channels in the considered simple MIMO scenario, and (10.84) follows from the Taylor Series of expansion of $\log(1+x)$ around $x = 0$, which implies $\log(1+x) = x - \frac{x^2}{2} + o(x^2)$.

The achievability proof is now completed.

10.2.3.2 Converse

For converse bound, the maximal second-order moment of all distributions under the covert constraint is considered. First, the distribution of any code is defined as $P_{\mathbb{X}}$ to incorporate all random coding schemes (including deterministic coding as a special case), and regard each codeword as a matrix generated by $P_{\mathbb{X}}$. The average second moment of the distribution is expressed as $\frac{1}{mn}\mathsf{tr}(\mathsf{cCov}(\mathbb{X}) \otimes \mathsf{rCov}(\mathbb{X}))$, where $\mathsf{cCov}(\cdot)$, $\mathsf{rCov}(\cdot)$ denote the column covariance matrix and row covariance matrix of matrices, respectively. Below two distributions of the codes for covert communication over the MIMO AWGN channels are introduced.

Let $P_{\mathsf{X}}^{\mathrm{rnd}}$ be any distribution of X under given second moment constraint

$$\frac{1}{mn}\mathsf{tr}(\mathsf{cCov}(\mathsf{X}) \otimes \mathsf{rCov}(\mathsf{X})) = \rho, \tag{10.86}$$

and the induced output distribution of the MIMO AWGN channel is denoted as $P_{\mathbb{Y}}^{\mathrm{rnd}}$. Let $P_{\mathsf{X}}^{\mathrm{G}}$ be the Gaussian distribution with i.i.d. coordinates, and each coordinate has zero mean and variance ρ, the induced output distribution of the MIMO AWGN channel (with background noise $\mathcal{N}(0, 1)$) is $P_{\mathbb{Y}}^{\mathrm{G}}$.

For these distributions with the same parameter ρ, the following fact is an extension of [14, Theorem 9] to the MIMO scenario. For any randomly generated code with fixed second moment condition (10.86), Gaussian random codes $P_{\mathsf{X}}^{\mathrm{G}}$ minimize KLD between the induced output distribution and the distribution of the noise at the adversary. The equivalent statement is as follows. Any input distribution which satisfies the covert constraint, then the second moment should be less than the parameter ρ in (10.86) of the equation

$$D(P_{\mathbb{Y}}^{\mathrm{G}} \| P_{\mathbb{Y}_{\mathrm{w}}|0}^{n\times m}) = \delta. \tag{10.87}$$

For any codebook, the second moment is equivalent to the average power since it is assumed that the codewords are equiprobable. By solving Eq. (10.87), the variance ρ of $P_{\mathbb{Y}}^{\mathrm{G}}$ is obtained, then for any other distribution with variance ρ_{mn}, let $\rho \le \rho_{mn}$ to satisfy the covert constraint $D(P_{\mathbb{Y}}^{\mathrm{rnd}} \| P_{\mathbb{Y}_{\mathrm{w}}|0}^{n\times m}) \le \delta$. In other words, Gaussian random codes will have the largest average power among all random codes under covert constraint. Since at this point ρ_{mn} is the average power of Gaussian random codes, the following result is obtained:

$$M(n, \varepsilon, \delta) \le M_{\mathrm{a}}(n, \varepsilon, \rho). \tag{10.88}$$

Thus, a converse bound under covert constraint is obtained in terms of a converse bound under average power constraint.

As ρ is the solution of Eq. (10.87), calculating KLD obtains

$$D(P_{\mathbb{Y}}^{\mathrm{G}} \| P_{\mathbb{Y}_{\mathrm{w}}|0}^{n\times m}) = E_{P_{\mathbb{Y}}^{\mathrm{G}}}\left[\log \frac{\prod_{i\in[n]}\prod_{j\in[m]}\frac{1}{\sqrt{2\pi(1+\rho)}}\exp\left(-\frac{y_{ij}^2}{2(1+\rho)}\right)}{\prod_{i\in[n]}\prod_{j\in[m]}\frac{1}{\sqrt{2\pi}}\exp\left(-\frac{y_{ij}^2}{2}\right)}\right] \tag{10.89}$$

$$= E_{P_{\mathbb{Y}}^{\mathrm{G}}}\left[mn\cdot\log \frac{\frac{1}{\sqrt{2\pi(1+\rho)}}\exp\left(-\frac{y^2}{2(1+\rho)}\right)}{\frac{1}{\sqrt{2\pi}}\exp\left(-\frac{y^2}{2}\right)}\right] \tag{10.90}$$

$$= -\frac{mn}{2}\log(1+\rho) + E_{P_{\mathbb{Y}}^{\mathrm{G}}}\left[mn\cdot\frac{y^2\rho}{2(1+\rho)}\right] \tag{10.91}$$

$$= -\frac{mn}{2}\log(1+\rho) + \frac{mn\rho}{2} \tag{10.92}$$

$$= \frac{mn}{2}(\rho - \log(1+\rho)), \tag{10.93}$$

where (10.90) follows from the fact that $P_{\mathsf{X}}^{\mathrm{G}}$ are the Gaussian distribution with i.i.d. coordinates and (10.92) follows from the output with the distribution $\mathcal{N}(0, 1+\rho)$.

Similarly, from Lemma 10.1, a lower bound of (10.93) will lead to an upper bound of ρ:

$$\rho_{\mathrm{nec}} = \sqrt{\frac{4\delta\eta}{mn}}. \tag{10.94}$$

Let $\eta = 1 + \frac{1}{n}$. It follows that

$$\log M(n,\varepsilon,\delta) \le \log M_{\mathrm{a}}(n,\varepsilon,\rho) \tag{10.95}$$

$$\le \log M_{\mathrm{a}}(n,\varepsilon,\rho_{\mathrm{nec}}) \tag{10.96}$$

$$\le \log\left[\frac{1}{1-\frac{n}{n+1}}M\left(n,\varepsilon,\frac{(n+1)\rho_{\mathrm{nec}}}{n}\right)\right] \tag{10.97}$$

$$\le \frac{mn}{2}\log\left(1+\sqrt{\frac{4\delta\eta}{mn}}\right) \tag{10.98}$$

$$\le \sqrt{mn\eta\delta} \tag{10.99}$$

$$\le \sqrt{mn\delta}\left(1+\frac{1}{2n}\right) \tag{10.100}$$

$$= \sqrt{mn\delta} + \frac{\sqrt{m\delta}}{2\sqrt{n}}, \tag{10.101}$$

where (10.95) follows from the result in (10.88), (10.96) follows from the result in (10.94), (10.97) follows from [19, Eq. (196)], (10.98) from by substituting the power bound into $\log M = \frac{mn}{2} \log(1 + \text{SNR})$ [16, Eq. 9.17], (10.99) follows since $\log(1+x) \leq x$ for any $x \in \mathbb{R}_+$, and (10.100) follows since $\sqrt{\eta} = \sqrt{1 + \frac{1}{n}} \leq 1 + \frac{1}{2n}$ for any $n \in \mathbb{N}$.

10.3 Massive MIMO

10.3.1 Problem Formulation

10.3.1.1 The Sparse Channel Model for Massive MIMO

Figure 10.5 plots the system model for millimeter-wave (mmWave) massive MIMO communication via hybrid beamforming architecture, a confidential signal is transmitted from a legitimate user, Alice, to another legitimate user, Bob, while a potential adversary, Willie, attempts to detect the occurrence of communication. Assuming that Alice, Bob, and Willie have N_a, N_b, and N_w antennas, respectively, and the number of radio frequency (RF) chains at each user is denoted as M_a, M_b, and M_w. For massive MIMO systems, the number of antennas at each user is greater than or equal to the number of RF chains, i.e., $N_\ddagger \geq M_\ddagger$, where $\ddagger \in$ a, b, w. Furthermore, suppose that Alice employs hybrid beamforming, consisting of a digital precoder $\mathbf{W}$ and an analog precoder $\mathbf{U}_\ddagger$.

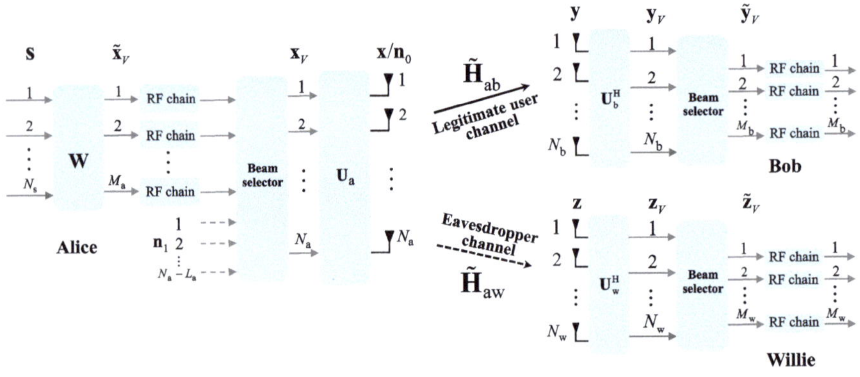

Fig. 10.5 System model for covert transmission with a transmitter communicating with a legitimate receiver in the existence of a potential adversary via mmWave massive MIMO hybrid beamforming architecture

To model the channel from Alice to Bob, the analog precoder at Alice is [20]

$$\mathbf{U}_{\mathrm{a}} = \frac{1}{\sqrt{N_{\mathrm{a}}}} \left[\mathbf{u}_{\mathrm{a}}(\theta_{\mathrm{a},1}), \mathbf{u}_{\mathrm{a}}(\theta_{\mathrm{a},2}), \ldots, \mathbf{u}_{\mathrm{a}}(\theta_{\mathrm{a},N_{\mathrm{a}}})\right], \tag{10.102}$$

for each $i \in [N_{\mathrm{a}}]$, the parameter $\mathbf{u}_{\mathrm{a}}(\theta_{\mathrm{a},i}) = \left[1, e^{-j2\pi\vartheta_{\mathrm{a},i}}, \ldots, e^{-j2\pi(N_{\mathrm{a}}-1)\vartheta_{\mathrm{a},i}}\right]^{\mathrm{T}}$ is the array steering vector, and $\theta_{\mathrm{a},i} = \arcsin\left(\frac{\lambda\vartheta_{\mathrm{a},i}}{d}\right)$ is the angle of departure (AoD) with respect to uniformly spaced spatial angles $\vartheta_{\mathrm{a},i} = \frac{(i-1-(N_{\mathrm{a}}-1)/2)}{N_{\mathrm{a}}}$, the antenna spacing d and the wavelength λ. Similarly, the analog precoders for Bob $\mathbf{U}_{\mathrm{b}}$ and $\mathbf{U}_{\mathrm{w}}$ are obtained by replacing the a as b and w, and the index i as j (and k) in $\mathbf{U}_{\mathrm{a}}$, respectively.

This way, the block fading channel from Alice to Bob satisfies [20, 21]

$$\tilde{\mathbf{H}}_{\mathrm{ab}} = \frac{1}{\sqrt{N_{\mathrm{b}}N_{\mathrm{a}}}} \sum_{j\in[N_{\mathrm{b}}]} \sum_{i\in[N_{\mathrm{a}}]} \mathbf{H}_{\mathrm{ab}}(j,i)\mathbf{u}_{\mathrm{b}}(\bar{\theta}_{\mathrm{b},j})\mathbf{u}_{\mathrm{a}}^{\mathrm{H}}(\bar{\theta}_{\mathrm{a},i}) \tag{10.103}$$

$$= \mathbf{U}_{\mathrm{b}}\mathbf{H}_{\mathrm{ab}}\mathbf{U}_{\mathrm{a}}^{\mathrm{H}}, \tag{10.104}$$

where $\mathbf{H}_{\mathrm{ab}} \in \mathbb{C}^{N_{\mathrm{b}}\times N_{\mathrm{a}}}$ is the equivalent virtual channel of physical channel $\tilde{\mathbf{H}}_{\mathrm{ab}}$ [21]. Since $\mathbf{U}_{\mathrm{a}}$, $\mathbf{U}_{\mathrm{b}}$, and $\mathbf{U}_{\mathrm{w}}$ are unitary discrete Fourier transform (DFT) matrices, the channels between Alice and Bob, as well as Willie, satisfy

$$\mathbf{H}_{\mathrm{ab}} = \mathbf{U}_{\mathrm{b}}^{\mathrm{H}}\tilde{\mathbf{H}}_{\mathrm{ab}}\mathbf{U}_{\mathrm{a}} \tag{10.105}$$

$$\mathbf{H}_{\mathrm{aw}} = \mathbf{U}_{\mathrm{w}}^{\mathrm{H}}\tilde{\mathbf{H}}_{\mathrm{aw}}\mathbf{U}_{\mathrm{a}}. \tag{10.106}$$

Note that $\mathbf{H}_{\mathrm{ab}}$ and $\mathbf{H}_{\mathrm{aw}}$ are generally different, unless Bob and Willie are located at the same position, in which case covert communication would be impossible.

As illustrated in Fig. 10.6, the sparsity of the mmWave massive MIMO channel [20] suggests that the channel matrix typically contains only a limited number of significant non-zero coefficients, referred to as dominant beams, while the remaining coefficients are zero, termed non-dominant beams. Taking into account the channel sparsity, the channel matrix $\mathbf{H}_{\mathrm{ab}}$ can be represented in a low-dimensional virtual space, which is given as follows [21]:

$$\mathbf{H}_{\mathrm{ab}}^{\mathrm{l}} = [\mathbf{H}_{\mathrm{ab}}(j,i)]_{(j,i)\in\mathcal{E}} \in \mathbb{C}^{L_{\mathrm{b}}\times L_{\mathrm{a}}}, \tag{10.107}$$

where $\mathcal{E} = \{(j,i) \mid j \in \mathcal{E}_{\mathrm{b}}, i \in \mathcal{E}_{\mathrm{a}}\}$ denotes the selection mask containing the indices of the dominant beams, and $\mathcal{E}_{\mathrm{a}}$, $\mathcal{E}_{\mathrm{b}}$ represent the sets of indices for the dominant receive and transmit beams, respectively. Similarly, the low-dimensional virtual representation of $\mathbf{H}_{\mathrm{aw}}$ is given by

$$\mathbf{H}_{\mathrm{aw}}^{\mathrm{l}} = [\mathbf{H}_{\mathrm{aw}}(k,i)](k,i) \in \mathcal{F} \in \mathbb{C}^{L_{\mathrm{w}}\times L_{\mathrm{a}}}, \tag{10.108}$$

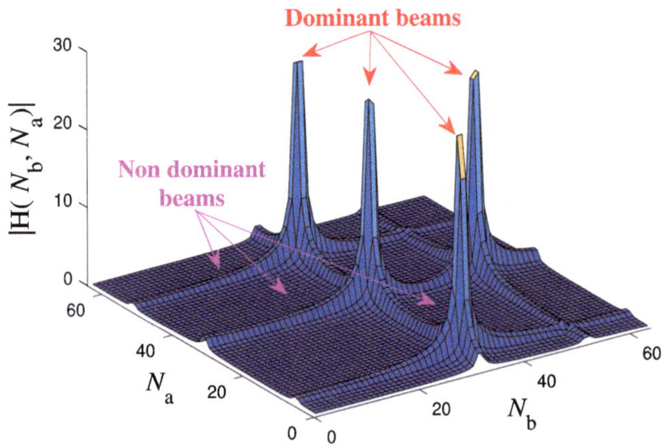

Fig. 10.6 Illustration of dominant beams and non-dominant beams of the virtual channel

where $\mathcal{F} = \{(k, i) \mid k \in \mathcal{F}_{\mathrm{w}}, i \in \mathcal{F}_{\mathrm{a}}\}$. For simplicity, let L_{a}, L_{b}, and L_{w} represent the number of dominant beams from the perspectives of Alice, Bob, and Willie, respectively. Note that $M_{\mathrm{a}} \leq L_{\mathrm{a}} \leq N_{\mathrm{a}}$, $M_{\mathrm{b}} \leq L_{\mathrm{b}} \leq N_{\mathrm{b}}$, and $M_{\mathrm{w}} \leq L_{\mathrm{w}} \leq N_{\mathrm{w}}$.

Inspired by [13, 22], the sparsity of a channel generally includes the angle sparsity and the power sparsity, respectively. The angle sparsity is

$$\rho = \frac{L_{\mathrm{a}}}{N_{\mathrm{a}}}, \ \rho \in \left[\frac{M_{\mathrm{a}}}{N_{\mathrm{a}}}, 1\right) \tag{10.109}$$

Furthermore, let G_{d} and G_{n} denote the gains of dominant beams and non-dominant beams, respectively. Thus the power sparsity is defined as

$$\eta = \frac{\|G_{\mathrm{n}}\|_2^2}{\|G_{\mathrm{d}}\|_2^2}, \ \eta \in (0, 1). \tag{10.110}$$

Using ρ and η, the sparsity metric for the massive MIMO channel can be defined as [13]

$$\chi(\rho, \eta) = \rho + \eta - \rho\eta. \tag{10.111}$$

Note that the sparsity metric $\chi(\rho, \eta) \in \left[\frac{M_{\mathrm{a}}}{N_{\mathrm{a}}}, 1\right)$ decreases in the angle sparsity ρ and decreases in the power sparsity η, respectively.

10.3.1.2 Covert Transmission Scheme

As illustrated in Fig. 10.5, Alice aims to securely transmit a confidential signal $\mathbf{s} \in \mathbb{C}^{N_s \times 1}$ to Bob, while simultaneously preventing Willie from detecting any legitimate communication activity. To achieve covert transmission, a jamming-aided scheme based on the framework proposed in [13] is employed in this section. Alice's transmission process is modeled under two hypotheses: H_0 and H_1.

Under hypothesis H_0, Alice does not transmit any confidential signal and instead sends a jamming signal $\mathbf{n}_0$ across all beams using N_a transmit antennas. The jamming signal $\mathbf{n}_0$ is drawn from a complex Gaussian distribution, $\mathcal{CN}\left(0, \frac{Q_n}{N_a}\mathbf{I}N\mathrm{a}\right)$, where Q_n represents the power of the jamming signal. In this case, the transmission only serves to mask any legitimate communication, making it difficult for Willie to detect any activity.

In contrast, under hypothesis H_1, Alice transmits the confidential signal by utilizing a subset of dominant beams, while simultaneously injecting a jamming signal $\mathbf{n}_1$ into the non-dominant beams. The jamming signal $\mathbf{n}_1$ follows a complex Gaussian distribution $\mathcal{CN}\left(0, \frac{(1-\phi)Q_t}{N_a - L_a}\mathbf{I}N\mathrm{a} - L_\mathrm{a}\right)$, where Q_t represents Alice's total transmit power and ϕ indicates the fraction of power allocated for the confidential signal. The remaining power is used for jamming in the non-dominant beams. This strategy ensures that the legitimate transmission remains concealed, even when the jamming signal is present.

When Alice transmits an i.i.d. signal under hypothesis H_0, our proposed jamming-aided covert transmission scheme aligns with the concept of stealthy communication discussed in [23]. Moreover, this scheme can be considered a special case of the covert communication model in [24], where the transmitter and the jammer are assumed to be the same entity.

Throughout the chapter, consistent with previous studies [20, 22], assume that full CSI of the channel $\mathbf{H}_{ab}$ is available only at Bob, and the adversary Willie only knows $\mathbf{H}_{aw}$. Furthermore, assume that a limited feedback from Bob to Alice is available for transmit beam selection. Note that existing studies on mmWave systems and reconfigurable antennas usually assume full CSI at the receiver or even at both the transmitter and the receiver, e.g., [20, 22, 25]. Furthermore, the estimation method for CSI has been studied in a number of papers such as [26, 27] for mmWave channels. Since Alice does not have full CSI, the digital precoder at Alice is given by $\mathbf{W} = \sqrt{\frac{\phi Q_t}{M_a}}\mathbf{I}_{M_a}$ [20, 22] and [28, Chapter 8] with equal power allocation between the N_s data streams with the transmit power constraint, and the beam selector can be easily realized by switches in practice.

Figure 10.7 illustrates the distribution of confidential signals and artificial noise in the equivalent virtual channel. This part specifies in detail how Alice transmits the confidential signal $\mathbf{s}$ under H_1. Firstly, using the precoding matrix $\mathbf{W}$, Alice generates the precoded signal $\tilde{\mathbf{x}}_V = \mathbf{W}\mathbf{s} \in \mathbb{C}^{M_a \times 1}$. Subsequently, the precoded signal $\tilde{\mathbf{x}}_V$ goes through the beam selector to generate the signal vector $\mathbf{x}_V \in \mathbb{C}^{N_a \times 1}$.

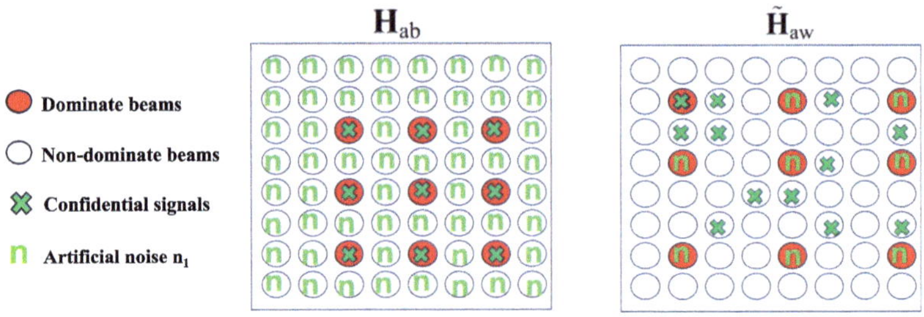

Fig. 10.7 Illustration of the distribution of confidential signals and artificial noise in the dominant beam and non-dominant beam of the legitimate user and eavesdropper channels

This way, the transmitted signal $\mathbf{x}_V$ of Alice is

$$[\mathbf{x}_\mathrm{V}(i)] = \begin{cases} \tilde{\mathbf{x}}_\mathrm{V}, \ i \in \bar{\mathcal{E}}_\mathrm{a}, \\ \mathbf{n}_1, \ i \notin \bar{\mathcal{E}}_\mathrm{a}, \\ 0, \quad i \in \mathcal{E}_\mathrm{a}, i \notin \bar{\mathcal{E}}_\mathrm{a}, \end{cases} \tag{10.112}$$

where i is the index of the dominant transmit beams in $\mathcal{E}_\mathrm{a}$ defined in Eq. (10.107), and $\bar{\mathcal{E}}_\mathrm{a}$ with size M_a is a subset of $\mathcal{E}_\mathrm{a}$, which is randomly selected from $\mathcal{E}_\mathrm{a}$.

After passing through the transmit array response matrix $\mathbf{U}_\mathrm{a}$, the transmit signal $\mathbf{x}$ is given by

$$\mathbf{x} = \mathbf{U}_\mathrm{a}\mathbf{x}_\mathrm{V}. \tag{10.113}$$

This way, Bob's received signal is

$$\mathbf{y} = \tilde{\mathbf{H}}_\mathrm{ab}\mathbf{x} + \mathbf{n}_\mathrm{b}, \tag{10.114}$$

where $\tilde{\mathbf{H}}_\mathrm{ab}$ is the fading channel between Alice and Bob in Eq. (10.104), and the additive noise at Bob satisfies $\mathbf{n}_\mathrm{b} \sim \mathcal{CN}(0, \sigma_\mathrm{b}^2\mathbf{I}_{N_\mathrm{b}})$. Using the analog decoder $\mathbf{U}_\mathrm{b}^\mathrm{H}$, the signal for beam selection is

$$\mathbf{y}_\mathrm{V} = \mathbf{U}_\mathrm{b}^\mathrm{H}\mathbf{y} = \mathbf{H}_\mathrm{ab}\mathbf{x}_\mathrm{V} + \mathbf{n}_\mathrm{b,V}. \tag{10.115}$$

Since $\mathbf{U}_\mathrm{b}^\mathrm{H}$ is a unitary matrix, the matrix $\mathbf{H}_\mathrm{ab}$ is the equivalent virtual channel between Alice and Bob in Eq. (10.106), and we have $\mathbf{n}_\mathrm{b,V} \triangleq A_\mathrm{b}^\mathrm{H}\mathbf{n}_\mathrm{b} \sim \mathcal{CN}(0, \sigma_\mathrm{b}^2\mathbf{I}_{N_\mathrm{b}})$. For the beam selection, a selection mask $\bar{\mathcal{E}}_\mathrm{b} \subseteq \mathcal{E}_\mathrm{b}$ is adopted to randomly choose M_b beams from the L_b dominant receive beams. Hence the signal for detection is

$$\tilde{\mathbf{y}}_\mathrm{V} = [\mathbf{y}_\mathrm{V}(j)]_{j \in \bar{\mathcal{E}}_\mathrm{b}} = \bar{\mathbf{H}}_\mathrm{ab}\tilde{\mathbf{x}}_\mathrm{V} + \hat{\mathbf{H}}_\mathrm{ab}\mathbf{n}_1 + \tilde{\mathbf{n}}_\mathrm{b,V}, \tag{10.116}$$

where $\bar{\mathbf{H}}_{\mathrm{ab}} = [\mathbf{H}_{\mathrm{ab}}(j,i)]_{i\in\bar{\mathcal{E}}_{\mathrm{a}}, j\in\bar{\mathcal{E}}_{\mathrm{b}}} \in \mathbb{C}^{M_{\mathrm{b}}\times M_{\mathrm{a}}}$ and $\hat{\mathbf{H}}_{\mathrm{ab}} = [\mathbf{H}_{\mathrm{ab}}(j,i)]_{i\notin\bar{\mathcal{E}}_{\mathrm{a}}, j\in\bar{\mathcal{E}}_{\mathrm{b}}} \in \mathbb{C}^{M_{\mathrm{b}}\times(N_{\mathrm{a}}-L_{\mathrm{a}})}$ are the submatrices of $\mathbf{H}_{\mathrm{ab}}$, and $\tilde{\mathbf{n}}_{\mathrm{b,V}} = [\mathbf{n}_{\mathrm{b,V}}(j)]_{j\in\bar{\mathcal{E}}_{\mathrm{b}}} \in \mathbb{C}^{M_{\mathrm{b}}\times 1}$ is the noise power in detection.

10.3.1.3 Willie's Detection of Legitimate Communication

Depending on whether Alice transmits a confidential signal or not, Willie's received signal $\mathbf{z}$ is

$$\mathbf{z} = \begin{cases} \tilde{\mathbf{H}}_{\mathrm{aw}}\mathbf{n}_0 + \mathbf{n}_{\mathrm{w}}, & \mathrm{H}_0, \\ \tilde{\mathbf{H}}_{\mathrm{aw}}\mathbf{x} + \mathbf{n}_{\mathrm{w}}, & \mathrm{H}_1, \end{cases} \tag{10.117}$$

where $\tilde{\mathbf{H}}_{\mathrm{aw}}$ is the fading channel between Alice and Willie, and $\mathbf{n}_{\mathrm{w}} \sim \mathcal{CN}(0, \sigma_{\mathrm{w}}^2\mathbf{I}_{N_{\mathrm{w}}})$ is the additive noise at Willie. Using the receive analog decoder $\mathbf{U}_{\mathrm{w}}^{\mathrm{H}}$, it follows that

$$\mathbf{z}_{\mathrm{V}} = \mathbf{U}_{\mathrm{w}}^{\mathrm{H}}\mathbf{z} = \begin{cases} \mathbf{H}_{\mathrm{aw}}\mathbf{n}_{0,\mathrm{V}} + \mathbf{n}_{\mathrm{w,V}}, & \mathrm{H}_0, \\ \mathbf{H}_{\mathrm{aw}}\tilde{\mathbf{x}}_{\mathrm{V}} + \mathbf{n}_{\mathrm{w,V}}, & \mathrm{H}_1, \end{cases} \tag{10.118}$$

where the matrix $\mathbf{H}_{\mathrm{aw}}$ is the equivalent virtual channel between Alice and Bob.

Subsequently, a beam selector $\bar{\mathcal{F}}_{\mathrm{w}} \subseteq \mathcal{F}_{\mathrm{w}}$ is used to randomly choose M_{w} dominant receive beams from all beams $\mathcal{F}_{\mathrm{w}}$. With this operation, the output signal $\tilde{\mathbf{z}}_{\mathrm{V}}$ satisfies

$$\tilde{\mathbf{z}}_{\mathrm{V}} = \begin{cases} \bar{\mathbf{H}}_{\mathrm{aw}}\mathbf{n}_{0,\mathrm{V}} + \hat{\mathbf{H}}_{\mathrm{aw}}\mathbf{n}_{0,\mathrm{V}} + \tilde{\mathbf{n}}_{\mathrm{w,V}}, & \mathrm{H}_0 \\ \bar{\mathbf{H}}_{\mathrm{aw}}\tilde{\mathbf{x}}_{\mathrm{V}} + \hat{\mathbf{H}}_{\mathrm{aw}}\mathbf{n}_{1,\mathrm{V}} + \tilde{\mathbf{n}}_{\mathrm{w,V}}, & \mathrm{H}_1. \end{cases} \tag{10.119}$$

Analogously, $\bar{\mathbf{H}}_{\mathrm{aw}}$ and $\hat{\mathbf{H}}_{\mathrm{aw}}$ are the submatrices of the equivalent virtual channel matrix $\mathbf{H}_{\mathrm{aw}}$, and the details of parameters in Eq. (10.119) are given in Table 10.1. The proof of Eq. (10.119) is left as an exercise.

Note that Willie's received signal $\tilde{\mathbf{z}}_{\mathrm{V}}$ satisfies $\tilde{\mathbf{z}}_{\mathrm{V}} \sim f_0 = \mathcal{CN}(0, \boldsymbol{\Sigma}_0^{\mathrm{AN}})$ under H_0 and $\tilde{\mathbf{z}}_{\mathrm{V}} \sim f_1 = \mathcal{CN}(0, \boldsymbol{\Sigma}_1^{\mathrm{AN}})$ under H_1, where $\boldsymbol{\Sigma}_0^{\mathrm{AN}}$ and $\boldsymbol{\Sigma}_1^{\mathrm{AN}}$ satisfy

$$\boldsymbol{\Sigma}_0^{\mathrm{AN}} = \left(\left\|\bar{\mathbf{n}}_0^1\right\|_2^2 + \eta\left\|\bar{\mathbf{n}}_0^2\right\|_2^2 + \left\|\hat{\mathbf{n}}_0^1\right\|_2^2 + \eta\left\|\hat{\mathbf{n}}_0^2\right\|_2^2 + \sigma_{\mathrm{w}}^2\right)\mathbf{I}_{M_{\mathrm{w}}}, \tag{10.120}$$

$$\boldsymbol{\Sigma}_1^{\mathrm{AN}} = \left(\left\|\bar{\mathbf{x}}_{\mathrm{V}}^1\right\|_2^2 + \eta\left\|\bar{\mathbf{x}}_{\mathrm{V}}^2\right\|_2^2 + \left\|\hat{\mathbf{n}}_1^1\right\|_2^2 + \eta\left\|\hat{\mathbf{n}}_1^2\right\|_2^2 + \sigma_{\mathrm{w}}^2\right)\mathbf{I}_{M_{\mathrm{w}}}. \tag{10.121}$$

To detect whether Alice transmits a confidential signal to Bob or not, Willie uses binary hypothesis testing [15]. For this purpose, consider the Cartesian product

Table 10.1 Explanations of key matrices

Matrix Parameters	Definitions
$\bar{\mathbf{H}}_{\text{aw}}^1 \in \mathbb{C}^{M_{\text{w}} \times \frac{L_{\text{a}} M_{\text{a}}}{N_{\text{a}}}}$	the low-dimensional representation of $\bar{\mathbf{H}}_{\text{aw}}$ corresponds to the confidential signal $\tilde{\mathbf{z}}_V$ via the dominant beam
$\hat{\mathbf{H}}_{\text{aw}}^1 \in \mathbb{C}^{M_{\text{w}} \times \frac{L_{\text{a}}(N_{\text{a}}-L_{\text{a}})}{N_{\text{a}}}}$	the low-dimensional representation of $\bar{\mathbf{H}}_{\text{aw}}$ corresponds to the AN $\mathbf{n}_1$ via the dominant beam
$\bar{\mathbf{H}}_{\text{aw}}^2 \in \mathbb{C}^{M_{\text{w}} \times \frac{(N_{\text{a}}-L_{\text{a}}) M_{\text{a}}}{N_{\text{a}}}}$	the low-dimensional representation of $\bar{\mathbf{H}}_{\text{aw}}$ corresponds to the confidential signal $\tilde{\mathbf{z}}_V$ via the non-dominant beam
$\hat{\mathbf{H}}_{\text{aw}}^2 \in \mathbb{C}^{M_{\text{w}} \times \frac{(N_{\text{a}}-L_{\text{a}})^2}{N_{\text{a}}}}$	the low-dimensional representation of $\bar{\mathbf{H}}_{\text{aw}}$ corresponds to the AN $\mathbf{n}_1$ via the non-dominant beam

of the transmitted and received signals at Alice, Bob, and Willie. Specifically, to transmit a confidential signal, Alice transmits a sequence of n symbols $\mathbf{x}^n = (\mathbf{x}_1, \ldots, \mathbf{x}_n)$ where for each $i \in [n]$, $\mathbf{x}_i$ is generated as described in Eq. (10.113). This way, Willie's detection signals are $\tilde{\mathbf{z}}_{\text{V}}^n = (\tilde{\mathbf{z}}_{\text{V}1}, \ldots, \tilde{\mathbf{z}}_{\text{V}n})$, where for each $i \in [n]$, $\tilde{\mathbf{z}}_{\text{V}}$ is obtained as in Eq. (10.119). Then, to detect the occurrence of legitimate communication, Willie needs to determine whether $\tilde{z}_{\text{V}}^n$ is generated i.i.d. from f_0 or f_1, i.e., whether $\tilde{z}_{\text{V}}^n \sim \mathbb{P}_0^{\text{AN}^n} = f_0^n$ or $\tilde{z}_{\text{V}}^n \sim \mathbb{P}_1^{\text{AN}^n} = f_1^n$. Willie uses the log likelihood ratio test to make the decision. There are two types of errors: false alarm and miss detection. False alarm occurs when Willie decides that Alice is transmitting while Alice does not and its probability is denoted by α. Miss detection occurs when Willie decides Alice is not transmitting while Alice transmits a confidential signal, and its probability is denoted by β. In order to stop Willie from detecting legitimate communication between Alice and Bob, we need to keep the sum error probability $\text{P}_{\text{e}} := \alpha + \beta$ close to 1, which is the error probability when one makes a random guess.

10.3.2 Covertness Measure and Discussions

In this section, we use both the TVD [29, 30] and KLD as the covertness measures since they are closely related to the performance of the likelihood ratio test. We then derive bounds of both covertness measures and analyze the effect of the transmit power, the jamming power, and the channel sparsity on our derived upper bounds. These results build the foundation for the derivation of the covert transmission rates in the next section.

10.3.2.1 TVD Covertness Measure

From [15, Theorem 13.1.1], the TVD between $\mathbb{P}_0^{\mathrm{AN}^n}$ and $\mathbb{P}_1^{\mathrm{AN}^n}$ is related with $\mathrm{P_e}$ is

$$\mathrm{P_e} = 1 - \mathrm{TV}\left(\mathbb{P}_0^{\mathrm{AN}^n}, \mathbb{P}_1^{\mathrm{AN}^n}\right), \tag{10.122}$$

where $\mathrm{TV}\left(\mathbb{P}_0^{\mathrm{AN}^n}, \mathbb{P}_1^{\mathrm{AN}^n}\right)$ is the TVD between two distributions $\mathbb{P}_0^{\mathrm{AN}^n}$ and $\mathbb{P}_1^{\mathrm{AN}^n}$, i.e.,

$$\mathrm{TV}(\mathbb{P}_0^{\mathrm{AN}^n}, \mathbb{P}_1^{\mathrm{AN}^n}) = \frac{1}{2}\| f_0^n - f_1^n \|_1 \tag{10.123}$$

$$= \frac{1}{2}\int_{\mathcal{Z}^n} \left| \prod_{i\in[n]} f_0(z_i) - \prod_{i\in[n]} f_1(z_i)\right| \mathrm{d}z_{\mathrm{V}}^n. \tag{10.124}$$

To ensure the covertness, it is important to keep the error probability $\mathrm{P_e}$ of Willie to be close to one and keep $\mathrm{TV}(\mathbb{P}_0^{\mathrm{AN}^n}, \mathbb{P}_1^{\mathrm{AN}^n})$ as small as possible. To do so, we need to calculate the TVD exactly or approximately. However, it is intractable to obtain an exact formula of the TVD [9, 30] for multidimensional distributions. As a compromise, we give the following upper bound on the TVD:

$$\mathrm{TV}\left(\mathbb{P}_0^{\mathrm{AN}^n}, \mathbb{P}_1^{\mathrm{AN}^n}\right) \le \frac{\sqrt{n}\left(2+\sqrt{6}\right)}{2} \left\| \left(\mathbf{\Sigma}_1^{\mathrm{AN}}\right)^{-\frac{1}{2}} \mathbf{\Sigma}_0^{\mathrm{AN}} \left(\mathbf{\Sigma}_1^{\mathrm{AN}}\right)^{-\frac{1}{2}} - \mathbf{I} \right\|_{\mathrm{F}} \tag{10.125}$$

$$= \frac{\left(2+\sqrt{6}\right)\sqrt{nM_{\mathrm{w}}}}{2} \left| \frac{(\zeta Q_{\mathrm{n}} - Q_{\mathrm{t}})\,\chi}{\sigma_{\mathrm{w}}^2 + Q_{\mathrm{t}}\chi} \right| \tag{10.126}$$

$$* =: f_{\mathrm{TV}}\left(Q_{\mathrm{t}}, Q_{\mathrm{n}}, \chi\right), \tag{10.127}$$

where $\|\cdot\|_{\mathrm{F}}$ denotes the F norm, $\zeta = 1 + \frac{M_{\mathrm{a}}-L_{\mathrm{a}}}{N_{\mathrm{a}}}$, and χ is the channel sparsity defined in Eq. (10.111). The proof of Eq. (10.126) is left as an exercise.

Lemma 10.2 *The upper bound $f_{\mathrm{TV}}(Q_{\mathrm{t}}, Q_{\mathrm{n}}, \chi)$ on the TVD has the following properties.*

1. *The function $f_{\mathrm{TV}}(Q_{\mathrm{t}}, Q_{\mathrm{n}}, \chi)$ decreases with Q_{t} when $Q_{\mathrm{t}} \in (0, \zeta Q_{\mathrm{n}})$ and increases with Q_{t} when $Q_{\mathrm{t}} \in (\zeta Q_{\mathrm{n}}, Q_{\max})$.*
2. *The function $f_{\mathrm{TV}}(Q_{\mathrm{t}}, Q_{\mathrm{n}}, \chi)$ decreases with Q_{n} in $(0, \frac{Q_{\mathrm{t}}}{\zeta})$, and then increases with Q_{n} in $(\frac{Q_{\mathrm{t}}}{\zeta}, Q_{\max})$.*
3. *The function $f_{\mathrm{TV}}(Q_{\mathrm{t}}, Q_{\mathrm{n}}, \chi)$ increases with χ in $(\frac{M_{\mathrm{a}}}{N_{\mathrm{a}}}, 1)$.*

The results in Lemma 10.2 demonstrate the impact of various factors on the upper bound of the TVD covertness measure. Specifically, the analysis considers the transmit power Q_{t} of Alice for sending a confidential signal and a jamming signal under hypothesis H_1, the transmit power Q_n of Alice for sending only a jamming signal under hypothesis H_0, and the channel sparsity. It is shown that for a given jamming power Q_n and constant ζ, Alice can achieve an optimal transmit power of ζQ_n to minimize the upper bound of the TVD covertness measure, $f_\mathrm{TV}(Q_\mathrm{t}, Q_\mathrm{n}, \chi)$. Furthermore, for fixed values of Q_t and ζ, there exists an optimal jamming power Q_n that minimizes $f_\mathrm{TV}(Q_\mathrm{t}, Q_\mathrm{n}, \chi)$. Finally, the covertness of legitimate communication, as measured by the TVD, improves as the channel sparsity level χ increases.

When Alice does not send a jamming signal to confuse Willie, corresponding to $\phi = 1$ and $Q_\mathrm{n} = 0$, the upper bound on the $\mathrm{TV}(\mathbb{P}_0^n, \mathbb{P}_1^n)$ is simplified to

$$\mathrm{TV}(\mathbb{P}_0^n, \mathbb{P}_1^n) = \frac{(2+\sqrt{6})\sqrt{nM_\mathrm{w}}\chi Q_\mathrm{t}}{2(\sigma_\mathrm{w}^2 + \chi Q_\mathrm{t})} \tag{10.128}$$

$$=: f_\mathrm{TV}(Q_\mathrm{t}, \chi). \tag{10.129}$$

10.3.2.2 KLD Covertness Measure

It follows from Pinsker's inequality [15, Lemma 11.6.1] that

$$\mathrm{TV}\left(\mathbb{P}_0^{\mathrm{AN}^n}, \mathbb{P}_1^{\mathrm{AN}^n}\right) \le \sqrt{\frac{1}{2}\mathrm{D}(\mathbb{P}_0^{\mathrm{AN}^n} || \mathbb{P}_1^{\mathrm{AN}^n})}, \tag{10.130}$$

where $\mathrm{D}(\mathbb{P}_0^{\mathrm{AN}^n} || \mathbb{P}_1^{\mathrm{AN}^n})$ is the KLD of distributions $\mathbb{P}_0^{\mathrm{AN}^n}$ and $\mathbb{P}_1^{\mathrm{AN}^n}$. Note that the KLD has also been widely adopted to measure the detection performance of covert communications and enables us to obtain analytical results. Specifically, the closed-form expression of the KLD is given as follows:

$$\mathrm{D}(\mathbb{P}_0^{\mathrm{AN}^n} || \mathbb{P}_1^{\mathrm{AN}^n}) = n\int_{\mathcal{Z}} f_0(z) \ln \frac{f_0(z)}{f_1(z)} \mathrm{d}\tilde{z}_\mathrm{V} \tag{10.131}$$

$$= nM_\mathrm{w}\left[\ln\left(\frac{\sigma_\mathrm{w}^2 + \chi Q_\mathrm{t}}{\sigma_\mathrm{w}^2 + \zeta\chi Q_\mathrm{n}}\right) - \frac{(Q_\mathrm{t} - \zeta Q_\mathrm{n})\chi}{\sigma_\mathrm{w}^2 + \chi Q_\mathrm{t}}\right] \tag{10.132}$$

$$=: f_\mathrm{KL}(Q_\mathrm{t}, Q_\mathrm{n}, \chi). \tag{10.133}$$

The proof of Eq. (10.132) is left as an exercise.

Lemma 10.3 *The function $f_\mathrm{KL}(Q_\mathrm{t}, Q_\mathrm{n}, \chi)$ has the following properties.*

1. *The function $f_\mathrm{KL}(Q_\mathrm{t}, Q_\mathrm{n}, \chi)$ decreases with Q_t when $Q_\mathrm{t} \in (0, \zeta Q_\mathrm{n})$ and increases with Q_t when $Q_\mathrm{t} \in (\zeta Q_\mathrm{n}, Q_\mathrm{max})$.*
2. *The function $f_\mathrm{KL}(Q_\mathrm{t}, Q_\mathrm{n}, \chi)$ decreases with Q_n in $\left(0, \frac{Q_\mathrm{t}}{\zeta}\right)$, and then increases with Q_n in $\left(\frac{Q_\mathrm{t}}{\zeta}, Q_\mathrm{max}\right)$.*

3. *The function* $f_{\text{KL}}\left(Q_{\text{t}}, Q_{\text{n}}, \chi\right)$ *increases with* χ *in* $\left(\frac{M_{\text{a}}}{N_{\text{a}}}, 1\right)$.

Note that under the KLD covertness measure, the same comments of Lemma 10.2 also apply since both $f_{\text{KL}}\left(Q_{\text{t}}, Q_{\text{n}}, \chi\right)$ and $f_{\text{TV}}\left(Q_{\text{t}}, Q_{\text{n}}, \chi\right)$ have the same monotonicity properties with respect to its parameters, despite the significant differences of the formulas and the proof steps.

Analogously, when Alice does not send a jamming signal to confuse Willie, corresponding to $\phi = 1$ and $Q_{\text{n}} = 0$, the KLD $\text{D}(\mathbb{P}_0^n||\mathbb{P}_1^n)$ is simplified to

$$\text{D}(\mathbb{P}_0^n||\mathbb{P}_1^n) = nM_{\text{w}}\left[\ln\left(1 + \frac{\chi Q_{\text{t}}}{\sigma_{\text{w}}^2}\right) - \frac{\chi Q_{\text{t}}}{\sigma_{\text{w}}^2 + \chi Q_{\text{t}}}\right] \tag{10.134}$$

$$=: f_{\text{KL}}\left(Q_{\text{t}}, \chi\right). \tag{10.135}$$

10.3.2.3 The Effect of Jamming on Covertness Measure

This section analyzes the effect of the jamming power on the upper bounds of both covertness measures, revealing that there exists an optimal jamming power Q_{n} for Alice to effectively confuse Willie when no confidential signal is transmitted.

Lemma 10.4 *The following conclusions hold:*

1. *The power of the jamming signal* $\mathbf{n}_1$ *when Alice transmits a confidential signal under* H_1 *does not affect the upper bounds of both the TVD and the KLD covertness measures.*
2. *The power of jamming signal* $\mathbf{n}_0$ *under* H_0 *when Alice is not transmitting a confidential signal affects bounds of both covertness measures, and the optimal jamming power for the jamming signal* $\mathbf{n}_0$ *that minimizes upper bounds for both covertness measures is* $Q_{\text{n}} = \frac{Q_{\text{t}}}{\varsigma}$.

10.3.3 Covert Transmission Rate

Using bounds on the covertness measures in Sect. 10.3.2, the optimization problems can be established to maximize the covert transmission rate. The objective function is the ergodic capacity, and the constraints are on the covertness of the legal communication via a bound on either $f_{\text{TV}}(\cdot)$ or $f_{\text{KL}}(\cdot)$.

10.3.3.1 Covert Transmission Rate Without Jamming

When Alice does not transmit a jamming signal, corresponding to $\phi = 1$ and $Q_{\text{n}} = 0$. It follows from [22] that under block fading, with the transmit power Q_{t}, the transmission rate of legitimate communication is

$$R(Q_{\rm t}) = M_{\rm a} \log_2 \left(1 + \frac{M_{\rm b} Q_{\rm t}}{M_{\rm a} \sigma_{\rm b}^2} \right). \tag{10.136}$$

To satisfy the covertness of communication, the transmit power $Q_{\rm t}$ needs to be constrained. By constraining both covertness measures by some positive number δ, the δ-covert transmission rate is obtained via the following optimization problem:

$$\mathrm{P}_1: \quad \max_{Q_{\rm t} \in (0, Q_{\max}]} \quad R(Q_{\rm t}) \tag{10.137a}$$

$$\text{s.t.} \quad f_{\dagger}(Q_{\rm t}, \chi) \leq \delta, \tag{10.137b}$$

where $\dagger \in \{\text{TV}, \text{KL}\}$ denotes the type of the covertness measure. Let $R_{\dagger}$ be the solution of the optimization problem P_1. Using monotonicity of $R(Q_{\rm t})$ and $f_{\dagger}(Q_{\rm t}, \chi)$ with respect to $Q_{\rm t}$, P_1 can be solved with the following theorem.

Theorem 10.1 The solution to the optimization problem P_1 is

$$R_{\dagger} = M_{\rm a} \log_2 \left(1 + \frac{M_{\rm b} Q_{\dagger}}{M_{\rm a} \sigma_{\rm b}^2} \right), \tag{10.138}$$

where $Q_{\dagger}$ is the solution to $f_{\dagger}(Q_{\rm t}, \chi) = \delta$.

Proof Note that the transmission rate $R(Q_{\rm t})$ increases monotonously with the transmit power $Q_{\rm t}$. To improve the covert transmission rate, the transmit power $Q_{\rm t}$ should be as large as possible. However, the larger the transmit power, the looser the covertness guarantee. Thus, under each covertness measure, there is a maximal transmit power $Q_{\rm t}$. Since $f_{\rm TV}(Q_{\rm t}, \chi)$ monotonously increases with $Q_{\rm t}$, to satisfy the covertness constraint δ, the maximal transmit power is

$$Q_{\rm TV} = \frac{2\sigma_{\rm w}^2 \delta}{\chi \left[\left(2 + \sqrt{6} \right) \sqrt{n M_{\rm w}} - 2\delta \right]}. \tag{10.139}$$

This way, the maximal covert transmission rate on the TVD measure is obtained.

Analogously, when the KLD is the covertness measure, the covertness constraint is $f_{\rm KL}(Q_{\rm t}, \chi) \leq \delta$. Thus, we have

$$n M_{\rm w} \left[\ln \left(1 + \frac{\chi Q_{\rm t}}{\sigma_{\rm w}^2} \right) - \frac{\chi Q_{\rm t}}{\chi Q_{\rm t} + \sigma_{\rm w}^2} \right] = \delta, \tag{10.140}$$

which is equivalent to

$$\ln\left(\frac{\sigma_{\mathrm{w}}^2+\chi Q_{\mathrm{t}}}{\sigma_{\mathrm{w}}^2}\right)+\frac{\sigma_{\mathrm{w}}^2}{\chi Q_{\mathrm{t}}+\sigma_{\mathrm{w}}^2}=\frac{\delta}{nM_{\mathrm{w}}}+1. \tag{10.141}$$

Let $\xi=\frac{\sigma_{\mathrm{w}}^2}{Q_{\mathrm{t}}\chi+\sigma_{\mathrm{w}}^2}$, and $T=\frac{\delta}{nM_{\mathrm{w}}}+1$, Eq. (10.141) is equivalent to

$$-\ln\xi+\xi=T. \tag{10.142}$$

Since Eq. (10.142) is a transcendental equation, its root is

$$\xi=-W_{-1}\left(-e^{-T}\right), \tag{10.143}$$

where $W_{-1}(\cdot)$ is the Lambert W function [31]. Note that given $\ell\in\mathbb{C}$, the Lambert W function satisfies $W(\ell)e^{W(\ell)}=\ell$. Thus, the solution to Eq. (10.143) gives the optimal transmit power

$$Q_{\mathrm{KL}}=\frac{-\sigma_{\mathrm{w}}^2}{\chi}\left[W_{-1}\left(-e^{-\frac{\delta}{nM_{\mathrm{w}}}-1}\right)^{-1}+1\right] \tag{10.144}$$

Hence, the maximal covert transmission rate on the KLD measure is obtained. □

10.3.3.2 Covert Transmission Rate with Jamming

When Alice transmits jamming signals, from [22], with the transmit power Q_{t}, the transmission rate of legitimate communication is

$$R^{\mathrm{AN}}(Q_{\mathrm{t}})=M_{\mathrm{a}}\log_2\left[1+\frac{M_{\mathrm{b}}\phi Q_{\mathrm{t}}}{M_{\mathrm{a}}\sigma_{\mathrm{b}}^2}\left(1-\frac{f(\mu,\upsilon)}{4\mu\upsilon}\right)\right], \tag{10.145}$$

where $\mu:=\frac{(1-\phi)\eta Q_{\mathrm{t}}}{\sigma_{\mathrm{b}}^2}$, $\upsilon:=\frac{M_{\mathrm{b}}}{N_{\mathrm{a}}-L_{\mathrm{a}}}$, and $f(\mu,\upsilon)=\left(\sqrt{\mu(1+\sqrt{\upsilon})^2+1}-\sqrt{\mu(1-\sqrt{\upsilon})^2+1}\right)^2$.

Similarly to the case without jamming, with a covertness constraint $\delta\in\mathbb{R}_+$, the maximal transmission rate can be obtained via the following optimization problem:

$$\mathrm{P}_2:\quad \max_{Q_{\mathrm{t}}\in(0,Q_{\max}]}\quad R^{\mathrm{AN}}(Q_{\mathrm{t}}) \tag{10.146a}$$

$$\text{s.t.}\quad f_{\dagger}(Q_{\mathrm{t}},Q_{\mathrm{n}},\chi)\le\delta, \tag{10.146b}$$

$$0<Q_{\mathrm{n}}\le Q_{\max}. \tag{10.146c}$$

Let $R_{\dagger}^{\mathrm{AN}}$ be the solution of the optimization problem P_2. Using the monotonicity of $R^{\mathrm{AN}}(Q_{\mathrm{t}})$ with respect to Q_{t} and the monotonicity of $f_{\dagger}(Q_{\mathrm{t}}, Q_{\mathrm{n}}, \chi)$ with respect to Q_{t} and Q_{n}, the solution to P_2 and the following result can be obtained.

Theorem 10.2 The solution to the optimization problem P_2 is

$$R_{\dagger}^{\mathrm{AN}} = M_{\mathrm{a}} \log_2 \left[1 + \frac{M_{\mathrm{b}} \phi Q_{\dagger}^{\mathrm{AN}}}{M_{\mathrm{a}} \sigma_{\mathrm{b}}^2} \left(1 - \frac{f(\mu, \upsilon)}{4\mu\upsilon} \right) \right], \tag{10.147}$$

where $Q_{\dagger}^{\mathrm{AN}}$ is the solution to $f_{\dagger}(Q_{\mathrm{t}}, Q_{\mathrm{n}}, \chi) = \delta$.

Proof Note that the transmission rate $R^{\mathrm{AN}}(Q_{\mathrm{t}})$ is observed to increase monotonically with the transmit power Q_{t}. To improve the covert transmission rate, the transmit power Q_{t} should be maximized. However, due to the covertness constraint, the transmit power Q_{t} cannot exceed a certain limit. Therefore, the values of Q_{t} and Q_{n} should be optimized carefully. From Lemma 10.4, it is known that the optimal jamming power is given by $Q_{\mathrm{n}} = \frac{Q_{\mathrm{t}}}{\zeta}$, which is no less than the transmit power Q_{t}, since $\zeta \leq 1$ according to its definition in Eq. (10.127) and the condition that $M_{\mathrm{a}} \leq L_{\mathrm{a}}$. Thus, under the covertness constraint, it is concluded that the optimal transmit power $Q_{\dagger}^{\mathrm{AN}}$ and jamming power Q_{n} satisfy the relation $Q_{\dagger}^{\mathrm{AN}} \leq Q_{\mathrm{n}} \leq Q_{\max}$.

Using the upper bound of TVD covertness measure, i.e., $f_{\dagger}(Q_{\mathrm{t}}, Q_{\mathrm{n}}, \chi) = f_{\mathrm{TV}}(Q_{\mathrm{t}}, Q_{\mathrm{n}}, \chi)$. To satisfy the covertness constraint δ and maximize the covert transmission rate, the optimal transmit power is obtained as follows when $Q_{\mathrm{n}} = Q_{\max}$

$$Q_{\mathrm{TV}}^{\mathrm{AN}} = \frac{2\sigma_{\mathrm{w}}^2 \delta + \left(2 + \sqrt{6}\right) \sqrt{n M_{\mathrm{w}}} \chi \zeta Q_{\max}}{\left[\left(2 + \sqrt{6}\right) \sqrt{n M_{\mathrm{w}}} - \delta\right] \chi}. \tag{10.148}$$

Similarly, when the KLD is the covertness measure, i.e., $f_{\dagger}(Q_{\mathrm{t}}, Q_{\mathrm{n}}, \chi) = f_{\mathrm{KL}}(Q_{\mathrm{t}}, Q_{\mathrm{n}}, \chi)$. To satisfy the covertness constraint δ and maximize the covert transmission rate, the optimal transmit power is obtained as follows when $Q_{\mathrm{n}} = Q_{\max}$

$$Q_{\mathrm{KL}}^{\mathrm{AN}} = \frac{-1}{\chi} \left[\left(\sigma_{\mathrm{w}}^2 + \chi \zeta Q_{\max} \right) W_{-1} \left(-e^{-\frac{\delta}{n M_{\mathrm{w}}} - 1} \right)^{-1} + \sigma_{\mathrm{w}}^2 \right]. \tag{10.149}$$

□

Finally, the case where a suboptimal transmit power $Q_{\mathrm{t}} < Q_{\dagger}^{\mathrm{AN}}$ is used by the transmitter is commented on. In this case, for any transmit power Q_{t}, the optimal jamming power $Q_{\mathrm{n}}^{\dagger}$ required to satisfy the covertness constraint under the TVD and

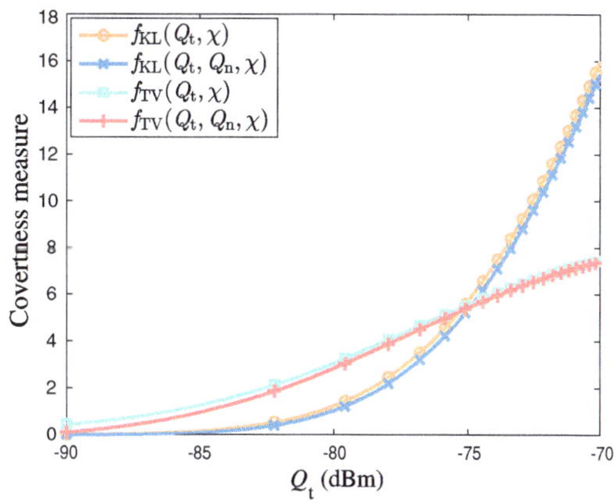

Fig. 10.8 Numerical plots of the covertness measure versus the transmit power Q_t, where $M_a = M_w = 16$, $L_a = 48$, $\eta = 0.25$, $n = 1$, $\sigma_w^2 = -80$ dBm, $Q_n = -90$ dBm

KLD covertness measures is

$$Q_n^{TV} = \frac{\left(2+\sqrt{6}\right)\sqrt{nM_w}Q_t\chi - 2\left(\chi Q_t + \sigma_w^2\right)\delta}{\left(2+\sqrt{6}\right)\sqrt{nM_w}\zeta\chi}, \tag{10.150}$$

$$Q_n^{KL} = \frac{-1}{\zeta\chi}\left[\left(\sigma_w^2 + \chi Q_t\right)W_{-1}\left(-e^{-\frac{\delta}{nM_w}-1}\right) + \sigma_w^2\right]. \tag{10.151}$$

10.3.4 Numerical Results and Discussions

A series of numerical examples are presented to validate the results derived in Theorems 10.1 and 10.2. Specifically, the bounds on the TVD and KLD as covertness measures are first examined in Sect. 10.3.2. Then, in Sect. 10.3.3, an analysis of the covert transmission rates is provided. Furthermore, the impact of various factors on both the covertness measures and the corresponding covert transmission rates is explored, including channel sparsity, the number of RF chains, the number of dominant beams, the transmit power of the confidential signal, and the power allocated for jamming.

Figure 10.8 plots the bounds on the TVD and KLD covertness measures, as defined in Eqs. (10.127), (10.129), (10.133), and (10.135), for both scenarios with and without jamming, as a function of the transmit power Q_t of the confidential signal. From the figure, it is evident that when Alice employs a jamming signal, both covertness measures are smaller compared to the scenario without jamming. This suggests that the presence of jamming enhances the covertness of legitimate communication. For both cases, it is observed that the KLD covertness measure initially decreases with Q_t but increases after surpassing a critical transmit power. This behavior indicates that the TVD covertness measure is stronger at lower

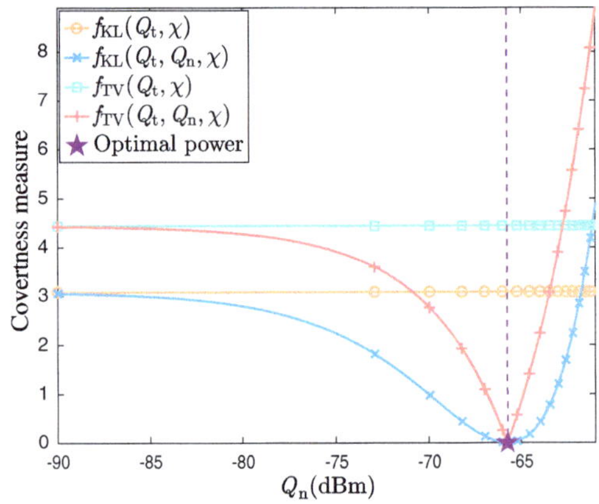

Fig. 10.9 Numerical plots of the covertness measure versus the jamming power Q_n with a fixed transmit power Q_t, where $M_b = M_w = 16,\ L_a = 96,\ n = 1,\ \chi = 0.1,\ \sigma_w^2 = -80$ dBm, $Q_t = -70$ dBm

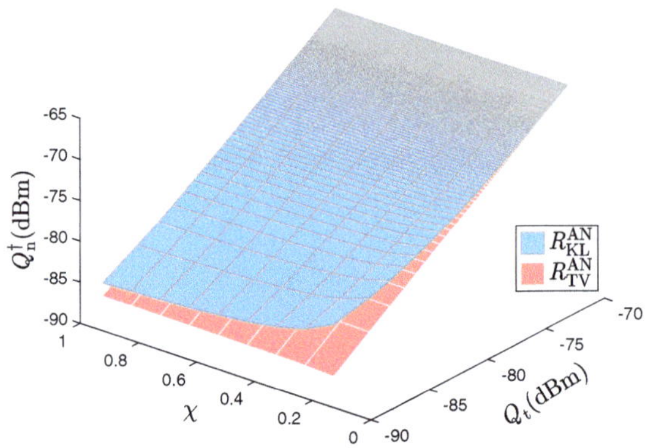

Fig. 10.10 Numerical plots of jamming power Q_n versus the channel sparsity χ and the transmit power Q_t with different covertness measure, where $M_b = M_w = 16,\ n = 1,\ \delta = 0.01,\ \sigma_w^2 = -80$ dBm

transmit power values. This phenomenon aligns with Pinsker's inequality, as shown in Eq. (10.130).

Figure 10.9 plots the bounds on both covertness measures as a function of the jamming power Q_n. It is observed that for small values of jamming power Q_n, the covertness function $f_\dagger\,(Q_t, Q_n, \chi)$ with jamming is consistently smaller than the corresponding covertness function $f_\dagger\,(Q_t, \chi)$ without jamming. However, for large values of Q_n, the covertness function $f_\dagger\,(Q_t, Q_n, \chi)$ becomes larger than $f_\dagger\,(Q_t, \chi)$. This indicates the existence of an optimal jamming power that minimizes the covertness, given the covert communication constraint, channel sparsity, and the transmit power Q_t. Notably, the bounds on both covertness measures attain their minimum value at the same jamming power, specifically at $Q_n = \frac{Q_t}{\zeta}$, as shown in Lemma 10.4.

Figure 10.10 plots the optimal jamming power $Q_n^\dagger$, as derived in Eqs. (10.150) and (10.151), versus the channel sparsity χ and transmit power Q_t for both covertness measures. From the figure, it is observed that the optimal jamming power increases with the transmit power Q_t and decreases as the channel sparsity χ increases. These trends align with the theoretical results presented earlier.

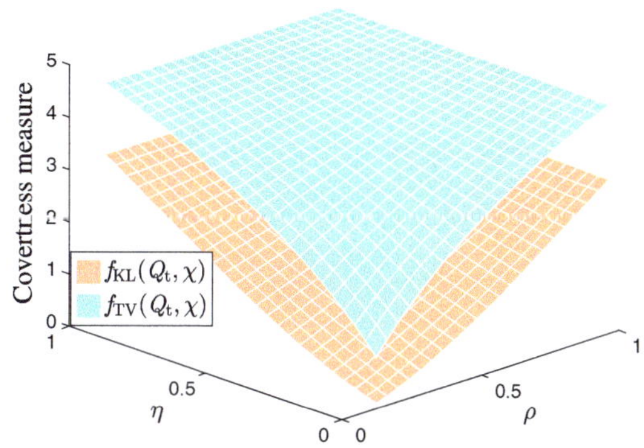

Fig. 10.11 Numerical plots of the covertness measure versus the channel sparsity parameters ρ and η when jamming is not utilized, where $M_{\mathrm{a}} = M_{\mathrm{w}} = 16$, $n = 1$, $\sigma_{\mathrm{w}}^2 = -80$ dBm, $Q_{\mathrm{t}} =$ 80 dBm

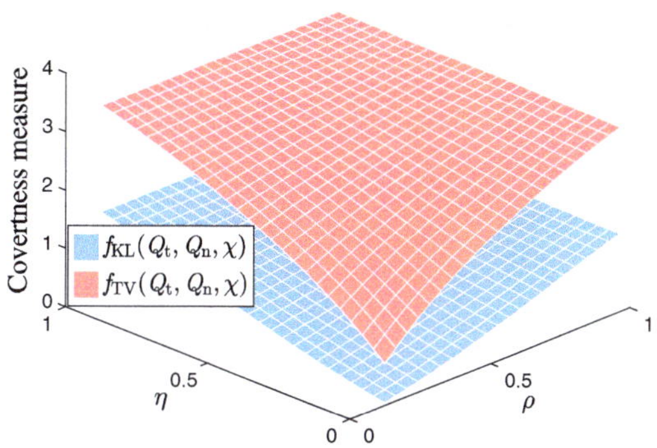

Fig. 10.12 Numerical plots of the covertness measure versus the channel sparsity parameters ρ and η when jamming is utilized, where $M_{\mathrm{a}} = M_{\mathrm{w}} = 16$, $n = 1$, $\sigma_{\mathrm{w}}^2 = -80$ dBm, $Q_{\mathrm{t}} = -80$ dBm, $Q_{\mathrm{n}} = -85$ dBm

Bounds on both covertness measures as functions of the channel sparsity parameters ρ and η are next presented. These bounds are shown for the case without jamming in Fig. 10.11 and for the case with jamming in Fig. 10.12. The transmit power Q_{t} is chosen to be relatively small so that smaller values of the KLD measure are obtained. From both figures, it is observed that the covertness functions $f_{\dagger}(Q_{\mathrm{t}}, \chi)$ and $f_{\dagger}(Q_{\mathrm{t}}, Q_{\mathrm{n}}, \chi)$ increase as either sparsity parameter increases. This suggests that a greater channel sparsity results in a stronger covertness guarantee. The underlying reasons for this behavior are detailed in Lemmas 10.2 and 10.3, where the effects on the TVD and KLD covertness measures are explained.

In Figs. 10.13 and 10.14, the results of Theorems 10.1 and 10.2 are illustrated, and the impact of channel sparsity on covert transmission rates is examined by plotting the covert transmission rates under both covertness measures as functions of the channel sparsity parameters ρ and η. These are shown for the case without jamming in Fig. 10.13 and for the case with jamming in Fig. 10.14. From both figures, it is observed that the covert transmission rates $R_{\dagger}$ and $R_{\dagger}^{\mathrm{AN}}$ increase as either sparsity parameter decreases, indicating that sparser channels result in higher covert transmission rates. Since a stronger guarantee is provided by the TVD

Fig. 10.13 Numerical plots of the covert transmission rate versus the channel sparsity parameters ρ and η when jamming is not utilized, where $M_a = M_w = M_b = 16$, $n = 500$, $\phi = 0.9$, $\delta = 0.01$, $\sigma_b^2 = -80$ dBm, $\sigma_w^2 = -80$ dBm

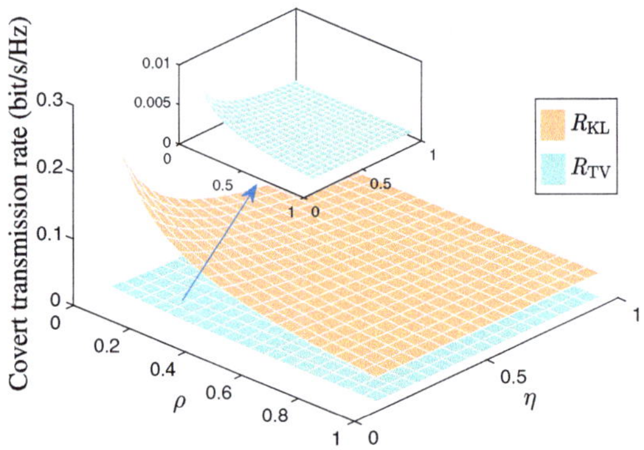

Fig. 10.14 Numerical plots of the covert transmission rate versus the channel sparsity parameters ρ and η when jamming is utilized, where $M_a = M_w = M_b = 16$, $\phi = 0.9$, $n = 500$, $\delta = 0.01$, $\sigma_b^2 = -80$ dBm, $\sigma_w^2 = -80$ dBm, $Q_n = -60$ dBm

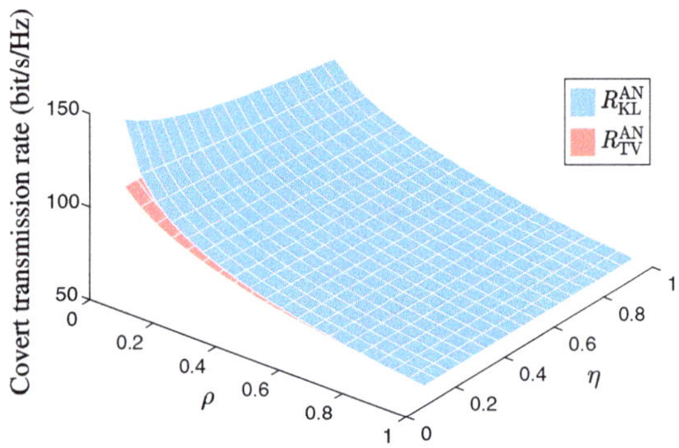

Fig. 10.15 Numerical plots of the covert transmission rate versus the detection probability δ and channel uses n, where $M_a = M_w = M_b = L_a = 16$, $\eta = 0.25$, $\phi = 0.9$, $\delta = 0.01$, $\sigma_b^2 = -90$ dBm, $\sigma_w^2 = -90$ dBm, $Q_n = -65$ dBm

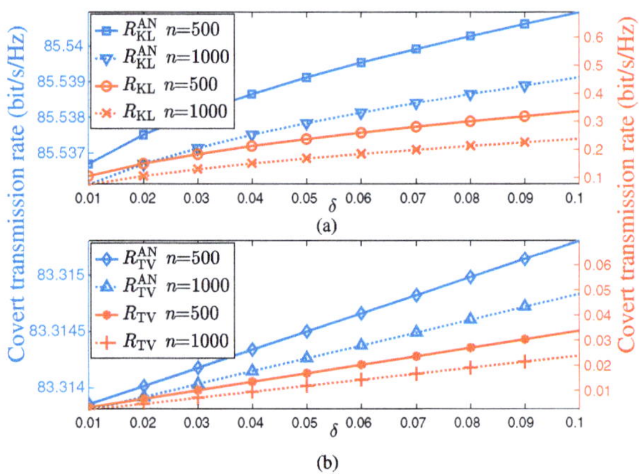

covertness measure compared to the KLD, the corresponding covert transmission rates R_{TV} and R_{TV}^{AN} are lower than R_{KL} and R_{KL}^{AN} under the KLD covertness measure, particularly when jamming is not employed. Furthermore, the jamming-aided transmission scheme significantly enhances the covert transmission rates.

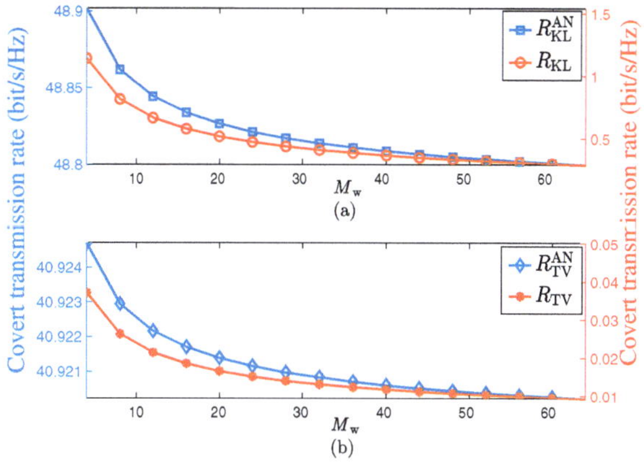

Fig. 10.16 Numerical plots of the covert transmission rate versus the number of RF chains M_w at Willie, where $L_a = 64$, $M_b = M_a = 16$, $\phi = 0.9$, $\eta = 0.25$, $\delta = 0.01$, $n = 500$, $\sigma_b^2 = -90$ dBm, $\sigma_w^2 = -80$ dBm, $Q_n = -80$ dBm

In Fig. 10.15 plots the covert transmission rates under both covertness measures as a function of the tolerable covertness constraint δ, for different numbers of channel uses (blocklength) n, considering both the cases without and with jamming. The following observations can be made. First, the jamming-aided transmission significantly enhances the covert transmission rate. Since the KLD is a looser covertness measure, it results in higher covert transmission rates. Second, the covert transmission rate increases as either the covertness constraint δ increases or the number of channel uses n decreases. This is because a larger δ or a smaller n corresponds to a more relaxed covert constraint, allowing for higher optimal transmit power.

Figure 10.16 plots the covert transmission rates under both covertness measures as a function of the number of RF chains M_w at Willie, for both the cases with and without jamming. It is observed that the covert transmission rate decreases as the number of RF chains M_w increases. This trend is consistent with the theoretical results in Eqs. (10.127), (10.129), (10.133), and (10.135), which indicate that both covertness measures $f_\dagger(Q_t, \chi)$ and $f_\dagger(Q_t, Q_n, \chi)$ increase with M_w. This phenomenon suggests that a more powerful adversary, represented by a larger number of RF chains at Willie, leads to a reduction in the covert transmission rate.

10.4 Chapter Summary

This chapter studies covert communication over single-antenna AWGN channels, MIMO AWGN channels, and massive MIMO channels. For single-antenna systems, the square root law establishes that the covert transmission rate has a zero first-order term and a second-order term scaling inversely with the square root of the blocklength. This result highlights the theoretical limit for covert communication, where maintaining covertness imposes a strict constraint on the transmission rate and requires extremely low transmission power to avoid detection by malicious users. In

MIMO AWGN channels, the square root law is extended to account for the spatial degrees of freedom due to multiple antennas. The covert transmission rate achieves a gain proportional to the square root of the number of antennas, demonstrating a significant improvement compared to single-antenna systems. However, the gain remains sub-linear, reflecting the inherent limitations of covert communication in terms of the square root law. Finally, this chapter presents a comprehensive analysis of covert communication over mmWave massive MIMO channels with spatial sparsity. A metric for channel sparsity is introduced and analyzed on how sparsity and system parameters impact the performance of covert communication. A key finding is that increasing channel sparsity and using artificial noise significantly improve covert transmission rates. Furthermore, the covert rate increases with more RF chains for the legitimate user and decreases with more RF chains for the adversary. The study of massive MIMO covert communication highlights the effectiveness of friendly jamming in enhancing covert communication, providing insights for the design of next-generation secure communication systems using mmWave MIMO technology.

10.5 Exercises

10.1 What is covert communication? What are the metrics used to evaluate the covertness?

10.2 What is the role of multi-antenna in covert communication?

10.3 Prove the inequality in (10.21) that $\log(1+x) \geq x - \frac{x^2}{2}$.

10.4 Prove Lemma 10.1 concerning the proof for the MIMO AWGN channel.

10.5 Plot the covert transmission rate versus the blocklength n when the covertness level $\delta = 0.8$ and the number of antennas $m = 20$.

10.6 Why are mmWave massive MIMO channels sparse? How does one evaluate the sparsity of mmWave massive MIMO channels?

10.7 Justify the results on calculating covertness measures in Eq. (10.126) under TVD and in Eq. (10.132) under KLD, respectively.

10.8 Plot the curve that shows the relationship between the covert transmission rate and the number of RF chains of the legitimate transmitter and the dominant beams of the channel based on Theorems 10.1 and 10.2, both with and without the use of jamming.

References

1. Y. Liu, H.-H. Chen, L. Wang, Physical layer security for next generation wireless networks: theories, technologies, and challenges. IEEE Commun. Surveys Tuts. **19**(1), 347–376 (2017)
2. X. Chen, J. An, Z. Xiong, C. Xing, N. Zhao, F.R. Yu, A. Nallanathan, Covert communications: a comprehensive survey. IEEE Commun. Surveys Tuts. **25**(2), 1173–1198 (2023)
3. C.E. Shannon, Communication theory of secrecy systems. Bell Labs Tech. J. **28**(4), 656–715 (1949)
4. A.D. Wyner, The wire-tap channel. Bell Syst. Tech. J. **54**(8), 1355–1387 (1975)
5. I. Csiszár, J. Korner, Physical channels with confidential messages. IEEE Trans. Inf. Theory **24**(3), 339–348 (1978)
6. U.M. Maurer, Secret key agreement by public discussion from common information. IEEE Trans. Inf. Theory. **39**(3), 733–742 (1993)
7. R. Ahlswede, I. Csiszar, Common randomness in information theory and cryptography. I. Secret sharing. IEEE Trans. Inf. Theory **39**(4), 1121–1132 (1993)
8. M. Bloch, J. Barros. *Physical-Layer Security: From Information Theory to Security Engineering*, 1st edn. (Cambridge University Press, Cambridge, 2011)
9. B.A. Bash, D. Goeckel, D. Towsley, Limits of reliable communication with low probability of detection on AWGN channels. IEEE J. Sel. Areas Commun. **31**(9), 1921–1930 (2013)
10. L. Wang, G.W. Wornell, L. Zheng, Fundamental limits of communication with low probability of detection. IEEE Trans. Inf. Theory **62**(6), 3493–3503 (2016)
11. M.R. Bloch, Covert communication over noisy channels: a resolvability perspective. IEEE Trans. Inf. Theory **62**(5), 2334–2354 (2016)
12. P.H. Che, M. Bakshi S. Jaggi, Reliable deniable communication: hiding messages in noise, in *IEEE ISIT* (2013), pp. 2945–2949
13. L. Bai, J. Xu, L. Zhou, Covert communication for spatially sparse mmWave massive MIMO channels. IEEE Trans. Commun. **71**(3), 1615–1630 (2023)
14. X. Yu, S. Wei, S. Huang, X. Zhang, The second order asymptotics of covert communication over AWGN channels (2023). arXiv: 2305.17924
15. E. Lehmann, J. Romano, *Testing Statistical Hypotheses*, 3rd edn. (Springer, New York, 2005)
16. T.M. Cover, J.A. Thomas, *Elements of Information Theory* (Wiley, New York, 1991)
17. Q. Zhang, M.R. Bloch, M. Bakshi, S. Jaggi, Undetectable radios: covert communication under spectral mask constraints, in *IEEE ISIT* (2019), pp. 992–996
18. Q. Zhang, *On Covert and Stealthy Communication* (The Chinese Univ. of Hong Kong, Hong Kong, 2019)
19. Y. Polyanskiy, H.V. Poor, S. Verdú, Channel coding rate in the finite block length regime. IEEE Trans. Inf. Theory **56**(5), 2307–2359 (2010)
20. B. He, H. Jafarkhani, Low-complexity reconfigurable MIMO for millimeter wave communications. IEEE Trans. Commun. **66**(11), 5278–5291 (2018)
21. A.M. Sayeed, Deconstructing multiantenna fading channels. IEEE Trans. Signal Process. **50**(10), 2563–2579 (2002)
22. J. Xu, W. Xu, D.W.K. Ng, A.L. Swindlehurst, Secure communication for spatially sparse millimeter-wave massive MIMO channels via hybrid precoding. IEEE Trans. Commun. **68**(2), 887–901 (2020)
23. J. Hou, G. Kramer, Effective secrecy: reliability, confusion and stealth, in *IEEE ISIT* (2014), pp. 601–605
24. H. ZivariFard, M.R. Bloch, A. Nosratinia, Covert communication via non-causal cribbing from a cooperative jammer, in *IEEE ISIT* (2021), pp. 202–207
25. O. El Ayach, S. Rajagopal, S. Abu-Surra, Z. Pi, R.W. Heath Jr., Spatially sparse precoding in millimeter wave MIMO systems. IEEE Trans. Wireless Commun. **13**(3), 1499–1513 (2014)
26. A. Alkhateeb, O. El Ayach, G. Leus, R.W. Heath Jr., Channel estimation and hybrid precoding for millimeter wave cellular systems. IEEE J. Sel. Topics Signal Process. **8**(5), 831–846 (2014)

27. M. Kokshoorn, H. Chen, P. Wang, Y. Li, B. Vucetic, Millimeter wave MIMO channel estimation using overlapped beam patterns and rate adaptation. IEEE Trans. Signal Process. **65**(3), 601–616 (2017)
28. D. Tse, P. Viswanath, *Fundamentals of Wireless Communication* (Cambridge University Press, Cambridge, 2005)
29. M. Tahmasbi, M.R. Bloch, First- and second-order asymptotics in covert communication. IEEE Trans. Inf. Theory **65**(4), 2190–2212 (2019)
30. S.-Y. Wang, M.R. Bloch, Covert MIMO communications under variational distance constraint. IEEE Trans. Inf. Forensics Security **16**, 4605–4620 (2021)
31. I. Chatzigeorgiou, Bounds on the Lambert function and their application to the outage analysis of user cooperation. IEEE Commun. Lett. **17**(8), 1505–1508 (2013)

MIX
Papier aus verantwortungsvollen Quellen
Paper from responsible sources
FSC® C105338

If you have any concerns about our products,
you can contact us on
ProductSafety@springernature.com

In case Publisher is established outside the EU,
the EU authorized representative is:
Springer Nature Customer Service Center GmbH
Europaplatz 3, 69115 Heidelberg, Germany

Printed by Libri Plureos GmbH
in Hamburg, Germany